LANTIC OCEAN · *TECUMSEH FORMATION*, PENNSYLVANIAN, WEEPING WATER, CASS CO., NEBRASKA, U.S.A. · *NEGLI CREEK LIMESTONE*, MISSISSIPPIAN, PEACH HILL, PERRY CO., INDIANA, U.S.A. · *WHITEHORSE FORMATION*, TRIASSIC, LLAMA MOUNTAIN, ALBERTA, CANADA · *SILURIAN*, VISBY, GOTLAND, SWEDEN · *PLIENSBACHIAN*, JURASSIC, STE. BAUME, BOUCHES DU RHÔNE, FRANCE · *MANLIUS FORMATION*, DEVONIAN, HELDERBERG MOUNTAINS, RENSSELAER CO., NEW YORK, U.S.A. · *CRETACEOUS*, TIRUCHCHIRAPPALLI, TIRUCHCHIRAPPALLI DISTRICT, MADRAS, INDIA · *HUNGRY HOLLOW FORMATION*, DEVONIAN, WEST WILLIAMS TOWNSHIP, MIDDLESEX CO., ONTARIO, CANADA · *DENTON CLAY MEMBER*, CRETACEOUS, FORT WORTH, TARRANT CO., TEXAS, U.S.A. · *BEERSHEVA FORMATION*, JURASSIC, MAKHTESH RAMON, ISRAEL · *BRASSFIELD LIMESTONE*, SILURIAN, JEFFERSON CO., INDIANA, U.S.A. · *GERSTER FORMA*[illegible] PERMIAN, LEACH MOUNTAINS, ELKO CO., NEVADA, [illegible] LOWER ORDOVICIAN, BORNHOLM, DENMARK · *B*[illegible] *FORMATION*, MISSISSIPPIAN, HARDIN CO., KENT[illegible], U.S.A. · *LOWER JURASSIC*, COMANA VALLEY, PE[illegible]I MOUNTAINS, ROMANIA · *JURASSIC*, POLIWNA-ON-THE-

Alan Stanley Horowitz · Paul Edwin Potter

Introductory Petrography of Fossils

With 100 Plates and 28 Figures

Photomicrography by George R. Ringer

Springer-Verlag
New York · Heidelberg · Berlin 1971

Alan Stanley Horowitz, Curator of Paleontology
Paul Edwin Potter, Professor of Geology
Indiana University, Bloomington, IN 47401/U.S.A.

ISBN 0-387-05275-5 Springer-Verlag New York Heidelberg Berlin
ISBN 3-540-05275-5 Springer-Verlag Berlin Heidelberg New York

Universitätsdruckerei H. Stürtz AG Würzburg

Shells to bits,
Bits to dust,
Aragonite to calcite,
The microscope's a must.

H. & P.

Preface

This is a book for beginners. Not geological beginners, because an introductory course in paleontology and some knowledge of the petrographic microscope is assumed, but for beginners in the study of the petrography of fossil constituents in sedimentary rocks. Fossils are studied for various reasons: 1) to provide chronologic (time) frameworks, 2) to delineate rock units and ancient environments, or 3) to understand the past development (evolution) of living plants and animals. All of these uses may be attained through petrographic studies of thin sections of fossils embedded in sedimentary rocks. Some knowledge of the appearance of fossils in thin section is also fundamental for general stratigraphic studies, biofacies analyses, and is even useful in studying some metamorphic rocks. Commonly, fossils are essential for the delineation of carbonate rock types (facies or biofacies). We have written this book for sedimentary petrologists and stratigraphers, who routinely encounter fossils as part of their studies but who are not specialists in paleontology, and for students who are seeking a brief review and an introduction to the literature of the petrography of fossiliferous sedimentary rocks.

Although experienced paleontologists may be appalled by the many generalized statements on size, shape, and principal fossil characters recited herein, we counter that we have had some success in introducing non-paleontologically oriented geologists to the use and identification of fossil constituents without using excessive paleontological terminology and detailed systematics. While the variability of shape and form within major fossil groups is a mark of their evolutionary success, it is an apparently bewildering chaos to the uninitiated. We have tried to orient the user toward the major shapes and features of fossils as viewed in thin section in order to simplify identification and utilization of major fossil groups in petrographic studies.

The thin section study of carbonate rocks was long slighted, but is now a key part of stratigraphic and sedimentological studies. Certainly, the composition of skeletal debris is frequently difficult or impossible to obtain by any other technique. Although our review would have been most useful two decades ago when modern emphasis on carbonate petrographic studies began, the amount of information then available inhibited a review of the type presented herein. Much of the information then available on the petrography of fossils was summarized by J. H. Johnson in his "An Introduction to the Study of Organic Limestones" in 1951. More recently O. P. Majewske has prepared a summary, "Recognition of Invertebrate Fossil Fragments in Rocks and Thin Sections" published in the E. J. Brill International Sedimentary Petrographical Series. Our work likewise will be superceded as extensive studies and additional summaries of the shell microstructure of various groups of fossils become available. In particular, the scanning electron microscope provides two orders of magnitude better resolution than we discuss. Eventually atlases of electron microscopy of fossil shell microstructure will be compiled comparable to those for the optical microscope. However, we do not believe the optical microscope will be displaced because it is convenient, provides much information at small cost, and reveals textural relations visible only by means of transmitted light.

Horowitz began his studies of the petrography of fossils under the auspices of the Marathon Oil Company, and thanks are due R. D. Russell for providing the opportunity, J. H. Johnson for presenting the initial instruction, and L. C. Pray and J. L. Wray for many hours of friendly counseling. Potter became interested via the teaching of sedimentary petrology. It seemed only natural to pool our complementary interests. For his part Potter learned a lot about limestones and paleontologists. Horowitz thought it would be fun to do a joint project with Potter. The ratio of fun to work was rather small. Of course, one is not supposed to say such things in print because it tarnishes the image, so much admired in scientific circles, of scientific work conducted for compelling logical reasons. Consequently, for public consumption, this book was prepared in order to provide a convenient introductory review to the study of fossils in thin sections. As such, it provides an example of the blending of the fields of paleontology and sedimentary petrography.

This book is a product of much assistance by numerous people. We have been most pleased at the response from geologists all over the world to our request for samples of carbonate rocks. Both the plates and text have greatly benefited from the cooperation of those listed below: Robert L. Anstey, Michigan State University (samples, text review); Christina L. Balk, New Mexico Institute of Mining and Technology (samples); Maxwell R. Banks, University of Tasmania, Australia (samples); Roger L. Batten, American Museum of Natural History (text review); James W. Baxter, Illinois State Geological Survey (samples, thin sections); Frank W. Beales, University of Toronto, Canada (thin sections, text review); Harold J. Bissell, Brigham Young University (samples); Richard S. Boardman, U. S. National Museum (text review); Arthur J. Boucot, Oregon State University (text review); Sydney D. Bowers, Arabian American Oil Company,

Saudi Arabia (samples); M. Braun, Geological Survey of Israel (samples); William P. Brosgé, U. S. Geological Survey (samples); Colin Bull, Ohio State University (samples); S. K. Chanda, Jadavpur University, India (samples); Jean Charollais, Université de Genèva, Switzerland (samples); Alan H. Cheetham, U. S. National Museum (text review); Roberto Colacicchi, Piazza Università, Perugia, Italia (samples); Earle R. Cressman, U. S. Geological Survey (thin sections); Bruno D'Argenio, Università di Napoli, Italia (samples); Nelson M. DaSilva, Petróleo Brasileiro S. A., Brazil (samples); David L. Dilcher, Indiana University (text review); J. Robert Dodd, Indiana University (samples, text review); Mircea Dumitriu, Geological Committee, Romania (samples); William H. Easton, University of Southern California (text review); Graham F. Elliott, British Museum (Natural History), England (thin sections); Paulo M. de Figueiredo, Universidade do Rio Grande do Sul, Brazil (samples); Robert H. Finks, Queens College City College of New York (text review); Alfred G. Fischer, Princeton University (samples); Donald W. Fisher, New York State Museum and Science Service (samples); Robert L. Folk, University of Texas (samples); Robert W. Frey, University of Georgia (text review); K. L. Gauri, University of Louisville (text review); S. Ghosh, Hindustan Minerals and Natural History Supply Company, India (samples); D. W. Gibson, Geological Survey of Canada (samples); E. D. Gill and H. E. Wilkinson, National Museum of Victoria, Australia (samples); Robert N. Ginsberg, University of Miami (Florida) (samples); F. F. Gray, Sunlite Oil, Ltd., Canada (samples); Carlos G. Gutiérrez, Santa Maria de la Paz y Anexas, S. A., México (samples); Walter Häntzschel, Geologische Staatsinstitut, Hamburg, Deutschland (text review); Lawrence Hardie, The Johns Hopkins University (samples); Kotora Hatai, Tohoku University, Japan (samples); Donald E. Hattin, Indiana University (samples, thin sections); Monin ul Hoque, University of Libya, Libya (samples); B. F. Howell, Princeton University (text review); Stephanie V. Hrabar, University of Cincinnati (samples); John W. Huddle, U. S. Geological Survey (text review); Donald Hyers, Texaco, Incorporated (samples); L. V. Illing, V. C. Illing and Partners, England (samples); Arie Janssens, Ohio Geological Survey (samples); Erle G. Kauffman, U. S. National Museum (samples, text review); Marshall Kay, Columbia University (samples); Porter M. Kier, U. S. National Museum (text review); Harry S. Ladd, U. S. Geological Survey (samples); N. Gary Lane, University of California, Los Angeles (samples); Aurèle La Rocque, Ohio State University (samples); Alfred R. Loeblich, Jr., Chevron Research Company (text review); Earle F. McBride, University of Texas (samples); M. J. McCarthy, University of Natal,

Republic of South Africa (samples); MICHAEL MACLANE, Indiana University (samples); R. MANNIL, Academy of Sciences of the Estonian S. S. R. (samples); F. M. MARTIN, Universidad de Madrid, España (samples); HARUO NAGAHAMA, Geological Survey of Japan (samples); HENRY F. NELSON, Mobil Research and Development Corporation (samples); M. G. N. MASCARENHAS NETO, Direcção Provincial dos Serviços de Geologia e Minas, Portuguese Angola (samples); MATTHEW H. NITECKI, Field Museum of Natural History (text review); MASAMICHI OTA, Akiyoshi-dai Science Museum, Japan (samples); A. R. PALMER, State University of New York, Stony Brook (samples, text review); DEAN PENNINGTON, Occidental of Libya, Inc. (samples); THOMAS G. PERRY, Indiana University (text review); JOHN POJETA, Jr., U. S. Geological Survey (text review); GILBERT POMMIER and M. J. THOUVENIN, Cie Français des Pétroles, France (samples); REX T. PRIDER and D. RHODES, University of Western Australia, Australia (samples); WAYNE PRYOR, University of Cincinnati (samples); RAJKA RADOIČIĆ, Zavod za Geol. i Geof. Istraživanja, Yugoslavia (samples); R. F. REBOILE, Direcção dos Serviços de Geologia e Minas, Moçambique (samples); JÜRGEN REMANE, Universitè de Neuchâtel, La Suisse (thin sections); CARL B. REXROAD, Indiana Geological Survey (samples, text review); J. KEITH RIGBY, Brigham Young University (thin sections, text review); JUAN CARLOS RIGGI, Yacimiento Petróleo Fomento, Argentina (samples); D. N. RIMAL, Nepal Geological Survey (samples); JOSE MA RIOS GARCIA, Empresa National Minera del Sahara, S. A. España (samples); PERRY O. ROEHL, Union Oil Company of California (samples); J. P. ROSSOUW, Geological Survey Republic of South Africa (samples); JOSEPH ST. JEAN, Jr., University of North Carolina (text review); HANS SCHMINCKE, Institut für Mineralogie und Petrologie, Bochum, Deutschland (samples); SUPRIYA SENGUPTA, Indian Statistical Institute, India (samples); CHARLES T. SIEMERS, Indiana University (samples); GEORGE SNYDER, U. S. Geological Survey (thin sections); I. G. SOHN, U. S. Geological Survey (text review); NILS SPJELDNAES, Aarhus University, Denmark (samples); R. THORSTEINSSON and J. W. KERR, Geological Survey of Canada (samples); P. N. VARFOLOMEEV, Central Scientific Geological and Prospecting Museum, U. S. S. R. (samples); HARRY R. WARMAN, Iranian Oil Exploration and Production Consortium, Iran (samples); J. W. WELLS, Cornell University (text review); PHILIP R. WOLCOTT, Creole Petroleum Corporation, Venezuela (samples); JOHN L. WRAY, Marathon Oil Company (samples, text review); ELLIS YOCHELSON, U. S. Geological Survey (samples, text review); FREDERICK G. YOUNG, McGill University, Canada (samples); RAINER ZANGERL, Chicago Museum of Natural History (text review); W. ZIMMERLE, Deutsche Erdöl-Aktiengesellschaft, Deutschland (samples).

In addition to the samples requested specifically for this book, we have utilized collections reposited by the students and staff in the Department of Geology at Indiana University.

We are grateful to many geologists, cited and uncited, who have provided us with bibliographic assistance. Because our expertise is limited, we have availed ourselves of a wide array of paleontologists to assist in eliminating egregious errors from the fossil text. However, we have not always adopted their advice.

All the photomicrographs were taken by Mr. George R. Ringer of the Indiana Department of Geology and the Indiana Geological Survey. We are indebted to him for the skill with which he prepared the negatives and prints. Most of the thin sections were prepared by Mr. Leonard Neal, preparator in the Geology Department, Indiana University. Garry Anderson prepared some special thin sections of conodonts. We obtained some thin sections on loan, as indicated above in the acknowledgements, and are indebted to these workers for permission to photograph the borrowed slides.

Miss Ellen Freeman of the Department of Geology Library and Mrs. Naomi Lawlis and her staff at the Interlibrary Loan Office of Indiana University have provided much needed support in obtaining difficult citations and works not otherwise available to us.

We received considerable assistance in the "book keeping" aspects of this project: 1) cataloging samples, 2) numbering of specimens, negatives, and prints, and 3) filing correspondence, negatives, prints and samples, from Mrs. Merilee Darko, Kenneth Heiner, Gary O'Neal, Mrs. George Ringer, and Peggy Staser. The plates were prepared by Ann Pavlick. Mrs. Janice Perry and Mrs. Thea Brown typed various versions of the manuscript and we thank them for their patience.

Many of the line drawings are Horowitz primitives (they don't look too bad after a couple of beers) and were prepared in final form by Ann Pavlick of Indiana University.

Funds for this project were obtained from the following:

American International Oil Company, Chicago, Illinois;
Atlantic Richfield Company, Dallas, Texas;
Home Oil Company, Calgary, Alberta, Canada;
Indiana University Foundation;
Marathon Oil Company, Denver, Colorado;
Mobil Research and Development Corporation, Dallas, Texas;
Murphy Oil Corporation, El Dorado, Arkansas;
Pan American Petroleum Corporation, Tulsa, Oklahoma;
Union Oil Company, Brea, California;
Sale of the syllabus "Geology of Sand and Sandstones" by Pettijohn, *et al.*, 1965.

The support of the above companies and the Indiana University Foundation has made our project possible. Royalties from this book will go initially to the J. J. Galloway Fund of the Indiana University Foundation, which will further work in paleontology and stratigraphy of the students and staff in the Department of Geology at Indiana University.

Finally, we are indebted to the Department of Geology at Indiana University, for its support.

Bloomington
May 15, 1970

ALAN STANLEY HOROWITZ
PAUL EDWIN POTTER

Table of Contents

An instructional set of fifty black and white 2 × 2 slides selected from the plates can be purchased from Springer-Verlag Berlin · Heidelberg · New York

1. Introduction

The text consists of four chapters. The first discusses the organization of the book and explains some details of its composition and construction. The second provides a rationale and overview of carbonate petrography and the place of biotic constituents in the field of carbonate petrography. The third chapter presents a group by group discussion of the thin section characters of biotic constituents. Most of the biotic groups are discussed under four headings: skeletal architecture, skeletal microstructure, distribution, and comparisons. Skeletal architecture refers to the external or internal geometry of the shells or skeletons. Skeletal microstructure refers to the internal character of the shell or skeletal wall. A few groups, such as the arthropods and mollusks, contain an additional introductory section. The fourth chapter contains an introduction to the plates and plate descriptions.

The taxonomic level of the groups selected for discussion is not uniform. Most of the groups of animals considered herein are phyla, the highest rank generally used in the animal kingdom. However, the tintinnines and foraminifers are orders, and the radiolarians are a subclass of one-celled organisms (Protozoa). Some classes within phyla are discussed in sections on mollusks and arthropods. The algal subdivisions are generally families. The selection of these groups has been influenced by the ease of recognition of their general characters and their occurrence in the geologic record. An experienced paleontologist could determine much finer taxonomic divisions than we have attempted. On the other hand, a few groups do not lend themselves very well to detailed discrimination by means of thin sections. This is true at present for most echinodermal and trilobite debris.

The chronologic ranges of the various fossil groups usually are taken from HARLAND *et al.* (1967), which is the latest authoritative one-volume summary. Fig. 28 illustrates ranges and taxonomic abundances for the major fossil groups discussed in the text.

While the second chapter might be read for the pleasure of a short, coherent account of a field of study, the third is obviously for reference only. The third chapter also contains some comments on the mineralogy of fossil skeletons and their post-depositional modifications. Each chapter has its own bibliography, and we call attention to the historically arranged selected bibliography on carbonate rock studies in Chapter 2. A short annotated list of microfacies monographs is also presented in Chapter 3. Finally, the last—and most important—chapter consists of plates that show the common biotic constituents in rocks of Cambrian to Recent age from all over the world.

References Cited

HARLAND, W. B., *et al.*, 1967, The fossil record, a symposium with documentation: London, Geological Society of London, 827 p.

2. Getting Started in Carbonate Petrography: Methodology and Applications

Introduction

An introductory text on the petrography of fossils would be incomplete without mention of its uses as well as informing the reader how to begin the study of carbonate rocks in thin section.

The study of fossils in thin section is used primarily in chronostratigraphy, systematic paleontology, and for petrographic mapping of biofacies, one of the best ways to identify ancient environments of carbonate deposition. Although other uses come to mind, for example, in skeletal diagenesis or porosity mapping in a carbonate reservoir, we believe that for many years to come a principal use of "fossil petrography" will be its role in helping sedimentologists identify ancient depositional environments.

This chapter is a brief introduction to the key ideas associated with the study of carbonates as well as an annotated list of the more important papers, the latter provides a history of development that starts with the first petrographic study of a limestone by HENRY SORBY in 1851. Our presentation begins with carbonate petrography—for surely it is the essential foundation of almost all carbonate studies—and then considers the role played by the other aspects of a deposit, particularly its size, shape, and relation to bounding units. Largely excluded, however, are the chemical aspects of carbonates—not because we regard them as unimportant—but because they lie mostly outside of a petrographic treatment of fossils. The study of dolomite and dolomitic textures is also largely excluded. There are, however, a number of sources where one can now find broad, balanced coverage of carbonates. Comprehensive coverage including petrography as well as chemical aspects is available in the two volumes of "Carbonate Rocks" edited by G.V. CHILINGAR *et al.* (1967). BARRS' (1963) brief review of the petrology of carbonate rocks is strongly founded on modern processes and is still one of the best concise reviews available. Another source book is CAROZZI's (1960), "Microscopic Sedimentary Petrology" a part of which includes a qualitative petrography of carbonate rocks. Two chapters of "Petrology of the Sedimentary Rocks" by HATCH *et al.* (1965) are devoted to an elementary treatment of carbonates. The annotated bibliography of POTTER (1968) gives a helpful overview to the literature of the many different aspects of carbonate rocks.

Finally, a few words about objectives in this chapter. The following is not intended to be a summary of present knowledge nor to contain a digest of answers. Rather it is offered as an outline of methodology, an outline whose focus is on organizing the many different aspects of a carbonate body into a rational unity so that it can best be related to its neighbors and the basin in which it occurs.

In this age of sophisticated equipment and complicated preparation procedures, it is refreshing to note that the techniques needed for carbonate petrography are mostly simple ones. These include: acid etching and the making of peels, polished sections, and thin sections. Stains help distinguish dolomite from calcite and the various calcites from one another on such preparations. The use of the X-ray diffractometer to determine carbonate mineralogy on bulk samples is also commonplace. These techniques are covered in the methods chapter of "Carbonate Rocks" by CHILINGAR *et al.* (1967), GERMAN MÜLLER's (1967) "Methods in Sedimentary Petrology", and appropriate sections of KUMMEL and RAUP's (1964)

"Handbook of Paleontological Techniques". The Journal of Sedimentary Petrology frequently contains notes on techniques.

Essential Elements of Carbonate Petrography

Background

A carbonate rock consists of 50 percent or more carbonate minerals, which are most commonly calcite and dolomite (Fig. 1). Most of the calcite of ancient limestones is low-Mg calcite and only in Recent carbonate muds and sands does one find much of either high-Mg calcite or aragonite (CHAVE, 1954). If the mineral dolomite predominates over calcite, the rock is a dolomite or dolostone, but if not, it is a limestone. Here we are primarily concerned with limestones. Non-carbonate impurities exist in all but a very few carbonates and include all the minerals that are stable on the surface of the earth. Common minerals found in limestones include anhydrite, gypsum, chert, glauconite, and phosphate as well as terrigenous quartz and some feldspar. Other accessories may be a few heavy minerals or perhaps volcanic debris, if the limestone accumulated on a shallow marine platform near a volcanic complex. With increasing amounts of impurities, carbonates may grade into terrigenous sands or muds, into chemically precipitated evaporites, or even into sands rich in volcaniclastic detritus.

Texture and composition are the two fundamental pillars of all petrography. By texture is meant the size, sorting, shape, and arrangement (packing and fabric) of a rock's components. As pointed out by PETTIJOHN (1957, p. 13), textural terms are very

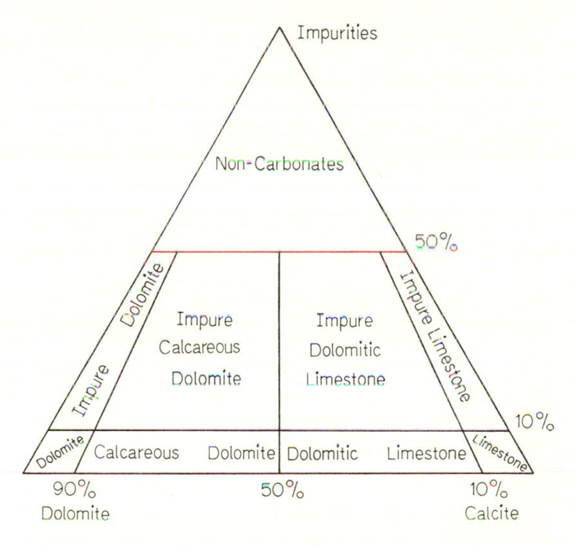

Fig. 1. Compositional triangle and terminology for carbonate rocks. An appropriate compositional term should be substituted for the word "impure" where possible (LEIGHTON and PENDEXTER, 1962, Fig. 2, with permission of the American Association of Petroleum Geologists).

Table 1. *Classification of Carbonate Rocks According to Depositional Texture* (modified from DUNHAM, 1962, Table 1)

<table>
<tr><th colspan="5">Depositional Texture Recognizable</th><th rowspan="5">Depositional Texture Not Recognizable</th></tr>
<tr><td colspan="4">Original Components Not Bound Together During Deposition</td><td rowspan="4">Original components bound together during deposition—as shown by intergrown skeletal matter, lamination contrary to gravity, or sediment-floored cavities that are roofed over by organic or questionably organic matter and are too large to be interstices</td></tr>
<tr><td colspan="3">Contains mud (particles of clay and fine silt size)</td><td rowspan="3">Lacks mud and is grain-supported</td></tr>
<tr><td colspan="2">Mud-supported</td><td rowspan="2">Grain-supported</td></tr>
<tr><td>Less than 10 percent grains</td><td>More than 10 percent grains</td></tr>
<tr><td>Mudstone</td><td>Wackestone</td><td>Packstone</td><td>Grainstone</td><td>Boundstone</td><td>Crystalline Carbonate</td></tr>
</table>

largely *geometric* ones and as such contrast with compositional terms (Table 1). In the terrigenous sands and muds as well as in igneous and metamorphic rocks, composition is specified mostly by mineralogy alone, but in the carbonates and, especially in the limestones, one or two carbonate minerals prevail so that *grain type* is much more important—is the grain an oolith (ooid), of skeletal origin, or a fragment of semiconsolidated lime mud derived from within the basin?

In describing carbonates it is always best to separate textural from compositional terminology because the same well-sorted, medium-grained carbonate sand can have, for example, a framework of mostly algae and foraminifers or a mixture of echinodermal plates and ooliths. Although this seems to imply the independence of composition and texture, grain type is in reality *dependent* on grain size—in short, texture and composition are no more independent in carbonate than in terrigenous sands (Fig. 2). This is so for two reasons: first, most organisms have a characteristic size and, secondly, after death most skeletal debris disintegrates into segments depending on the details of its architecture (FORCE, 1969). Hydraulic sorting after death helps segregate this debris. Consequently, a strong correlation exists between grain type and size in most limestones.

Textural Components

Detrital mud, chemical precipitates (cement), framework (detrital grains and *in situ* organisms) plus pores are the fundamental building blocks of all sedimentary rocks. Thus one can, in fact, say that the goal of a sedimentary petrography is to give a unique geologic interpretation to each of the many different combinations of these four components. Such an interpretation is almost always an exercise in writing a careful and particular geologic history—how did a particular sand form, what relations does it have to its associated lithologies, how does it fit into the basin's evolution, and what does it tell us about its post depositional events? But to move toward such a goal with confidence, a good understanding and clear definition of these four fundamental components is essential.

Framework.—Discrete, detrital particles of a sediment that form its **framework** or **skeleton** are called **grains.** Commonly these are of sand size (between 62 and 2000 microns) but they may also be silt. One may define the lower size limit of grains functionally in terms of its framework role or arbitrarily in terms of its size. For most sands 62 microns is an acceptable lower limit but 30 microns has been suggested (LEIGHTON and PENDEXTER, 1962, p. 35) and in siltstones even 20 microns might be used. The precise lower limit depends on the sediment and on the ability of the investigator to discern its details—whether one uses a hand lens, a binocular or a petrographic or a scanning electron microscope.

In situ **organic structures** constitute another important framework element. These are the corals, bryozoans and algae, which form an important clan of limestones (boundstones) whose framework is bound together by secreted calcium carbonate.

The rigid framework of either grains or *in situ* organisms makes possible the **pore system** of a rock. It is the pore system that permits a carbonate rock to store and transmit fluids. Detrital mud and later cementation may completely or partially fill the pore system. The initial porosity of a carbonate sand may also be greatly reduced by **solution packing** through stylolitic solution at points of grain contact (Pls. 32-1, 83, 89, 96). On the other hand, later selective solution may enhance porosity. CHOQUETTE and PRAY (1970) provide a very complete discussion of the nomenclature and classification of porosity in carbonate rocks.

Lime mud.—FOLK (1959, p. 8) first brought to the attention of most petrographers the microcrystalline calcite ooze that now is almost universally called **micrite.** As currently used, the term micrite applies to consolidated or unconsolidated carbonate mud. Micrite is the carbonate analog to the argillaceous mud that one finds between the framework grains of many terrigenous sands. As with argillaceous mud, petrographers have not been able to agree on a single upper size limit for micrite—should it be 62, 30, 20 or perhaps even 4 microns?

Again, the method of observation plays an essential role in the decision. Thirty microns might be the boundary for a binocular study but 4 microns might be best for a careful thin section study. To complicate matters, an electron microscope study of ancient micrites found most particles to lie between 0.5 and 2 microns (FLÜGEL *et al.*, 1968, Fig. 2). At present petrographers do not agree upon an upper size limit for micrite.

Micrite may be absent, may partially fill the void space between the framework, or may be the principal rock component so that if grains are present they are **mud supported** rather than **grain supported.** DUNHAM (1962) recognized this distinction as a major one in his classification of limestones which he assigned to three major groups: mud supported, grain supported, and boundstones (Table 1). This classification is simple and meaningful to apply and suggests the distinction made by DOTT (1964) between the clean washed arenites and the muddy wackes of terrigenous sandstones.

Studies with the electron microscope of modern carbonate muds show them to consist of discrete acicular (needlelike) crystals of aragonite or calcite up to 4 microns in length and of finely divided fossil debris. Because of the small size of its components, it is difficult to determine how much of a lime mud is a chemical precipitate from sea water supersaturated with respect to either calcite or aragonite, how much is of direct biologic origin, and how much comes from the disintegration of fossil debris. In thin section many carbonate muds have a clotted texture, which may be the result of indistinct pelletal boundaries (Pls. 19–1, 81, 86). MATTHEWS (1966) gave a good account of the properties and inferred origin of lime mud off Honduras, as did STOCKMAN *et al.* (1967) off Florida.

After deposition, lime mud generally tends to transform to a fine, stable fabric of equant grains, so that grains with unfavorable orientation and smaller size are eliminated. This diagenetic process is complex and not yet fully understood, fineness of grain being the main obstacle to progress based on the petrographic microscope. But without doubt micrite commonly becomes coarser grained with age. This alteration product has been called **microspar,** although the term **recrystallized calcite** has also been used (Pls. 5–3, 72, 94). Although size alone is not its only defining criterion, microspar has been defined as having diameters between 5 and 20 microns (CHILINGAR *et al.*, 1967, Table 3). STAUFFER (1962, p. 361–363) and FOLK (1965, p. 37–40) gave other criteria for its recognition. FOLK (1965) also furnished a detailed classification of the recrystallization fabrics of lime mud and

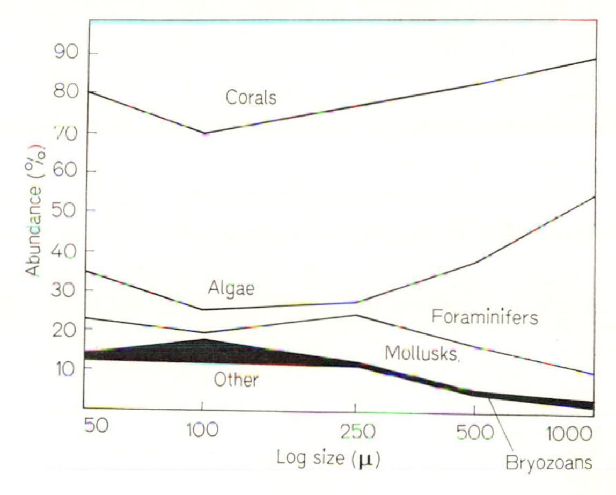

Fig. 2. Changing compositional abundance with grain size in an unconsolidated Recent sand, Andros Island, Bahamas (GOLDMAN, 1926, Tables 3 and 4).

proposed the term **neomorphism** to include such processes as inversion (aragonite→calcite), recrystallization (calcite→calcite), and strain recrystallization (strained→unstrained calcite). Figure 5 of FOLK (1965) presents a detailed code for description. CHILINGAR *et al.* (1967) also provided a valuable discussion of the diagenetic processes that transform lime mud into microspar, which can be confused with some fine-grained, chemically precipitated cements. When microspar is present, it should be tallied separately from micrite.

Cement.—Chemically precipitated cement may partially or completely fill the pore system of a sedimentary rock. Calcite is the chief cementing agent of carbonate rocks. Evaporitic minerals such as gypsum, anhyrite, and halite are mostly minor cementing agents. Quartz and chert are very rare cements except where the entire rock has been silicified. Unlike micrite, car-

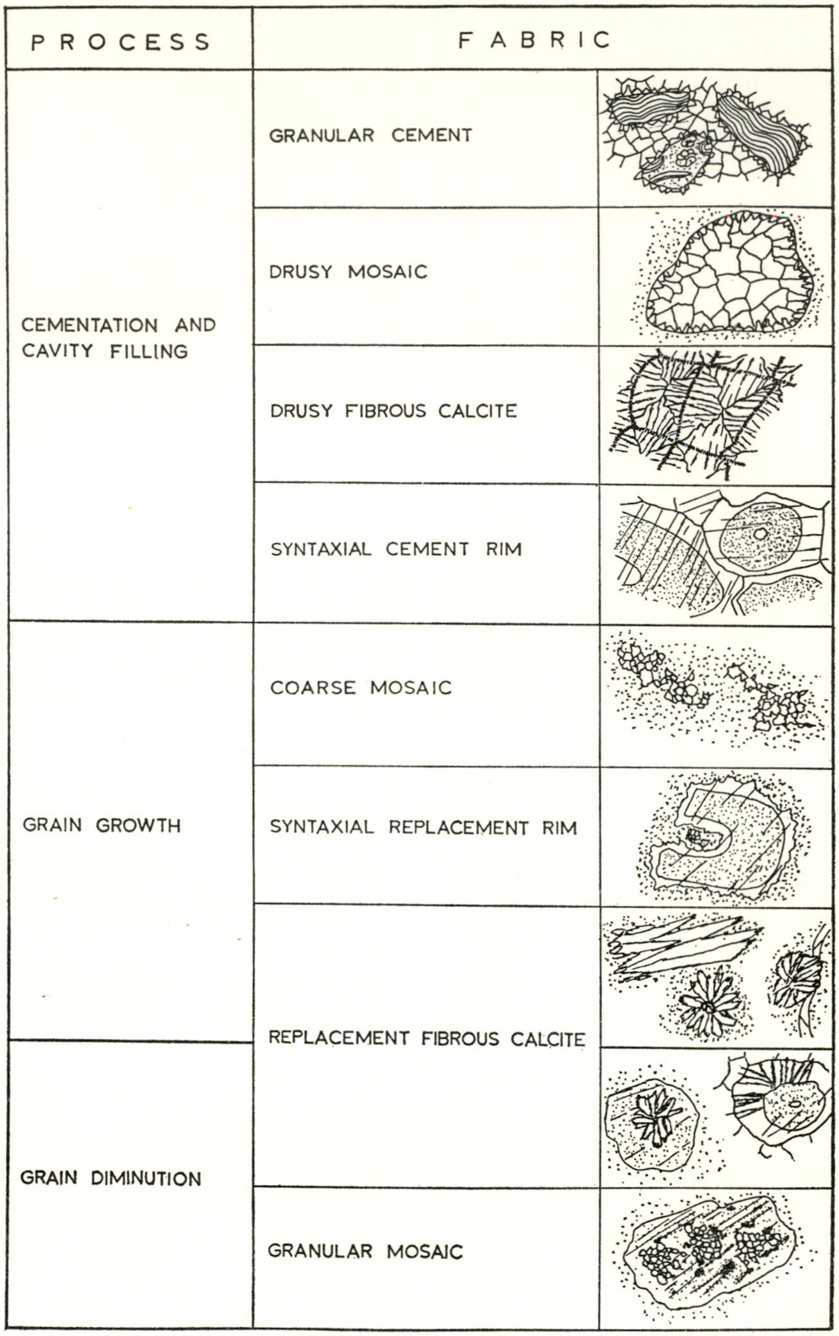

PROCESS	FABRIC
CEMENTATION AND CAVITY FILLING	GRANULAR CEMENT
	DRUSY MOSAIC
	DRUSY FIBROUS CALCITE
	SYNTAXIAL CEMENT RIM
GRAIN GROWTH	COARSE MOSAIC
	SYNTAXIAL REPLACEMENT RIM
	REPLACEMENT FIBROUS CALCITE
GRAIN DIMINUTION	
	GRANULAR MOSAIC

Fig. 3. Terminology of diagenetic fabrics (ORME and BROWN, 1963, Fig. 1).

bonate cement is clear and crystalline and may grow in optical continuity with some of the framework grains, particularly echinodermal plates and spines that are themselves single crystals (Pls. 10–2, 83). The clear, coarse-grained aspect of most of the carbonate cements has evoked the term **sparry calcite** (FOLK, 1959, p. 8) or simply **sparite** (CHILINGAR *et al.*, 1967, p. 320). As here used, sparry calcite is a descriptive term for translucent or transparent crystalline calcite or aragonite cement. Some petrographers use the term dolosparite for dolomitic cements of comparable texture.

Bathurst (1958) was the first to stress the textural complexities of sparry calcite and to recognize its many different fabrics and significance for the diagenetic history of a limestone. Later contributions have been made by Stauffer (1962), Orme and Brown (1963), Bathurst (1964), and Chilingar *et al.* (1967). Terminology as yet is not standardized so that the same fabric may be described by different terms. We find the terminology of Figure 3 helpful.

Granular cement (Pls. 24–1, 64, 80, 89) grows between framework elements, whereas **drusy cement** (Pls. 17–3, 33–4) fills cavities such as a brachiopod or pelecypod shell floored by mud. A "roof" formed by a shell may prevent complete infilling by mud, so that the resulting upper part of the void is later infilled by drusy cement. The resulting **geopetal** structure permits the determination of the up direction in a slide (Pls. 25–2, 29–2, 32–6, 61–5). Both granular and drusy cements generally have in common small clearly elongate or acicular crystals that rim the voids. Such crystals have grown on and with their c-axis perpendicular to a solid surface. Consequently, under crossed nicols one sees "marching men" pass the north-south cross hair as individual crystals go to extinction (Pl. 33–4). Away from the boundary, crystal size increases, and crystals become equant and anhedral so that the term **blocky cement** (Pls. 33–1, 45–5, 59–4) has been proposed (Dunham, 1969, p. 141). Drusy cement may also be composed of thin and acicular crystals so they appear to be superficially fibrous. Overgrowths in optical continuity with the host, are called **syntaxial rim cement** (Pls. 10–2, 50–3, 83).

All of the above terms describe cementation in the pore system. Unfortunately, diagenesis also produces spar that has recrystallized from finer grained micrite. **Coarse mosaic** and **syntaxial replacement rim** describe the resultant product (Fig. 3), although the term microspar is also used in a more general way. Strong evidence for a replacement origin includes: 1) regular grain size of the mosaic such that a systematic variation from small to large crystal size is missing, 2) crystals that truncate primary textures such as bedding and fossils, 3) "ghosts" of detrital grains in the mosaic, and 4) a granoblastic texture. The discussion by Stauffer (1962, p. 361–363) of criteria for recrystallized versus sparry calcite is still one of the best.

Grain diminution, shown by partially unreduced hosts and ghosts, is not so common.

The reader may well ask, "What good is all this terminology?" We answer by saying that because carbonates are more susceptible to diagenesis than terrigenous sandstones, they probably are capable of telling us more about their post-depositional history, for example, their proximity to later unconformities. Should a diagenetic history be fairly systematic and uniformly affect an appreciable volume of rock, its understanding and successful prediction might help us locate a stratigraphic trap. Certainly, the origin and distribution of the pore system of a carbonate rock is nearly always closely related to its diagenetic history (Choquette and Pray, 1970). As another example of the importance of cement and its fabric, Friedman (1969a) has suggested that carbonate cements possibly might tell us whether or not cementation was submarine or subaerial. Purdy's (1968) discussion of subsea, subaerial and subsurface diagenesis is an excellent, well-illustrated summary.

Types of Framework Grains

Inorganic and skeletal detrital grains, *in situ* organic structures, pellets, lumps, and coated grains are the six major types of framework grains in limestones (Table 2). Inorganic detrital grains may be intraclasts, partially or fully lithified lime mud eroded from nearby and redeposited, rock fragments of sand size carbonate and other terrigenous debris such as quartz or feldspar, and possibly a few other minerals that were ultimately derived from outside of the basin (Pl. 30–1, 30–3, 71). Sand size carbonate rock fragments derived from outside the basin are unusual and commonly only occur either in semiarid or arid climates or tectonic areas of high relief regardless of climate. If these conditions prevailed, they can form a significant fraction of terrig-

Table 2. *Grain Types in Limestones* (Modified from Leighton and Pendexter, 1962, Table 1)

Major Groups	Examples and Remarks
1. Detrital Inorganic Grains	Intraclasts, lithified rock fragments, quartz, feldspar, volcanic debris and heavy minerals
2. Detrital Skeletal Grains	Crinoids, mollusks, algae, brachiopods, foraminifers, trilobites, ostracodes, etc. May be whole or fragmented. Commonly transported after death
3. *In Situ* Skeletons	Corals, bryozoans, and algae, generally colonial or encrusting forms
4. Pellets	Probably mostly faecal material
5. Lumps	Algae lumps and composites such as grapestone
6. Coated Grains	Oolites, superficial oolites, pisolites plus algae and foraminiferal encrusted grains

enous sands. FOLK (1959, p. 4 and 7) proposed the terms **allochems** for transported grains that formed within the basin of deposition and **orthochems** for the carbonate that precipitated in a rock after its deposition.

Skeletal grains are fossil debris that may be whole or fragmental; commonly, the genetic interpretation of a limestone largely hinges on their recognition and on their associated macrofossils as well. Fragmental ***in situ* organic builders** such as corals and algae, differ from skeletal grains only in that they have not been transported and reworked by currents.

Pellets are silt-sized grains of micrite that are commonly ovoid in shape and lack internal laminations (Pls. 23–6, 54, 100). Perhaps the phrase, "well sorted, ellipsoidal, structureless grains of micrite" best describes them. Most are probably of faecal origin, but distinguishing some faecal pellets from small intraclasts can be difficult. FOLK (1962, p. 65) suggested that the two be separated arbitrarily by size, those smaller than 150 microns designated as pellets and those larger termed intraclasts.

Lumps pose another problem of definition. According to LEIGHTON and PENDEXTER (1962, p. 36), lumps are composite grains having surficial irregularities and believed to have formed by a process of aggregation on the sea floor and not in an animal's digestive system. The textural similarity between the lump and its associated sediment may aid their identification. Clear-cut operational definitions of lumps, pellets and intraclasts, definitions that are applicable throughout the geologic column and widely accepted by most petrographers, have not yet been formulated. Fortunately, lumps and intraclasts are commonly only minor constituents of most limestones, even though limestones formed almost entirely of lumps or intraclasts are known. Pelleted muds, however, are fairly widespread in the geologic column.

Coated grains are also fairly common and may form around any type of nucleus (Pls. 8–1, 51–4). **Ooliths** are spherical to elliptical grains less than 2 mm in size that consist of two or more coated layers. They may display concentric or radial structure (Pls. 40–2, 98–2). If the coats are less than the radius of the nucleus, the term **superficial oolith** (Pls. 51–4, 51–5) is used (LEIGHTON and PENDEXTER, 1962, p. 60). In practice, however, it may be simpler to define arbitrarily a superficial oolith as one having a thin coat, say less than 50 microns. Pisoliths differ from ooliths only in size, being larger than 2 mm.

Studying the Thin Section and Recording the Data

Given an understanding of and the ability to recognize the preceding textural components how does one start the study of a thin section or peel? One first scans it at low power (25×), obtains an approximate idea of its components and texture and then uses medium (80×) or high power (120×) to help identify all the framework elements (for example, foraminifers) and unravel all the details of its diagenetic history.

Qualitative observation is a vital element of all petrographic study as can be fully appreciated by consideration of the following questions. What kinds of components are present? Are coated grains present? Are the coated grains all of one type and size or not? Are post-diagenetic replacements selective as, for example, secondary pyrite that uses a specific host, for example, brachiopods? Is there evidence that some of the fossil debris has been reworked from another environment—perhaps mud-filled gastropods in a well-washed skeletal limestone cemented by sparry calcite (Pls. 24–1, 81, 91)? Is the micrite recrystallized? Has it replaced or attacked framework elements? To what extent has the original skeleton been modified (Pls. 39–3, 41–3, 43–1)? Is there a characteristic sequence of mineralogical replacements and cements? Many more questions such as these can be asked and practically all depend on qualitative observation. And the more astute the observer, the better he can answer them! Perceptive, qualitative observation can aid greatly in deciphering the diagenetic history of a rock as was clearly shown by DUNHAM's (1969) informative study of the distribution and significance of carbonate silt in a linear mound of Permian age in the Paradox Basin of New Mexico.

Certainly, we highly recommend careful observation as a prelude to quantitative point counts. But point counting is becoming more commonplace and is essential for later quantitative analysis of the data, possibly by computer. Along with point counting, careful attention should also be given to the mean size and range of individual components. This is commonly rewarding.

A standardized format for recording the data is essential. Table 3 gives a simple example of such a format. Here description and interpretation are clearly separated. But for most practical problems one can be much more specific and formulate a standardized data sheet that is specifically tailored to both the problem and tools one has in hand to solve it. An example of such a form has been published by KLOVAN (1964) in his study of the Devonian reefs in western Canada (Table 4). The objectives and rocks determine the format of the data sheet. One needs to reconnoiter fully a carbonate sequence before making such a data sheet. With some thought it may be possible to record the data directly on punch cards for subsequent computer analysis.

Table 3. *A Petrographic Report for Carbonates*

Name and Location	
Hand Specimen	
Color, grain size, induration, bedding, fossils (if any), and field name	
Thin Section Description	
Abstract:	Digest and condense all the petrography into 25 well-chosen words including rock name
Texture:	Broken or intact fossils plus DUNHAM's classification: packstone, wackestone, etc. Grain size and percent framework, cement, micrite, stylolites, and grain to grain relations such as solution packing. Fabric and cement. Amount and kind of porosity
Composition:	Kinds and amounts of allo- and orthochems plus terrigenous debris. Nature of cements and micrite
Interpretation	Draws on all of the preceding plus relations to associated sediments (Table 4). Petrographic name. Geomorphology and hydrography of depositional basin plus salinity, turbulence, and salient biology. Diagenesis

Classification

Classification of carbonate rocks, particularly limestones, has been a popular subject in the last 10 years, so much so that an entire memoir "Classification of Carbonate Rocks" edited by HAM (1962) is devoted to the subject. Subsequently, BISSELL and CHILINGAR (1967) have also reviewed the classification problem. MÜLLER-JUNGBLUTH and TOSCHEK (1969) have published a useful collection of classifications. Although the classification of carbonate rocks involves four fundamental building blocks (framework, cement, mud, and pores), they can be considered and arranged in many different ways so that the number of classifications are potentially limitless. Our purpose is not to recommend a particular classification but simply to point out to the beginner the steps needed to assign a name to a rock and briefly comment on its possible genetic significance.

One should first describe the over-all texture and then the composition of the framework elements of the rock. Thus an appraisal of its grain size should be made

Table 4. *Data Recording Sheet of* KLOVAN (1964, Table 6)

Well:	Unit Subsurface:	Thickness:
K.B.	Unit Subsea:	% of Unit

I. External Features

A–1 *Lithology*
- 1 pure ls.
- 2 sl. arg. ls.
- 3 v. arg. ls.
- 4 calc. sh.
- 5 dolomitic ls.
- 6 Dolomite

B–2 *Color*
- 7 white
- 8 grey
- 9 cream
- 10 buff
- 11 lt. brown
- 12 med. brown
- 13 dk. brown
- 14 black
- 15 green

C–3. *Impurities*
- 16 grn. sh. inclusions
- 17 blk. mud inclusions
- 18 arg. partings
- 19 dissem. arg.
- 20 18 and 19
- 21 no impurities

D–4. *Gross Texture*
- 22 laminite
- 23 sublithographic
- 24 calcarenite
- 25 bded. calcarenite
- 26 frag. (poikolitic)
- 27 crystalline
- 28 lam. shale
- 29 non-lam. shale
- 30 organically bound

II. Secondary Features

E–5. *Porosity*
- 31 poor
- 32 good
- 33 excellent

F–6. *Vugs*
- 34 none
- 35 few
- 36 abundant

G–7. *Stylolites*
- 37 none
- 38 grn. sh. residue
- 39 dk. sh. residue

III. Internal Features

H–8 *Mean Grain Size*
- 40 siltite
- 41 v. f. arenite
- 42 f. arenite
- 43 med. arenite
- 44 crse. arenite
- 45 v. crse. arenite

I *Nature of Grains*

		ab	r	np*
I–9	skeletal	46	47	48
I–10	nonskeletal	49	50	51
I–11	calcisphrs	52	53	54
I–12	crystals	55	56	57

J *Nature of > 2 mm Comp.*

		ab	r	np
J–13	skeletal	58	59	60
J–14	oncolites	61	62	63
J–15	lithoclasts	64	65	66

K–16 *Rounding of Grains*
- 67 poor
- 68 fair
- 69 well

L–17 *Packing of Grains*
- 70 in contact
- 71 floating

M–18 *Packing of > 2 mm Comp.*
- 72 in contact
- 73 floating

N–19 *Abundance of > 2 mm Comp.*
- 74 < 5%
- 75 > 5%

O–20 *Spar—Mud Ratio*
- 76 spar < mud
- 77 spar > mud

P–21 *Grain—Groundmass Ratio*
- 78 Grns : groundmass 1 : 9
- 79 Grns : groundmass 1 : 9–1 : 1
- 80 Grns : groundmass 1 : 1–9 : 1
- 81 Grns : groundmass 9 : 1

IV. Paleontologic Data

Q–22 *Est. % foss > 2 mm*
- 82 < 5
- 83 5–10
- 84 10–20
- 85 20–40
- 86 40–60
- 87 > 60

R–23 *Nature of Fossils*
- 88 unworn
- 89 sl. worn or frag.
- 90 well rounded

S–24 *Foss. Distribution*
- 91 scattered
- 92 in zones

V. Relative Abundance of Fossil Types

		ab	c	r	np
T–25	massive stroms	93	94	95	96
T–26	tabular stroms	97	98	99	100
T–27	Stachyodes	101	102	103	104
T–28	Amphipora	105	106	107	108
T–29	Solitary Rugosa	109	110	111	112
T–30	Colonial Rugosa	113	114	115	116
T–31	Tabulates	117	118	119	120
T–32	Brachiopods	121	122	123	124
T–33	Gastropods	125	126	127	128
T–34	Megalodon	129	130	131	132
T–35	Algal heads	133	134	135	136
T–36	sl. Algal coating	137	138	139	140
T–37	strong Algal coating	141	142	143	144
T–38	Algal-Strom. consortium	145	146	147	148
T–39	crinoids	149	150	151	152
T–40	forams	153	154	155	156

* **Note:** ab abundant, c common, r rare, np not present.

(Fig. 4). Secondly, one should consider the ratio of framework to mud. Is the rock mud or framework supported? If the latter, is it grain supported or does it have an *in situ* organic framework? Answers to these questions permit one to use the classification of DUNHAM (Table 1). His illustrations should be consulted to obtain an idea of how mud versus grain supported fabrics appear in thin section. DUNHAM uses mud to infer an approximate measure of current strength—the stronger the current the less likely mud could have been trapped between the framework. Most petrographers follow this interpretation. The third step is to specify the composition of the framework. What predominates? What are important accessory framework elements?

In assigning a name to a rock one lists first textural and then compositional adjectives as for example, "a medium-grained, echinodermal, brachiopodal grainstone" or a "coarse-grained, rudistid boundstone". One lists the major framework constituent before lesser ones.

Other classifications of limestones, such as those by FOLK (1962), LEIGHTON and PENDEXTER (1962) or BISSELL and CHILINGAR (1967), use terms other than wacke, grainstone and boundstone, but all have in common recognition of the importance of the framework-mud ratio and all give it the same "current strength" interpretation. Absence of mud commonly means infilling by later sparry calcite so that the term "sparite" means also washed, carbonate rock that is cemented by spar.

We recommend that the reader consider some of the above classifications and select the one that he prefers. But we also would like to remind him that detailed classifications are not necessary in order to understand and describe carbonates because simple, well-chosen descriptions will suffice for many problems. Moreover, variation of framework composition is commonly the essential clue to outlining the shape and other characteristics of a carbonate sand or reefal body as well as perhaps the basin in which they accumulated. Or one may be content to simply record the percentages of components.

Grade Size Scales (mm)	Wentworth (1922)	Leighton and Pendexter (1962)
8.0	Pebble Gravel	Breccias and Conglomerates
4.0	Granule Gravel	
2.0	Very Coarse Sand	Very Coarse Grained
1.0	Coarse Sand	Coarse Grained
0.5	Medium Sand	Medium Grained
0.25	Fine Sand	Fine Grained
0.12	Very Fine Sand	Very Fine Grained
0.06	Silt	Micro-Grained (Coarsely / Finely)
0.03		
0.016		
0.008		
0.004	Clay	Crypto-Grained
0.002		
0.001		

Fig. 4. Grade scales for sedimentary rocks (LEIGHTON and PENDEXTER, 1962, Fig. 3, with permission of the American Association of Petroleum Geologists).

The term micrite was proposed by FOLK (1959, p. 26) for a carbonate rock consisting mostly of lime mud. This is approximately the equivalent of DUNHAM's mudstone or of the earlier term calcilutite. Two types of micritic limestones are recognized—**microclastic** and **microcrystalline** (LEIGHTON and PENDEXTER, 1962, p. 37). Microclastic micrites have a dull luster, are muddy in appearance and may contain silt-sized quartz and some scattered fossil debris, whereas the microcrystalline ones have a high luster and a fabric of tightly interlocked calcite crystals. A thin section generally is needed to distinguish these two, especially at first. The electron microscope has greatly advanced our knowledge of the texture of micritic limestones (Fischer *et al.*, 1967; Flügel *et al.*, 1968).

Looking for "petrographic end members" is another aspect of the classification problem. If well-chosen, end members account for most of the petrographic variation. Their selection may be done subjectively, or factor or cluster analysis may be used. Cluster analysis (Bonham-Carter, 1965; Parks, 1966; Wishart, 1969) assigns groups of either variables or samples into hierarchial families. Factor or vector analysis (Imbrie and Purdy, 1962; Imbrie and Van Andel, 1964) aspires to answer four questions: 1) how many end members are present, 2) what is their composition, 3) what proportions of each sample belong to each end member, 4) and how much of the observed variance can be "explained" by each factor?

Mapping and the Fundamental Characteristics of Carbonate Bodies

Much of the focus of carbonate petrology is directed toward determining the depositional environment of a carbonate unit and discerning its subsequent diagenetic history, because together depositional environment and diagenesis determine reservoir characteristics—the kind, distribution, and amount of porosity.

The key to successful understanding of carbonate as well as in terrigenous sandstone bodies is systematic mapping—in the field or in the subsurface be it with the naked eye, a hand lens, a binocular or a petrographic microscope. Usually a combination is required for best results. The phrase, "Map it and it will come out right", perhaps sums up better than any other expression our own viewpoint. Some carbonate lithologies and their associated fossils may be uniquely tied to a single environment of deposition and thus ideally capable of identification in a hand specimen or even perhaps a thin section. In most cases, however, the most definitive and meaningful judgments require more information: the shape of the deposit, its size, its composition, its position in the basin and especially its relation to its bounding lithologies—those that proceeded and followed it in time as well as those that were seaward and landward of it (Table 5). These five factors are as important to the study of terrigenous sand bodies as they are for carbonate bodies. Indeed, they provide a comprehensive format for asking questions about any type of sedimentary body. And if we have sufficient information about them, we can nearly always obtain a reasonable interpretation. *Only careful systematic mapping provides this information.* Difficulties arise when only information on perhaps one or two of the factors rather than all five are available. Because we do not have information on these five fundamental characteristics, we have not attempted the interpretation of the limestones illustrated in our plates.

Because the diversity of carbonate debris is much greater than terrigenous debris, a carbonate body as a rule displays much greater variation in composition of its framework grains and skeletal material than do terrigenous sands. Secondly, because it is easier to distinguish by geophysical logs in the subsurface a terrigenous sand body from shale than to distinguish a carbonate sand from a micritic limestone, our knowledge of the geometry of carbonate sands is

Table 5. *Fundamental Aspects of Carbonate Bodies*

Geometry
Ribbons, sheets and mounds (linear, planar and circular map patterns)

Size
Length, width, and thickness

Composition
Mineralogy, petrology, megafossils, sedimentary structures and pore system

Internal Organization
Distribution of compositional features, bedding, and pore space

Position in Basin
Lateral (margin versus center of basin) and vertical (top, middle or bottom of basin)

Bounding Lithologies
Lateral (seaward and landward equivalents) and vertical neighbors (conformable or unconformable)

commonly less than that of terrigenous sands. In short, in most carbonate studies geophysical logs are inadequate to separate a particular limestone from associated limestones whereas most sandstones are treated as single entities, so that one simply need separate them from associated shales and limestones. And we might add that the geometry, size, and bounding lithologies of a carbonate body are as vital to understanding its origin as its composition. As a consequence, petrographic mapping has maximum significance for the carbonates—to delineate a biofacies from its neighbors, to help outline its size and shape, and to determine possible variations within it.

The level of mapping can vary greatly. For example, one might use a mixture of hand specimens, polished sections, and peels studied under the binocular microscope. Or mostly thin sections and only a few peels might be used. Any particular mix depends on what is available, outcrops, cores or cuttings, as well as the scope of the problem. A study of a unit extending over 40,000 km^2 will rely mostly on megascopic features, whereas the zonation of a small bioherm almost certainly requires microscopic study, particularly for the beginner. In almost any study the beginner and even the experienced sedimentologist will probably find a few thin sections useful to more carefully identify the constituents and correlate their appearance in thin section with peel or hand specimen.

LAPORTE's (1967) study of the Manlius Formation (Devonian) exposed for over 200 miles in New York State is a nice example of biofacies mapping employing a mixture of methods. He initially visited 47 outcrops, revisited 24 for more careful sampling and from them prepared 200 4 by 6 inch polished and etched sections. One or more large thin sections were cut from each of these and examined so that finally 12 outcrops were selected for careful petrographic and paleontologic study. This approach —first megascopic, then binocular and finally microscopic examination—not only best correlates thin and hand specimen information but is most efficient because it maximizes one's ability to examine, classify, and map as large an area as possible.

Factor analysis has been used to help identify mapping units with maximum petrographic contrast (GRIFFITH *et al.*, 1969).

Models and Cycles

A sedimentary model has been described as an "intellectual construct" which is commonly based on a prototype (POTTER and PETTIJOHN, 1963, p. 226). A graded bed, a Devonian reef, a carbonate mound, or a large sedimentary basin filled by carbonates may all be represented by models. The model integrates the components, explains how they are arranged with respect to one another, and commonly gives an explanation of the origin of both the arrangement and its fill. A model may be based solely on the ancient, solely on the modern, or on a combination of both. Paleocurrents play a vital role in sedimentary models. In some deep tectonic basins the current system may consist of episodic turbidity flows which transport calcareous debris into deep water. On the other hand, in a broad, shallow lagoon, the current system may be sluggish and supersaline. In either model, currents control oxygenation, transport nutrients, importantly affect the framework-mud ratio of a carbonate sand, the sorting of its framework, and significantly influence, via turbulence, the bottom fauna that contributes to the ultimate skeletal framework.

Of what value are such models? They offer the possibility of glittering universal generalization, improved prediction of lithologic distribution within a three dimensional rock body, and the possibility of improved understanding because they tend to better link process and product. As we already have seen, the model concept provides a good format for asking questions about sedimentation in general. In Table 6 we provide a few broad environmental groups of some carbonate studies that either explicitly contain a model or provide the data from which models are constructed.

EDIE (1958) apparently first published a general carbonate model based largely on his study of cratonic sedimentation in western Canada. His model (Fig. 5) is a

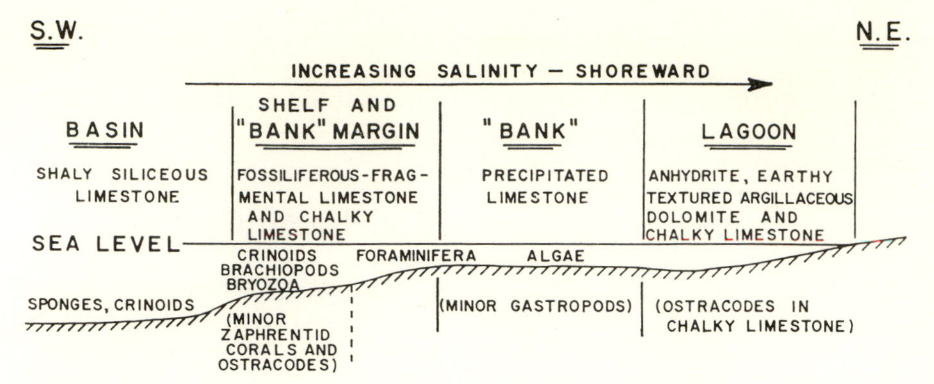

Fig. 5. The general carbonate model of EDIE (1958, Fig. 5, with permission of the American Association of Petroleum Geologists).

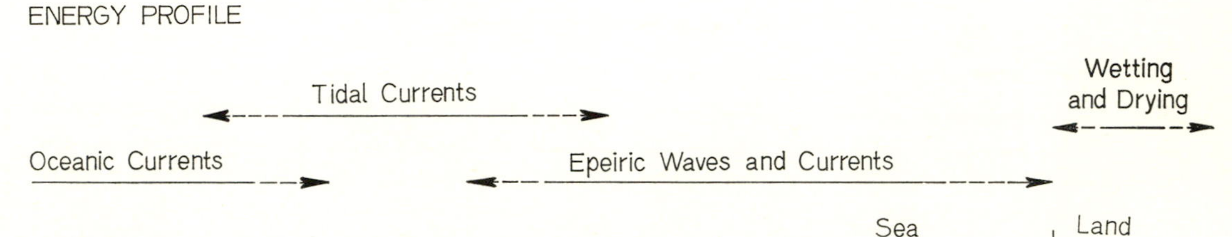

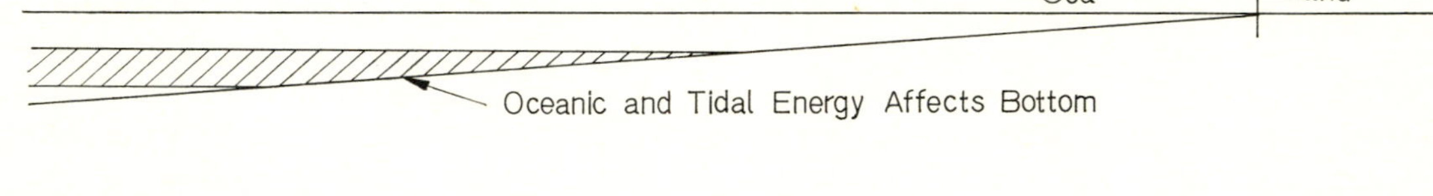

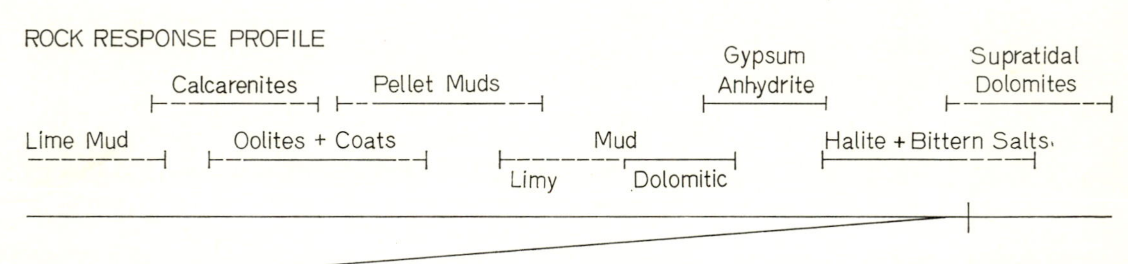

Fig. 6. Simplified energy and rock response profiles across a stable cratonic shelf having no terrigenous input (modified from SHAW, 1964, Figs. 8–1 and 8–2).

cross section, perpendicular to strandline, that shows the different major carbonate environments and their spatial relation to each other. The controlling environmental factor is salinity, which increases shoreward because of progressively restricted circulation. The model best applies to epeiric seas where shelfs are wide (perhaps hundreds of kilometers), have low slope (perhaps less than a decimeter per kilometer) and water is shallow (perhaps 5 meters or less). The essential feature of EDIE's model is that it spatially relates rock types to salinity and by implication permits one to distinguish seaward from landward deposition. It also implies tectonic stability and, broadly speaking, platform sedimentation. Subsequently, SHAW (1964, pp. 42–49) and IRWIN (1965) expanded on EDIE's model based on salinity and current strength (Fig. 6).

EDIE's model may be modified by three important factors: the state of organic evolution, climate, and whether or not the

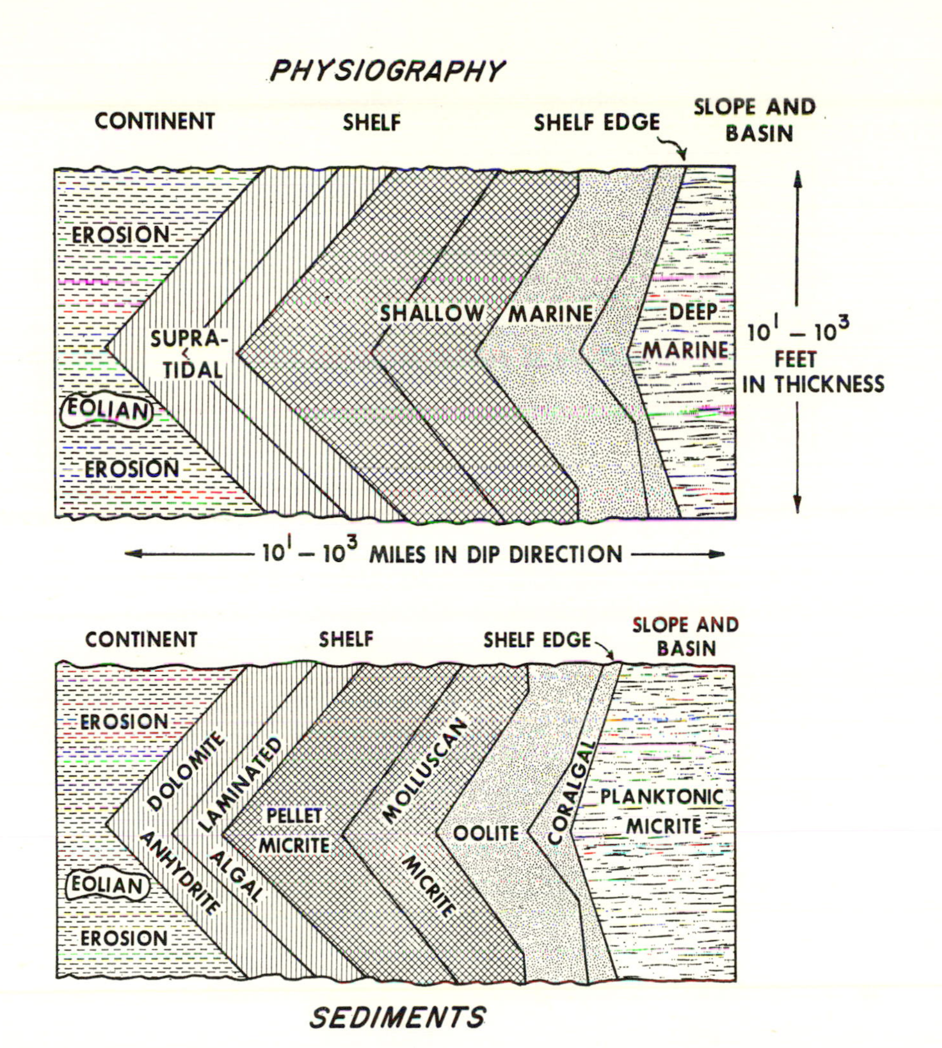

Fig. 7. Idealized distribution of physiography (environments) and sediments perpendicular to the paleoslope on a shelf. Note correlation between the two (COOGAN, 1969, Figs. 2 and 3).

transition from shallow to deepwater is abrupt or gradual.

Organic evolution controls the biotic types of the skeletal framework and, as a consequence, a Cretaceous sequence will show different biotic constituents than a Cambrian one. Climate plays a role because a saline or supersaline lagoon can develop only in an arid or semiarid climate; if there is a wet climate, excess rainfall will keep the lagoon well flushed and, as a consequence, a "primary", supratidal, dolomitic facies will not develop. The nature of the seaward sloping surface is equally important: if there is a sharp, well defined topographic break between shelf and basin, the shelf edge may be marked by a series of either marginal reefs or sands (oolitic and/or skeletal). Occasionally, the latter may supply detritus to turbidity currents that transport calcareous, shallow-water skeletal debris into a nearby deep basin. On the other hand, there may be no sharp shelf edge but instead a gradual and irregular increase in water depth. Under these conditions probably neither deep water turbidites nor a pronounced contrast between shelf facies on the one hand and a slope or basin facies on the other will develop. Lack of pronounced shelf edge may favor development of "random" reefs on many platforms.

Carbonate cycles are closely related to the model concept and are included in a general discussion of cyclothems by DUFF

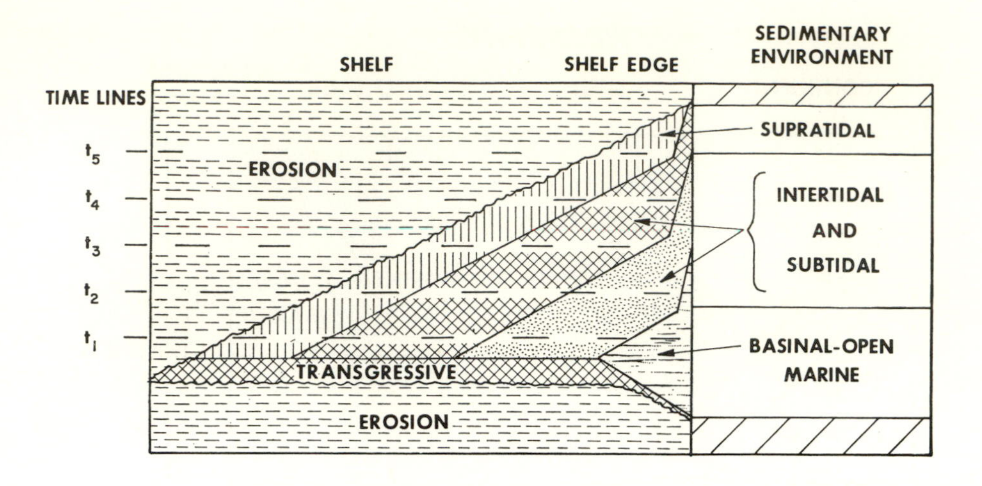

Fig. 8. Cyclic carbonate wedge on shelf showing typical flat base (COOGAN, 1969, Fig. 5).

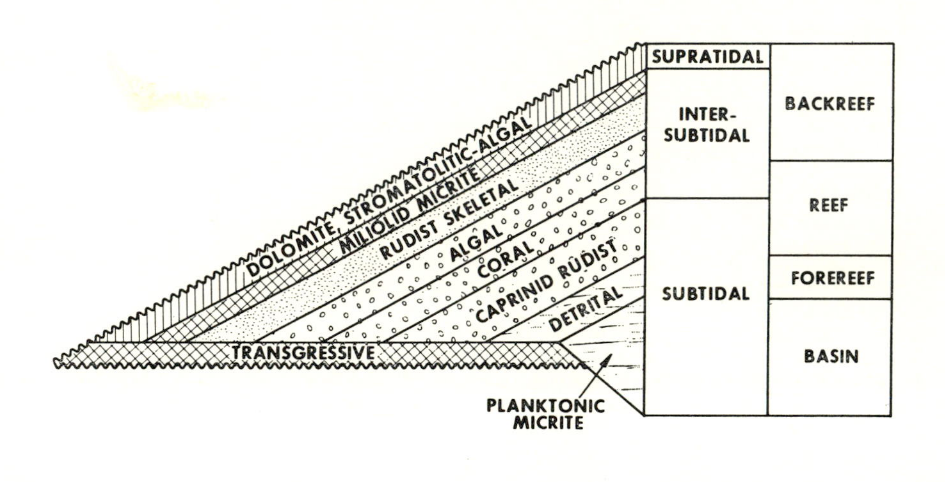

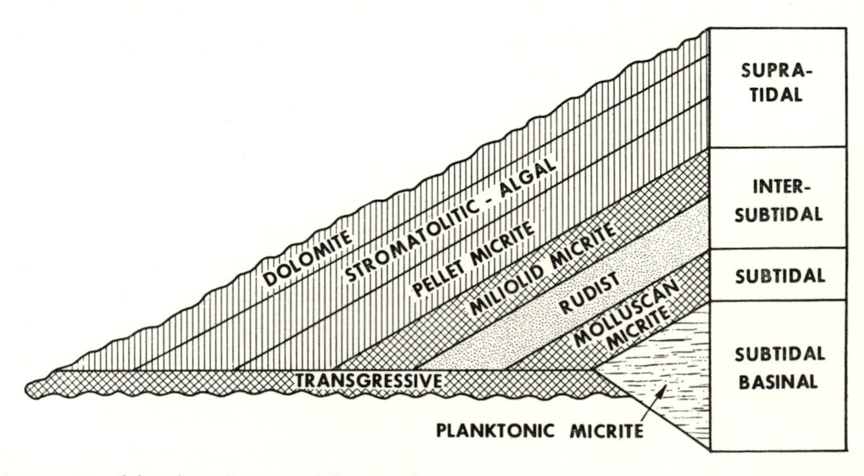

Fig. 9. Cretaceous reef (top) and nonreef (bottom) cycles (COOGAN, 1969, Figs. 7 and 11).

et al. (1967). COOGAN (1969) has perhaps best explained the rationale and use of carbonate cycles. His idealized modern cycle contains lithologies that span the shelf, slope and basin environments (Fig. 7). He considered planktonic mud to be slope and basinal. Above the basal unconformity are the sub- and intertidal coralgal (or reef) and oolitic sands, the molluscan and pellet muds, the supratidal stromatolitic-algae sediments and dolomite or gypsum. The cycle is completed with the progradation of supratidal sediments to the shelf edge and a drop in sea level exposing the shelf (producing a disconformity) or sea level may rise, beginning a new cycle.

According to COOGAN (1969, Fig. 5), a real cycle (Fig. 8) is likely to have a nearly flat base; has a thin, basal transgressive unit; thickens seaward; and is regressive so that it is strongly diachronous. Figure 9 shows two cycles in the Cretaceous, one for a reef and the other nonreef. Both show broad similarity to the modern. These and most other carbonate cycles of sedimentation apparently are a response to eustatic rise and fall of sea level. Although its recognition, documentation and explanation is only beginning in the study of carbonates, the cycle concept appears to be a practical and fruitful one that should find progressively more use in our attempts to interpret carbonate sequences.

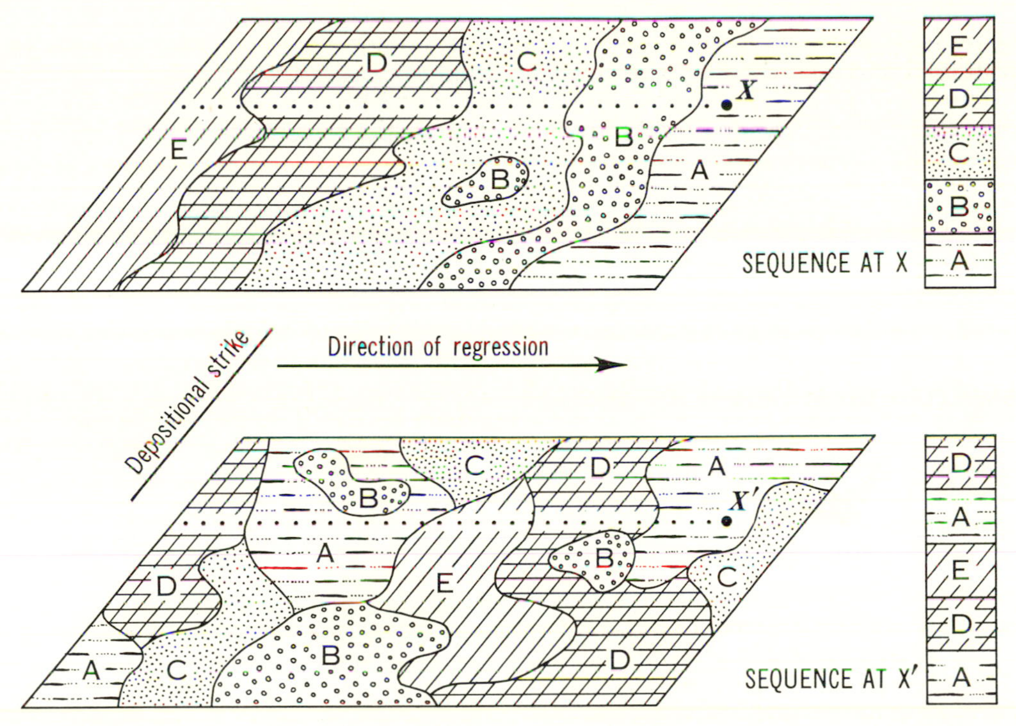

Fig. 10. Regular versus "crazy quilt" distribution of environments and resultant lithologies at X and X′ produced by regression from left to right (modified from POTTER and BLAKELY, 1968, Fig. 2). Orderly distribution of environments produces orderly vertical section but random pattern does not.

WALTHER'S Law (WALTHER, 1894, p.62) is relevant to our discussion of models, especially when models are used to help predict the occurrence of a particular lithology. WALTHER related the lateral migration of environments to vertical change of lithology in a stratigraphic section. He stated that vertical lithologic change in a section resulted from the lateral migration of different environments past it. SHAW (1964, p. 50) and VISHER (1965, pp. 41–42) also stressed the idea that vertical lithologic change at any point in a basin reflects the lateral migration of environments across it (Fig. 10). KAUFFMAN (1967, especially figs. 9–12) has illustrated this type of model in his presentation of rock types and faunal assemblages in the Cretaceous of the Western Interior of the United States. Effective use of WALTHER's Law requires 1) knowing the depositional strike (a line parallel to strandline), 2) remaining within the same cycle, 3) having fairly straight boundaries between the different depositional environments and 4) having environmental boundaries that do not change their orientation greatly during lateral migration. Depositional strike is perhaps the real key when using WALTHER's Law and the model concept because it tells one how to orient the model—a vital ingredient for its successful use.

How Carbonate Petrology Started

We believe an annotated selected bibliography of the literature of petrographic applications is valuable to the beginner, professional and teacher. We have searched the geologic literature for studies of carbonate rocks that have made major use of the petrographic microscope. The majority of the papers in our bibliography have integrated petrography with the other geological aspects of the deposit so that many comprehensive and integrated studies are included. As a consequence, we have excluded some important landmark papers that did not make much use of the petrographic microscope. For example, chemical studies and those that deal only with the megascopic aspects of a carbonate such as biofacies based on megafossils are excluded as are recent applications of the scanning electron microscope.

So that the reader may better appreciate the historical development of petrographic studies of carbonate rocks, we have arranged these references from oldest to youngest. We believe we have included the majority of the outstanding pre-World War II papers in our bibliography with the possible exception of some Russian papers. But for the papers following World War II, we necessarily have been selective and tried to include only the most interesting, informative, and some of the more unusual ones that we believe will provide maximum insight into what has been and can be done by carbonate petrography.

HENRY SORBY* founded carbonate petrography in 1851, when he examined in thin section a calcareous grit and studied its insoluble residue under the microscope. Two years later he published the first modal analysis of a rock, a fossiliferous limestone. In 1879 SORBY published the first comprehensive paper on carbonate petrology. As FOLK (1965) has pointed out, this paper truly establishes SORBY as the father of carbonate petrology. Six years later BORNEMANN (1885) published a comprehensive report including photomicrographs on the Muschelkalk of Germany. This work seems to be the first fully integrated study. The next major milestone was CAYEUX's (1897) report on the Mesozoic and Tertiary chalks and limestones of the Paris Basin. Many years passed before an advance comparable to that made by CAYEUX occurred even though it was but four years later when HOVELACQUE and KILIAN published the first microfacies atlas in 1900. Like most such books theirs deals with the limestones of a particular region and consists mostly of pictures with little accompanying text. In 1904 CULLIS's study of diagenetic textures in the Funafuti atoll appeared; his was the first paper to give major stress to cementation and diagenesis and, along with that of SKEATS (1903), the first to study a vertical profile.

From CULLIS' time until after World War II, carbonate petrology existed but did not prosper, although there were some outstanding studies during the period. Those by GOLDMAN (1926a and 1926b) in America, KAISIN (1927) in Belgium, ZIL'BERMINTS and MASLOV (1928) in Russia, SUJKOWSKI (1931) in Poland, BLACK (1933) in England, PIA (1933) in Germany, CAYEUX (1935) in France, and SANDER (1936) in Austria match in many respects some of the best of today. For example, GOLDMAN's study of a modern reef sand from the Bahamas wherein he made petrographic and chemical

* See selected annotated references for this and other historical references.

analyses of its different size fractions has few duplicates even today. The same could be said of his study of two samples of Mississippian and Ordovician limestones separated by an unconformity in Texas. Pre-World War II petrographic knowledge is summarized by HATCH *et al.* (1938) in the third edition of their "Petrology of Sedimentary Rocks". But in spite of the above and a few other studies, carbonate petrology remained an oddity practiced only by a few. Probably failure to use carbonate petrology in the petroleum industry was the main reason for its lack of popularity among geologists. Sandstone petrology fared better during this period so that even in the early 1900's heavy mineral studies were fairly common and in the late 1930's thin section studies of sandstones had their first serious beginnings and applications.

After World War II the use of carbonate petrology spread so that today it rivals and, in some respects, has more followers and surpasses the accomplishments of sandstone petrology. As FRIEDMAN (1969b, p. 1) pointed out, this is largely the result of the study of carbonate rocks by the petroleum industry, which either trained or supported many outstanding carbonate petrologists of today. Perhaps more important, the petroleum industry found after World War II that carbonate petrology was a discipline that "paid off". Although neither the first nor the last to offer a classification of carbonate rocks, ROBERT FOLK'S 1959 "Practical Petrographic Classification of Limestones" certainly can be taken as a sign that carbonate petrology had arrived and was ready for use in helping to unravel geologic processes and geologic history.

Table 6 classifies some of the selected annotated references according to major environment. Commonly subenvironments are recognized within major ones. We have not attempted to tabulate these subenvironments.

Table 6. *Environmental Grouping of Some Carbonate Studies*

Reefs

Atolls: SKEATS (1903), CULLIS (1904), EMERY *et al.* (1954), SCHLANGER (1964).

Cratons: HENSON (1950), LINK (1950), MASLOV (1950), BONET (1952), BERGENBACK and TERRIERE (1953), NEWELL *et al.* (1953), GINSBURG* (1956), MYERS *et al.* (1956), ANDRICHUK (1958), LECOMPTE (1958), LOZO *et al.* (1959), NEWELL *et al.* (1959), KORNICKER and BOYD* (1962), HOSKINS* (1963), SODERMAN and CAROZZI (1963), KLOVAN (1964), LUCCHI (1964), PITCHER (1964), TEXTORIS and CAROZZI (1964), HARBAUGH *et al.* (1965), NICHOLS (1965), PÜMPIN (1965), WOLF (1965), KREBS (1966), MURRAY (1966), STANLEY (1966), TEXTORIS and CAROZZI (1966), TOOMEY (1966), BONET* (1967), ROSSI (1967), FISCHBUCH (1968), JENIK and LERBEKMO (1968), LANGTON and CHIN (1968), LEAVITT (1968), OTA (1968), DUNHAM (1969b), LOGAN *et al.* (1969), ZANKE (1969)

Platforms (Includes supra- and intertidal, lagoonal and open shelf)
DIXON and VAUGHAN (1911), GILARD (1926), ZIL'BERMINTS and MASLOV (1928), BLACK* (1933), ILLING* (1954), BEALES (1956), EDIE (1958), DAETWYLER and KIDWELL* (1959), ILLING (1959), BEL'SKAGA (1960), FOLK (1962), LAPORTE (1962), STAUFFER (1962), PERKINS (1963), PURDY* (1963), TERRIERE (1963), FOLK and ROBLES* (1964), HARBAUGH and DERMIRMEN (1964), PESZAT (1964), BOSELLINI (1965), MURRAY (1965), PAYTON (1966), LAND *et al.* (1967), LAPORTE (1967), MONTY (1967), ROEHL (1967), VAN ANDEL and VEEVERS (1967), WILSON (1967), MACQUEEN and BAMBER (1968), WILSON (1958), WOBBER (1968), DUNHAM (1969a), KENDALL (1969), KENDALL and SKIPWITH* (1969), VEEVERS (1969)

Slope and Deep Basin
MEISCHNER (1962), COLACICCHI (1967), DOODGE (1967), GARRISON (1967), WILSON (1967b), FRIEDMAN (1969)

* Recent sediments.

Selected Chronologic, Annotated References

SORBY, H. C., 1851, On the microscopical structure of the calcareous grit of the Yorkshire coast: Geol. Soc. London Quart. Jour., v. 7, p. 1–6.

First use of a petrographic microscope to examine limestone in thin section and first use of insoluble residue technique.

— 1853, On the microscopical structure of some British Tertiary and post-Tertiary freshwater marls and limestones (abstract): Geol. Soc. London Quart. Jour., v. 9, p. 344–346.

Probably the first published modal analysis made from thin section study of carbonate rocks.

— 1879, The structure and origin of limestones: Geol. Soc. London Proc., v. 35, p. 56–95, 11 figs, 2 tables.

Decades in advance of his contemporaries, SORBY considered the general structure of various groups of shells, their mineralogy and relation to shell microstructure, their disintegration and replacement. Also described different types of carbonates of all ages in the British Isles. Reading this paper keeps one humble for here—in a single paper—are the foundations of carbonate petrology!

BORNEMANN, J. G., 1885, Beiträge zur Kenntniß des Muschelkalks, insbesondere der Schichtenfolge und der Gesteine des unteren Muschelkalks in Thüringen: Jahrb. Kgl. preuss. geol. Landesanstalt Bergakademie (1885), v. 6, p. 267–321, pls. 7–14, fig. A.

One of the first petrographic studies of a carbonate. Excellent plates with identifications. Six major rock types recognized. Petrography integrated with several vertical profiles. Faunal lists. Discusses classification of Muschelkalk in Thüringen. Classical paper.

CAYEUX, LUCIEN, 1897, Contribution à l'étude micrographique des terrains sédimentaires I. Étude de quelques dépôts siliceux secondaires et tertiaires du Bassin de Paris et de la Belgique II. Craie du Bassin de Paris: Lille, Le Bigot Frères, Imprimeurs-Editeurs, 589 p., 10 pls., 20 figs.

This volume summarizes the author's early work (starting in the 1880's) on the petrographic study of sedimentary rocks. In Part II the Cretaceous chalk is studied in detail concluding with an analysis of depositional environment (chapter 15). Interesting bibliography of 312 items arranged in sequential order. The first systematic petrography. Classical paper.

BLEICHER, M., 1898, Contribution à l'étude lithologique, microscopique et chimique des roches sédimentaires secondaires et tertiaries du Portugal: Communicações da Direcção dos Trabalhos geológicos de Portugal, v. 3, p. 251–289, 7 pls.

Early description of carbonates in Portugal, particularly dolomites. Only one reference.

HOVELACQUE, MAURICE, and KILIAN, C. C. W., 1900, Album de microphotographies des roches sédimentaires: Paris, Gauthier-Villars, Imprimeur-Librarie, 14 p., 69 pls.

Good quality photomicrographs at moderate to low magnifications of Jurassic, Triassic, and Cretaceous limestones of the French calcareous Alps. Virtually no text but two to six photomicrographs per plate, each photograph with a very brief description. Some identifications to species. Appears to be the first microfacies album, although the term is not used.

SKEATS, E. W., 1903, The chemical composition of limestones from upraised coral islands, with notes on their microscopical structures: Bull. Mus. Comp. Zoology (Harvard), v. 42 (Geol. Ser., v. 6), 126 p., 10 figs.

Chemical analyses and plots thereof against depth plus some petrography from a variety of tropical islands. Eight drawings of microscopic views. Good review of earlier studies.

CULLIS, C. G., 1904, The mineralogical changes observed in the cores of the Funifuti borings, *in* The atoll of Funafuti: London, Royal Soc. London, Rept. Coral Reef Comm., p. 392–420, pl. F, figs. 24–69.

An outstanding petrographical inquiry into the diagenesis and cementation of a vertical section based on a diamond drill core. Many beautiful and informative line drawings of thin sections at low to moderate power. Certainly a landmark in carbonate petrography and in the study of diagenesis.

JUKES-BROWN, A. J., and HILL, W., 1904, The Cretaceous rocks of Britain. 3. The Upper Chalk of England: Geol. Survey Great Britain Mem., 566 p., 79 figs.

Mainly a systematic description by counties of stratigraphy and paleontology as well as measured sections and faunal lists. Microscopical petrology, chemical position, hydrology and economic products of the chalk. Bibliography.

GAUB, FRIEDRICH, 1910, Die jurasischen Oolithe der schwäbischen Alb: Geol. u. Palaeont. Abh., n. ser., v. 13, 80 p., 10 pls.

Considers oolites and their nuclei and the Schwabischen Alps in particular. Many references and sixty photomicrographs.

DIXON, E. E. L., and VAUGHAN, ARTHUR, 1911, The Carboniferous succession in Gower (Glamorganshire), with notes on its fauna and conditions of deposition: Geol. Soc. London Quart. Jour., v. 67, p. 477–571, 4 pls., 10 figs., 5 tables.

Minimal petrography but very careful lithologic and faunal description define three sedimentary cycles in dolomitic, lagoonal carbonates and their lateral equivalents. Environmental analysis far ahead of its time. Probably the first paper to develop the model concept, although not explicitly. Compare with EDIE (1958).

Archangelsky, A. D., 1912, Verkhnemelovye otlozheniya vostoka Europeiskoi Rossiĭ (Upper Cretaceous deposits of eastern European Russia): Materali dlya geologiĭ Rossiĭ (Izd. Imperatorskago Mineralogischeskago Obshchestva), v. 25, 631 p., 10 pls., 20 figs.

A comprehensive monographic report in four parts in which petrography plays only a small role. There are 4 plates of photomicrographs with short descriptions. More extended are descriptions of fossils and their paleoecology (paleooceanography) and chronologic zonation. A good example of an early Russian carbonate study. Author aware of relevant non-Russian literature (see also von Bubnoff's long review *in* Neues Jahrb. Mineral. Geol. Palaont., Jahrg. 1914 (1), p. 456–471).

Blanchet, F., 1918, Étude micrographique des calcaires urgoniens: Univ. Grenoble lab. géologie fac. sci. Travaux, v. 9 (1916–1917), p. 29–86, 2 pls., 14 figs.

Four major parts: brief description of 200 thin sections, practical information on how to identify fossils in thin section, and petrographic characteristics of Cretaceous and Jurassic limestones in the Jura region. Largely descriptive.

Gilard, P., 1926, Recherche sur la constitution des craies du Limbourg: Acad. royale de Belgique Mém. cl. sci., 2nd ser., v. 8, fasc. 1, 73 p., 4 pls., 8 figs.

An integrated stratigraphic-petrographic study using chemical analyses, insoluble residues, and thin sections. Suggests a shallow water origin for the chalk but with strand line far away. Sparingly referenced.

Goldman, M. I., 1926a, Petrology of the contact of the Ordovician Ellenburger Limestone and the Mississippian limestone of Boone age in San Saba County, Texas, *in* Mississippian formations of San Saba County, Texas, Pt. IV: U.S. Geol. Survey Prof. Paper 146, p. 44–59, pls. 7–33, 1 fig.

One of the most remarkable carbonate petrographic studies ever made—and based on only 2 specimens! The starting point of the investigation was microstratigraphic—whether to assign a 2 cm layer to the Mississippian or Ordovician. Among topics treated are pelleted limestones, "echinoderm breccia", soft-sediment microfolding, envelopes around calcite fragments, and silicification. Concise interpretation. Excellent plates. Don't pass it by.

Goldman, M. I., 1926b, Proportions of detrital organic calcareous constituents and their chemical alteration in a reef sand from the Bahamas: Carnegie Inst. Washington Pub. 344 (Papers from the Dept. of Marine Biology, v. 23), p. 39–66, 1 fig., 13 tables.

An elaboration of Goldman's earlier 1918 studies in Publication 213, Carnegie Institute of Washington. Fifteen petrographic constituents recognized in different size grades down to fine sand. Chemical composition of constituents for different size fractions unusual. Diagenesis. Relative solubilities of calcium and magnesium carbonates. Careful description and interpretation by the master, the most versatile American sedimentary petrographer of his day, one who was equally at home studying ancient and modern carbonates as well as terrigeneous and chemical sediments.

Kaisin, Felix, 1927, Contribution à l'études des caractères lithologiques et du mode de formation des roches calcaires de Belgique: Acad. royale de Belgique Mém. cl. sci., ser. 2, v. 8, fasc. 5, 118 p., 20 pls., 1 fig.

A monograph in the Cayeux style. Begins with descriptions of Belgium limestones in stratigraphic order. Of most interest are chapters 4, 5, and 6 which contain carbonate petrography, including classification, diagenesis, and stratification. Over 200 references. The high quality plates contain 92 figures. An outstanding study for its day.

Zil'bermints, V. A., and Maslov, V. P., 1928, K litologiĭ kamennougol'nykh izvestnyakov donetskogo basseĭna (On the lithology of the Carboniferous limestones of the Donetz Basin): Moscow, Institut Pribladnoĭ Mineralogiĭ i Metallurgiĭ, Trudy no. 35, Nauchno-Teknicheskoe Upravlenie, no. 222, 215 p., 87 figs., 20 tables.

Systematic study of Carboniferous limestones in Donetz coal basin. Includes megascopic and microscopic examination, classification, and geographic and stratigraphic distribution of four major limestone types. Muddy limestones predominate. Many thin sections individually described and biotic constituents (mostly foraminifers) are identified to generic and specific level. English summary.

Sujkowski, Z., 1931, Petrografja kredy Polski. Kreda z glebokiego wiercenia w Lublinie w porównaniu z kreda niektórych innych obszarów Polski (Étude pétrographique du Crétacé de Pologne. La série de Lublin à sa comparaison avec la craie blanche): Sprawozdania polskiego Instytutu Geologicznego (Bull. Service géol. Pologne), v. 6, p. 483–628, pls. 6–13, 4 figs.

An early systematic petrographic study of a core (855 m) in Cretaceous carbonates. Detailed description of facies and constituents via lithology, thin sections, and study of residues. Has a plot of composition versus depth for all 855 m—the first of its kind? Origin of chalk is prime concern. Strongly influenced by Cayeux and aware of all the important literature of the day. Sixty-five references. Faunal lists. French resumé.

Pia, Julius, 1933, Die rezenten Kalksteine: Zeitschr. für Kristallographie, Mineralogie u. Petrographie. Abt. B: Mineralog. u. petrog. Mitt., n. ser., Ergänzungsband, 420 p., 4 pls., 22 figs., 64 tables.

All aspects of modern limestones: their chemistry, character of precipitating waters, lime muds and lime sands, and ecology of modern, shell-forming animals. Marine and fresh water environments. No thin section petrography but some chemistry. Many references.

BLACK, MAURICE, 1933, The algal sediments of Andros Island, Bahamas: Royal Soc. London Philos. Trans., ser. B., v. 222, p. 165–192, pls. 21–22, 16 figs., 1 table.

Ecology, description of some types of algal mats, discussion of lamination and its relationship to environment. Eight photomicrographs of algal sediments. Early modern carbonate sediment study. Classic.

CAYEUX, LUCIEN, 1935, Les roches sédimentaires de France. Roches carbonatées (calcaires et dolomies): Paris, Masson et Cie, 447 p., 26 pls., 9 figs.

Classic, descriptive petrography of limestones and dolomites. Begins with the modern and then turns to the many different, ancient lithologies. Treatment of lacustrine carbonates unusual. Volume contains the basis for all the later mapping applications. Superb plates. One hundred and one figures on the plates.

SANDER, BRUNO, 1936, Beiträge zur Kenntnis der Anlagerungsgefüge (Rhythmische Kalke und Dolomit aus der Trias); II. Südalpine Beispiele, Hauptdolomit, Allgemeines: Mineralog. u. petrog. Mitt., v. 48, p. 141–209, 10 figs., 16 tables (translated by E. B. KNOPF, 1951, Contributions to the study of depositional fabric: Tulsa, Am. Assoc. Petroleum Geologists).

Primarily concerned with the textural relations (fabrics) of Triassic limestones and dolomites in the Alps and as such is a milestone in the study of carbonates even through little mention is made of fossils. Compare with BOSELLINI (1965).

KRESTOVNIKOV, V., and TEODOROVICH, G. I., 1938, K petrographii paleozoĭskikh otlozheniĭ iuzhnogo urala (Petrography of the Paleozoic deposits of the southern Urals): Leningrad, Neftianyi Geologo-Razvedochnyi Institut, Trudy, ser. A., fasc. 93, 47 p., 8 pls., 13 tables.

Considers classification, grain size, and chemical analysis, along with mineralogical and biological constituents in a predominantly carbonate sequence. Fossil identifications (mostly foraminifers) to generic level.

HATCH, F. H., *et al.*, 1938, The petrology of the sedimentary rocks, 3rd ed.: London, Thomas Murby and Co., 383 p., 75 figs., 16 tables.

Chapter 8 (p. 152–197) treats limestones and dolomites. The first English language text to consider the petrology of carbonate rocks. Revised by M. BLACK, a carbonate specialist. Mostly historical interest.

HADDING, ASSAR, 1941, The pre-Quaternary sedimentary rocks of Sweden VI. Reef limestones: Lunds Univ. Årrsk. N.F., Avd. 2, v. 37, no. 10, 137 p., 88 figs.

General occurrence, form and size, building materials, diagenesis, relation to enclosing sediments and paleogeography of Silurian reefs. Descriptive, qualitative petrography well-illustrated.

CRICKMAY, G. W., 1945, Petrography of limestones, *in* Geology of Lau, Fiji: Bernice P. Bishop Mus. Bull. 181, p. 211–250, 4 figs., 6 tables.

Evaluation of the various fossil groups as rock makers in Cenozoic limestones, some of which are tuffaceous. Twelve line drawings of microscopic views of limestones. Many modal analyses. Diagenesis and cementation.

HENSON, F. R. S., 1950, Cretaceous and Tertiary reef formations and associated sediments in Middle East: Am. Assoc. Petroleum Geologists Bull., v. 34, p. 215–238, 14 figs., 1 table.

Under ideal conditions back-reef, reef and forereef (basinward) facies of a reef complex can be differentiated by petrographic and micropaleontologic criteria recognizable in thin sections and even in cuttings.

LINK, T. A., 1950, Theory of transgressive and regressive reef (bioherm) development and origin of oil: Am. Assoc. Petroleum Geologists Bull., v. 34, p. 263–294, 2 pls., 18 figs.

Not much petrography but essential stratigraphic framework of regression and transgression and their significance for reef development on the craton. Many informative diagrams. Compares Devonian reefs of Alberta to Permian reefs of Texas and New Mexico. Hydrocarbons and porosity within bioherms. "Big picture" thinking very well done.

MASLOV, V. P., 1950, Geologo-litologischeskoe issledovanie rifovich fatsiĭ ufimskogo plato (Geological and lithological investigation of the reef facies of the Ufimsk Plateau): Akad. Nauk SSSR, Institut Geologicheskiĭ Nauk, Trudy no. 118, Geol. ser. 42, 69 p., 24 pls., 26 figs., 1 table.

Distribution and types of biotic constituents of reefal and related Upper Paleozoic carbonates discussed at length. Many photomicrographs have identifications only to major fossil level (e.g. foraminifers). Concludes with a generalized facies map.

TEODOROVICH, G. I., 1950, Litologia karbontnych porod paleozoia Uralo-Volzhskoĭ oblasti (Lithology of Paleozoic carbonate rocks of the Ural-Volga region): Akad. Nauk SSSR, Inst. Nefti, 215 p., 93 figs.

Uses the carbonate rocks of a particular region to write a monograph of general interest: mineralogy, classification, origin, and lithic associations. Many photomicrographs with identified biotic debris. Comprehensive Russian bibliography.

BONET, FEDERICO, 1952, La facies Urgoniana del Cretácico Medio de la región de Tampico: Asoc. Mexicana de Geólogos Petroleros Bol., v. 4, p. 153–262, 50 figs.

Stratigraphy and petrography integrated with paleoecological comparison of coralline and rudistid reefs. Paleogeographic map and cross section summarize results.

BERGENBACK, R. E., and TERRIERE, R. T., 1953, Petrography and petrology of Scurry Reef, Scurry County, Texas: Am. Assoc. Petroleum Geologists Bull., v. 37, p. 1014–1029, 1 pl., 4 figs., 2 tables.

One of the early petrographic papers on reefs. Abundance of reef lithologies tabulated but not mapped.

NEWELL, N. D., *et al.*, 1953, The Permian reef complex of the Guadalupe Mountains region, Texas and New Mexico: San Francisco, W. H. Freeman and Co., 236 p., 32 pls., 85 figs.

Analysis and description from many points of view, well-done in 8 chapters. An early classic.

ILLING, L. V., 1954, Bahaman calcareous sands: Am. Assoc. Petroleum Geologists Bull., v. 38, p. 1–95, 9 pls., 13 figs., 7 tables.

Sedimentation, petrography, mineralogy and texture of modern detritus in the Bahamas. Some chemical data. A post-World War II classic.

EMERY, K. O., *et al.*, 1954, Geology of Bikini and nearby atolls: Part 1, Geology: U. S. Geol. Survey Prof. Paper 260–A, 265 p., 73 pls., 84 figs., 27 tables, 11 charts.

Modern carbonate sedimentation, geomorphology, and drilling to 2,556 feet yield expanded insight into the origin of an atoll famous for its bomb blasts. All limestones from top to bottom of drilled section are believed to be of shallow water origin. Recrystallization of aragonite to calcite in well cemented zones in subsurface is believed due to emergence. Detailed descriptions of cores and cuttings.

BEALES, F. W., 1956, Conditions of deposition of Palliser (Devonian) limestone of southwestern Alberta: Am. Assoc. Petroleum Geologists Bull., v. 40, p. 848–870, 3 pls.

An early petrographic study emphasizing comparison with the Bahamas. Relative abundance and facies pattern of microfacies. Diagenesis.

GINSBURG, R. N., 1956, Environmental relationships of grain size and constituent particles in south Florida carbonate sediments: Am. Assoc. Petroleum Geologists Bull., v. 40, p. 2384–2427, 10 figs., 9 tables.

Relates texture and composition to environment. One of the first of its kind and still instructive. Classic paper.

MYERS, D. A., *et al.*, 1956, Geology of the late Paleozoic Horseshoe atoll in west Texas: Univ. Texas Pub. 5607, 113 p., 18 pls., 10 figs., 10 tables.

Almost all aspects considered in this early study—internal fossil zonation excepted.

ANDRICHUK, J. M., 1958, Stratigraphy and facies analysis of Upper Devonian reefs in Leduc, Stettler and Redwater areas, Alberta: Am. Assoc. Petroleum Geologists Bull., v. 42, p. 1–93, 1 pl., 34 figs.

Little petrography but careful lithologic descriptions and mapping delineate reefs and associated facies. Pioneer paper.

EDIE, R. W., 1958, Mississippian sedimentation and oil fields in southeastern Saskatchewan, *in* GOODMAN, A. J., ed., Jurassic and Carboniferous of western Canada: Tulsa, Am. Assoc. Petroleum Geologists, John Andrew Allen Memorial volume, p. 331–363, 1 pl., 18 figs., 7 tables.

Four major and eight subenvironments recognized. Characteristics of basin versus shelf limestones discussed. First major use of cross section model for carbonates (see, however, the intellectual prototype of DIXON and VAUGHAN, 1911, Fig. 6). One of the best published examples combining petrology and regional stratigraphy for oil finding.

HADDING, ASSAR, 1958, The pre-Quaternary sedimentary rocks of Sweden VII. Cambrian and Ordovician limestones: Lunds Univ. Årssk. N.F., Avd. 2, v. 54, no. 5, 262 p., 193 figs.

Brief treatment of composition, structure, and texture followed by selected examples of Cambrian and Ordovician limestones. Numerous photomicrographs ×20. Mostly Swedish references.

LECOMPTE, MARIUS, 1958, Les récifs paléozoïques en Belgique: Geol. Rundschau, v. 47, p. 384–401, 7 figs.

Careful lithologic description details the internal zoning of a reef and its relation to off-reef sediments. Informative illustrations.

BARNES, V. E., *et al.*, 1959, Stratigraphy of the pre-Simpson Paleozoic subsurface rocks of Texas and southeast New Mexico: Univ. Texas Pub. 5924, 2 v., 836 p., 65 pls., 38 figs., 14 tables.

Comprehensive, very well-documented study of a thick section of Paleozoic carbonates and associated chemical sediments. Many photomicrographs. Petrography by R. L. FOLK.

DAETWYLER, C. C., and KIDWELL, A. L., 1959, The Gulf of Batabano, a modern carbonate basin, *in* Fifth World Petroleum Cong.: New York, Sec. I, paper 1, p. 1–21, 17 figs.

Eight carbonate shallow water facies recognized in a 30 by 20 mile basin. Water circulation more important than depth as a control on sedimentation. One of the first modern sediment studies that *mapped* constituents across a basin. Well-written, relevant description of the modern with the ancient in mind. Classic. Compare with NEWELL *et al.* (1959).

FOLK, R. L., 1959, Practical petrographic classification of limestones: Am. Assoc. Petroleum Geologists Bull., v. 43, p. 1–38, 41 figs., 2 tables.

Although neither the first limestone classification paper nor the last, probably the most significant in terms of its impact on sedimentary petrographers. Certainly a classic and still worth reading.

Illing, L. V., 1959, Deposition and diagenesis of some Upper Palaeozoic carbonate sediments in western Canada, *in* Fifth World Petroleum Cong.: New York, Sec. I, paper 2, p. 23–52, 3 pls., 8 figs.

Based on his experience with Devonian and Mississippian carbonates, author defines an ideal cycle and applies it to the Mississippian of the Moose Dome area. Much of general interest.

Lozo, F. E., *et al.*, 1959, Symposium on Edwards Limestone in central Texas: Univ. Texas Pub. 5905, 235 p., 40 pls., 23 figs., 3 tables.

Stratigraphy and detailed environmental analysis plus fossils as depth indicators and a review of oil production from the Edwards Limestone. Rudistid reef facies are well-described.

Newell, N. D., *et al.*, 1959, Organism communities and bottom facies, Great Bahama Bank: Bull. Am. Mus. Nat. History, v. 117, p. 181–228, pls. 58–69, 17 figs., 6 tables.

Correlates modern biofacies with topography and depositional environment. Good and plentiful illustrations in pioneer paper. Compare with Daetwyler and Kidwell (1959) and Purdy (1963).

Baxter, J. W., 1960, Salem Limestone in southwestern Illinois: Illinois Geol. Survey Circ. 284, 32 p., 3 pls., 6 figs., 3 tables.

A famous Mississippian limestone studied in outcrop in a county bordering the Mississippi River. Stratigraphy, vertical profiles of constituents, 29 chemical analyses, and brief inquiry into depositional history.

Bel'skaya, T. N., 1960, Pozdnedevonskoie more Kuznetskoi kotloviny, istoriia ego razvitiia naselenie i osadki (The late Devonian sea of the Kuznetz Basin, the history of its development, population, and sediments): Akad. Nauk SSSR, Trudy Paleontolocheskogo Institut, v. 82, 184 p., 17 pls., 54 figs., 4 tables.

A well-illustrated stratigraphic, sedimentary structure, petrologic, and paleontologic study of a 125 by 300 km basin. Numerous cross sections and diagrammatic facies maps.

Végh-Neubrandt, Elisabeth, 1960, Petrologische Untersuchung der Obertrias-Bildungen des Gerecsegebirges in Ungarn: Geologica Hungarica, Ser. Geol., v. 12, p. 1–132, 50 figs., 11 tables.

Four chapters including regional setting, sedimentary petrography, paleontology and stratigraphic problems, and analysis of depositional environment (mostly sedimentary cycles). A dominantly limestone-dolomite sequence studied by thin section, x-ray and chemical analysis. Hungarian, German subtitles and extended summary.

Rukhin, L. B., and Rukhina, E. V. L., 1961, Melovye otlozhenia Ferganskoĭ kotloviny (Cretaceous chalk deposits of the Fergana basin): Leningrad, Leningradskiĭ ordena Lenina gosudarstvennyi Universitet, 163 p., 70 figs., 8 tables.

Four well-illustrated chapters include stratigraphy, lithology and petrology, origin, and petroleum significance. Directional structures and paleogeographic maps. One page English summary.

Walpole, R. L., and Carozzi, A. V., 1961, Microfacies study of Rundle Group (Mississippian) of Front Ranges, central Alberta, Canada: Am. Assoc. Petroleum Geologists Bull. v. 45, p. 1810–1846, 1 pl., 25 figs., 6 tables.

Three sections and more than 800 thin sections showed 5 microfacies which are the basis for an ideal cycle. Bathymetry qualitatively estimated.

Folk, R. L., 1962, Petrography and origin of the Silurian Rochester and McKenzie Shales, Morgan County, West Virginia: Jour. Sed. Petrology, v. 32, p. 539–578, 5 pls., 10 figs.

Careful, perceptive, qualitative petrography on thin, unspectacular limestones and their associated shales in the Appalachian geosyncline. A distinctive style of petrography well worth studying.

Ham, W. E., ed., 1962, Classification of carbonate rocks—a symposium: Am. Assoc. Petroleum Geologists Mem. 1, 279 p.

Nine papers (plus a forward) by the leading carbonate petrographers of the day. A good many photomicrographs but texture and porosity stressed more than fossil identification.

Kornicker, L. S., and Boyd, D. W., 1962, Shallow-water geology and environments of Alacran reef complex, Campeche Bank, Mexico: Am. Assoc. Petroleum Geologists Bull., v. 46, p. 640–673, 34 figs., 4 tables.

Prime emphasis on processes and qualitative zonation of frame builders plus sediment contributors and binders. More ecology than petrology.

LaPorte, L. F., 1962, Paleoecology of the Cottonwood Limestone (Permian), northern Mid-Continent: Geol. Soc. America Bull., v. 73, p. 521–544, 4 pls., 5 figs., 4 tables.

Five distinct facies found in a thin (less than 7 feet) platform limestone. Study based on 45 outcrops, 40 chemical analyses, peels, and thin sections. Brief but helpful appendix.

Meischner, Klaus-Dieter, 1962, Rhenaer Kalk und Posidonienkalk im Kulm des nordöstlichen Rheinischen Schiefergebirges und der Kohlenkalk von Schreufa (Eder): Abh. hess. Landesamt Bodenforsch., v. 39, 47 p., 7 pls., 15 figs., 2 tables.

Episodic density currents transported calcareous debris into deep water distal to the terrigenous fill of a flysch basin. One of the earlier turbidite limestone papers.

Stauffer, K. W., 1962, Quantitative petrographic study of Paleozoic carbonate rocks, Caballo Mountains, New Mexico: Jour. Sed. Petrology: v. 32, p. 357–396, 38 figs.

Quantitative petrography of many variables on a single thin bed and on several vertical profiles. Integrated environmental analysis.

Fediaevsky, A., 1963, Méthode d'étude quantitative des microfaciès calcaires: Rev. Micropaléontologie, v. 6, p. 175–182, 3 figs.

Brief summary of petrographic description of limestones with examples of detailed vertical petrographic profiles.

Hoskin, C. M., 1963, Recent carbonate sedimentation on Alacran Reef, Yucatan Mexico: Nat. Acad. Sci., Nat. Res. Council Pub. 1089, 160p., 22 pls., 24 figs., 35 tables.

About 100 square miles of growing reef tract and recent carbonate sediments investigated by size analysis, thin section, x-ray and electron microscopy. Eleven major grain types recognized and comparison made with composition of 10 other reef tracts. Summary of sedimentary processes and a rich source of quantitative data.

Perkins, R. D., 1963, Petrology of the Jeffersonville Limestone (Middle Devonian) of southeastern Indiana: Geol. Soc. America Bull., v. 74, p. 1335–1354, 5 pls., 6 figs.

Qualitative petrography of some 14 sections. Four biozones recognized.

Purdy, E. G., 1963, Recent calcium carbonate facies of the Great Bahama Bank. 1. Petrography and reaction groups. 2. Sedimentary facies: Jour. Geology, v. 71, p. 334–355, 5 pls., 4 figs., 2 tables; p. 472–497, 1 pl., 4 figs., 7 tables.

Delineates major sedimentary provinces in modern sediments on Bahama Bank by systematic petrographic study of sand fraction. Uses reaction groups to reduce data.

Rich, Mark, 1963, Petrographic analysis of Bird Spring Group (Carboniferous-Permian) near Lee Canyon, Clark County, Nevada: Am. Assoc. Petroleum Geologists Bull., v. 47, p. 1657–1681, 5 pls., 2 figs., 3 tables.

Study of a 7,200 foot carbonate section in thin section plus insoluble residues—a vertical study of microfacies.

Soderman, J. W., and Carozzi, A. V., 1963, Petrography of algal bioherms in Burnt Bluff Group (Silurian), Wisconsin: Am. Assoc. Petroleum Geologists Bull., v. 47, p. 1682–1708, 2 pls., 20 figs., 2 tables.

Small, dolomitized bioherms consist of eight microfacies.

Terriere, R. T., 1963, Petrography and environmental analysis of some Pennsylvanian limestones from central Texas: U. S. Geol. Survey Prof. Paper 315-E, p. 79–126, pls. 31–38, 35 figs., 3 tables.

Limestones, a small part of the shaly Upper Pennsylvanian section, belong to four types. Detailed presentation of organic and inorganic constituents. Geologic history.

Bosellini, Alfonso, 1964, Stratigrafia, petrografia, e sedimentologia delle facies carbonatiche al limite permiano-trias nelle Dolomiti occidentali: Mem. Mus. Storia Naturale Venezia Tricentina, Anno 37–38 (1964–1965), v. 15, fasc. 2, 105 p., 57 figs.

Thorough study of the Permian-Triassic boundary in a carbonate sequence. Many photomicrographs. Eleven paragenetic events recognized. English abstract.

Folk, R. L., and Robles, Rogelio, 1964, Carbonate sands of Isla Perez, Alacran Reef complex, Yucatán: Jour. Geology, v. 72, p. 255–292, 4 pls., 19 figs., 2 tables.

Petrography of beach sands with emphasis on composition versus size, texture, and the sorting of carbonate debris. Suggests a "Sorby principle" for sizing of biotic debris.

Füchtbauer, Hans, 1964, Fazies, Porosität und Gasinhalt der Karbonatgesteine des norddeutschen Zechsteins: Zeitschr. deutsch. geol. Gesell. Jahrg., 1962, v. 114, p. 484–531, 3 pls., 19 figs., 3 tables.

Detailed petrography and lithology in a sparingly fossiliferous evaporite dolomite sequence. Algae, foraminifers, and ostracods are the principal fossils.

Harbaugh, J. W., and Demirmen, Ferruh, 1964, Application of factor analysis to petrologic variations of Americus Limestone (Lower Permian), Kansas and Oklahoma: Kansas Geol. Survey Spec. Distrib. Pub. 15, 41 p., 15 figs., 13 tables.

A 1 to 5 foot limestone is sampled with 97 hand specimens at 29 localities along a 300 mile outcrop. Q-mode factor analysis helps digest data. Unusual plots of vertical position of data in bed. Paleogeographic map integrates data. Compare with Toomey (1966).

Klovan, J. E., 1964, Facies analysis of the Redwater reef complex, Alberta, Canada: Bull. Canadian Petroleum Geology, v. 12, p. 1–100, 9 pls., 20 figs., 8 tables.

Geometry (maps and cross-sections) and clear informative petrography of a Devonian reef. Geographic and bathymetric zonation of reef building organisms. Seven lithologic facies recognized and 11 lithologies. Petrographic description forms recommended (Table 6). Sets the style for reef studies in western Canada.

Lucchi, F. R., 1964, Ricerche sedimentologiche sui lembi alloctoni della Val Marecchia: Giornale di Geologia, ser. 2a, v. 32, p. 545–626, 12 pls., 2 figs.

Using 170 thin sections from 12 sections, seven microfacies are recognized in Miocene carbonates in allochthonous blocks of the "argille scagliosa" of the Marecchia valley. Biostromes. Principal biotics are bryozoans, coralline algae, and foraminifers. Excellent plates.

MAMET, BERNARD, 1964, Sédimentologie des faciès "Marbles Noirs" du Paléozoïque Franco-Belge: Inst. royale Sci. Nat. Belgique Mém. 151, 131 p., 5 pls., 54 figs.

Good integration of paleontology and petrography with much quantitative data and 172 references. Five chapters: local and regional setting, petrographic classification, ecology of the black marbles, and chemical aspects.

PASSERINI, PIETRO, 1964, Il Monte Cetona (Provincia di Siena): Bull. Soc. Geol. Italiana, v. 83, p. 219–338, 58 figs.

Stratigraphy, sedimentary structures, microfacies, tectonics and geologic history of the Monte Cetona area from Triassic to present. Many photomicrographs. Some sandstone petrography. English abstract.

PESZAT, CZESLAW, 1964, Litologia jurajskich skal węglanowych między Tokarnią a Chmielnikiem (Lithology of the Jurassic carbonate rocks, southeastern margin of Holy Cross Mts., Poland): Acta Geol. Polonica, v. 14, p. 1–78, 6 pls., 10 figs., 12 tables.

Carbonate sedimentation in an epicontinental basin: chemical and petrographic analysis plus porosity and permeability. Oolitic limestones conspicuous. English summary.

PITCHER, MAX, 1964, Evolution of Chazyan (Ordovician) reefs of eastern United States and Canada: Bull. Canadian Petroleum Geology, v. 12, p. 632–691, 3 pls., 49 figs.

Small reefs and enclosing sediments studied in outcrop with principal emphasis on primary deposition rather than diagenesis. Bryozoans are the major constituents of the reefs. Portable core used to obtain samples from glaciated surfaces.

SCHLANGER, S. O., 1964, Petrology of the limestones of Guam: U. S. Geol. Survey Prof. Paper 403–D, 52 p., 21 pls., 5 figs., 11 tables.

Tertiary reef complex detailed by study of thin-sections, insoluble residues and chemical analyses. Very good photomicrographs. Volcanic debris admixed with skeletal material.

TEXTORIS, D. A., and CAROZZI, A. V., 1964, Petrography and evolution of Niagaran (Silurian) reefs, Indiana: Am. Assoc. Petroleum Geologists Bull., v. 48, p. 397–426, 2 pls., 24 figs.

Eight major microfacies were delimited via 800 thin sections from five selected reefs. Numerous vertical profiles and field relations are integrated into an "ideal" Niagaran reef.

BEHRENS, E. W., 1965, Environment reconstruction for a part of the Glen Rose Limestone, central Texas: Sedimentology, v. 4, p. 65–111, 10 pls., 11 figs., 8 tables.

One hundred and ninety-nine samples of a 50-foot thick limestone were studied and stratigraphically divided into five major facies based on 19 constituents. Facies and isopach maps summarize results. Thin sections, x-ray data, and insoluble residues supplied the data.

BLUCK, B. J., 1965, Sedimentation of Middle Devonian carbonates, southeastern Indiana: Jour. Sed. Petrology, v. 35, p. 656–682, 3 pls., 18 figs., 1 table.

Lacustrine, mudflat, lagoonal and off-shore environments recognized plus several microfacies. Quantitative petrographic profiles plus an interpretive, diagrammatic cross section.

BOSELLINI, ALPHONSO, 1965, Analisi petrografica della "Dolomia Principale" nel Gruppo di Sella (Regione Dolomitica): Mem. Geopaleontogiche, Univ. Ferrara, v. 1, fasc. 2, no. 3, p. 49–112, 5 pls., 37 figs.

A 300 meter section through the dolomite sequence studied by SANDER (1936). Seven petrographic types. Cycles considered intertidal rather than deep water. Good photomicrographs. English summary.

CARSS, B. W., and CAROZZI, A. V., 1965, Petrology of Upper Devonian pelletoidal limestones, Arrow Canyon Range, Clark County, Nevada: Sedimentology, v. 4, p. 197–224, 2 pls., 12 figs., 3 tables.

Twelve hundred feet of Devonian limestones examined by thin and polished sections (plus some x-ray studies) at 2-foot intervals. Seven microfacies were recognized based on eight variables. Two ideal cycles defined. Horizontal interpretation of microfacies is made and related to water depth. Identical style as in HEATH *et al.* (1967).

DEMIRMEN, FERRUH, and HARBAUGH, J. W., 1965, Petrography and origin of Permian McCloud Limestone of northern California: Jour. Sed. Petrology, v. 35, p. 136–154, 15 figs., 1 table.

A lense (100 to about 2500 feet thick and about 24 miles long) with graywackes and tuffs above and below. Small scale sedimentary structures are described plus allochems and depositional history. Geosynclinal setting makes deposit unusual.

HARBAUGH, J. W., *et al.*, 1965, Pennsylvanian marine banks in southeastern Kansas: Geol. Soc. America Assoc. Soc., Field Conf. Guidebook for Ann. Meeting, Kansas City, 54 p., 48 figs.

Stratigraphic and polished sections help define marine banks in the Desmoinesian and Missourian stages (Pennsylvanian) of Kansas. Isopach maps and cross sections. Includes a short, well-illustrated article by J. L. WRAY on algae.

MURRAY, J. W., 1965, Stratigraphy and carbonate petrology of the Waterways Formation, Judy Creek, Alberta Canada: Bull. Canadian Petroleum Geology, v. 13, p. 303–326, 5 pls., 4 figs., 2 tables.

The basin facies that envelops an oil producing, reef-fringed bank in west Alberta is studied by 70 thin sections, x-ray, and organic carbon determinations. See MURRAY (1966).

NICHOLS, R. A. H., 1965, Petrology of a Lower Carboniferous bryozoan limestone and adjacent limestones in North Wales, Great Britain: Jour. Sed. Petrology, v. 35, p. 887–899, 10 figs.

Detailed mapping and sampling of limestones from a small area plus qualitative petrography. Recognizes shoal and flank limestones and their varieties. Bryozoan limestones believed to have formed as prominences on the sea floor.

PEL, J., 1965, Étude du Givetien à sédimentation rythmique de la région de Hotton-Hampteau: Ann. Soc. Géol. Belgique, v. 88, p. 471–521, 10 pls., 2 figs.

Vertical profiles studied by 300 thin sections suggest an ideal cycle. Six megasequences. Clasticity indices help classify faunal variations. Two plates of photomicrographs and eight of vertical profiles. Petrography in the Carozzi style.

PÜMPIN, V. F., 1965, Riffsedimentologische Untersuchungen im Rauracien von St. Ursanne und Umgebung (Zentraler Schweizer Jura): Eclogae geol. Helvetia, v. 58, p. 799–876, 1 pl., 36 figs.

Bioherms (30 to 40 m thick) and associated calcarenites including algal-ball limestones, oolites, and chalky limestones. Numerous fossil lists and some chemical data. English summary.

WOLF, K. H., 1965, Petrogenesis and paleoenvironment of Devonian algal limestones of New South Wales: Sedimentology, v. 4, p. 113–178, 40 figs., 7 tables.

An algal reef complex superbly documented. Informative tables. Full discussion of paleoecology.

WOLF, K. H., and CONOLLY, J. R., 1965, Petrogenesis and paleoenvironment of limestone lenses in Upper Devonian red beds of New South Wales: Palaeogeography, Palaeoclimatology, and Palaeoecology, v. 1, p. 69–111, 18 pls., 6 figs., 4 tables.

Lenses of algal limestones in a fluvial or lacustrine red bed sequence. One of the few petrographic studies of a fresh-water limestone.

KREBS, WOLFGANG, 1966, Der Bau des oberdevonischen Langenaubach-Breitscheider Riffes und seine weitere Entwicklung im Unterkarbon (Rheinisches Schiefergebirge): Abh. Seckenbergischen Natur. Gesell. no. 511, 105 p., 13 pls., 18 figs., 3 tables.

A 1200 meter Upper Devonian reef complex dissected by means of petrology, sedimentary structures, and constituents. Reef complex divided into 10 microfacies (forereef, 4; core, 1; backreef, 5). Figure 74 is an excellent summary diagram. Many drawings of megascopic organic constituents.

MATTHEWS, R. K., 1966, Genesis of Recent lime mud in southern British Honduras: Jour. Sed. Petrology, v. 36, p. 428–454, 8 figs., 5 tables.

Petrographic study of Recent lime mud illustrates the problems of fossil identification with very small detritus. Mineralogical and strontium data.

MURRAY, J. W., 1966, An oil producing reef-fringed carbonate bank in the Upper Devonian Swan Hills Member, Judy Creek, Alberta: Bull. Canadian Petroleum Geology, v. 14, p. 1–103, 23 pls., 12 figs., 2 tables.

Nine major rock types defined by different combinations of organic framebuilders (massive and tabular stromatoporids plus some solenoporid algae and tabulate corals), skeletal and pelletal grains, micrite, sparry calcite, and void space. Tables 1 and 2 nicely define textural elements and rock types. Geometry and areal distribution of biofacies. Compare with KLOVAN (1964).

PAYTON, C. E., 1966, Petrology of the carbonate members of the Swope and Dennis Formations (Pennsylvanian), Missouri and Iowa: Jour. Sed. Petrology, v. 36, p. 576–601, 26 figs., 3 tables.

Integrated study including texture and crossbedding of some 200 miles of outcropping shelf limestones. Statistical analysis (Kendall coefficient of rank correlation) helps digest data gathered from 725 samples that were studied by polished and thin section, x-ray, and acid-insoluble residues. Four paleogeographic maps summarize results.

STANLEY, S. M., 1966, Paleoecology and diagenesis of Key Largo Limestone, Florida: Am. Assoc. Petroleum Geologists Bull., v. 50, p. 1927–1947, 1 pl., 12 figs., 1 table.

A homogeneous Pleistocene reef consists of about 30 percent framebuilders, the rest interstitial calcarenite. Oolites formed behind the reef in shallower water. Diagenetic history emphasized.

TEXTORIS, D. A., and CAROZZI, A. V., 1966, Petrography of a Cayugan (Silurian) stromatolite mound and associated facies (Ohio): Am. Assoc. Petroleum Geologists Bull., v. 50, p. 1375–1388, 1 pl., 4 figs., 2 tables.

Five closely spaced sections and 143 samples studied by polished surface and thin section reveal eight major microfacies. Explicit discussion and use of Walther's Law.

TOOMEY, D. F., 1966, Application of factor analysis to a facies of the Leavenworth Limestone (Pennsylvanian-Virgilian) of Kansas and environs: Kansas Geol. Survey Spec. Distrib. Pub. 27, 28 p., 13 figs., 6 tables.

A thin, widespread limestone is studied in 32 outcrop sections and comprehensively described. Resulting data digested by a variant of factor analysis called *vector analysis*. Compositional cross sections are unusual. A good paper to read for statistical facies mapping.

BISSELL, H. J., and CHILINGAR, G. V., 1967, Classification of sedimentary carbonate rocks, *in* G. V. CHILINGAR *et al.*, eds., Carbonate rocks (Developments in Sedimentology, v. 9A): Amsterdam, Elsevier Publishing Co., p. 87–168, 16 pls., 4 figs., 9 tables.

Compact, well-documented review with 96 photomicrographs, good bibliography and a

useful glossary of petrographic terminology relevant to carbonates.

Bonet, Federico, 1967, Biogeología subsuperficial del arrecife Alacranes, Yucatán: Univ. Nac. Auton. de México, Inst. Geología, Bol. 80, 192 p., 17 pls., 31 figs., 105 tables.

Ecology and sedimentation including unconsolidated sediments and immediately underlying consolidated ones. Diagenesis. Geologic history. Six of the plates contain photomicrographs.

Chilingar G. V., Bissell, H. J., and Wolf, K. H., 1967, Diagenesis of carbonate rocks, *in* G. V. Chilingar *et al.*, eds., Diagenesis in sediments (Developments in Sedimentology, v. 8): Amsterdam, Elsevier Publishing Co., p. 179–322, 24 pls., 21 figs., 12 tables.

Complete discussion with recognition of some 30 different diagenetic processes. Well-described plates illustrate many varieties of biotic debris and their diagenetic transformations. More than 500 references and a helpful glossary. Well-recommended.

Colacicchi, Roberto, 1967, Geologia della Marsica orientale: Geologica Romana, v. 6, p. 189–316, 72 figs.

Primarily a study of a shelf edge in a small part of central Italy (Marsica) from lower Lias to Tertiary. Recognition of turbidite limestones. Many excellent, well-described photomicrographs and some 250 references. Discusses tectonics and concludes by summarizing paleogeography. English summary.

Dooge, J., 1967, The stratigraphy of an upper Devonian carbonate-shale transition between the North and South Ram Rivers of the Canadian Rocky Mountains: Leidse Geol. Mededeel., v. 39, p. 1–53, 56 figs., 7 appendices.

Comprehensive study of a shallow shelf (carbonate) to deep water (shale) transition. Careful field and microscopic descriptions. Depositional history. Outstanding discussion of carbonate model (Chapter 9) and its worldwide application.

D'Argenio, Bruno, 1967, Geologia del Gruppo Taburno-Camposauro (Appennino Campano): Accad. Sci. Fische Matematiche (Napoli), ser. 3a, v. 6, 218 p., 39 figs., 19 pls.

Stratigraphy, tectonics, and geomorphology. Stratigraphy section has much data on sedimentation and microfacies. Table 2 is an unusual reconstruction (in a cross section) of Triassic and Cretaceous environments across central Italy (Adriatic to Mediterranean).

Eto, Joji, 1967, A lithofacies analysis of the lower portion of the Akiyoshi limestone group: Akiyoshi-dai Sci. Mus. Bull. 4, p. 7–42, 5 pls., 11 figs., 2 tables.

Petrography of tuffaceous fossiliferous limestones in a geosynclinal setting. Paleogeographical interpretations. Vertical petrographic profiles. Thirty photomicrographs. English abstract, captions, and summary.

Garrison, R. W., 1967, Pelagic limestones of the Oberalm Beds (Upper Jurassic—Lower Cretaceous), Austrian Alps: Bull. Canadian Petroleum Geology, v. 15, p. 21–49, 4 pls., 8 figs., 3 tables.

Very fine-grained limestones in a linear alpine trough studied by the optical and electron microscopes. Some allodapic or bioturbidite limestones also present.

Heath, C. P., *et al.*, 1967, Petrography of a carbonate transgressive-regressive sequence: The Bird Spring Group (Pennsylvanian), Arrow Canyon Range, Clark County, Nevada: Jour. Sed. Petrology, v. 37, p. 377–400, 10 figs.

A vertical profile (1846 ft.) of some 2095 samples yielded 11 microfacies and 78 complete cycles. Idealized and quartz rich cycles are illustrated. Bathymetric and horizontal interpretation of microfacies. Closely spaced samples are typical of Carozzi style.

Land, L. S., Mackenzie, F. T., and Gould, S. J., 1967, Pleistocene history of Bermuda: Geol. Soc. America Bull., v. 78, p. 993–1006, 4 pls., 5 figs., 4 tables.

Topics of general interest include: distinguishing eolianites from marine biocalcarenites, paleosoils, and diagenesis.

Laporte, L. F., 1967, Carbonate deposition near mean sealevel and resultant facies mosaic: Manlius Formation (Lower Devonian) of New York State: Am. Assoc. Petroleum Geologists Bull., v. 51, p. 73–101, 34 figs., 3 tables.

A study, based on detailed sampling of 24 sections along some 375 miles of outcrop, discriminates three major facies. Degree of submergence and water energy are considered the prime environmental factors that control major lithologies. Good illustrations of lithologies. Nice work on the outcrop.

Monty, C. L. V., 1967, Distribution and structure of Recent stromatolitic algal mats, eastern Andros Island, Bahamas: Ann. Soc. Géol. Belgique, v. 90 (Bull. 3), p. 55–100, 19 pls., 10 figs., 1 table.

The sedimentation, mineralogy and petrology of stromatolitic algal mats in marine intertidal, brackish intertidal, and supratidal settings. Excellent. Compare with Black (1933).

Murray, R. C., and Lucia, F. J., 1967, Cause and control of dolomite distribution by rock selectivity: Geol. Soc. America Bull., v. 78, p. 21–36, 5 pls., 7 figs.

Lithology, sedimentary structures, and petrography of carbonate muds and sands help explain selective dolomitization. Paleoecology plays an important role in clarifying dolomitization.

Roehl, P. O., 1967, Stony Mountain (Ordovician) and Interlake (Silurian) facies analogs of Recent low-energy marine and subaerial carbonates, Bahamas: Am. Assoc. Petroleum Geologists Bull., v. 51, p. 1979–2032, 48 figs.

The ancient appraised by careful comparison with the modern. Infratidal, intertidal, supratidal and subaerial diagenetic terrains are basis for comparing the ancient with modern (in the vicinity of Andros Island). Informative illustrations plus a brief appendix of petrographic types and their petrophysical characteristics. More sedimentation than petrography.

Rossi, Daniele, 1967, Dolomitizzazione delle formazioni anisiche e ladino-carniche delle Dolomiti: Memorie Mus. Tridentino di Sci. Nat. Trento, Anno 29–30, v. 16, fasc. 3, 89 p., 19 pls., 27 tables.

Dolomitization of middle Triassic reefs in the Italian Dolomites. Many chemical analyses and good photomicrographs. English summary (6 p.).

Van Andel, T. H., and Veevers, J. J., 1967, Morphology and sediments of theTimor Sea: Australian Dept. Nat. Devel., Bur. Min. Res., Geology and Geophysics Bull. 83, 173 p., 5 pls., 55 figs., 11 tables, 5 appendices.

An outstanding, comprehensive study of the morphology, structure, and setting of Timor Sea and adjacent areas, including sediments and depositional facies. Petrography includes texture, color, and $CaCO_3$ content as well as study of terrigeneous materials and detailed petrographic consideration of carbonates. A hierarchcial diagram and Q-mode factor analysis are used to find compositional assemblages and their distribution. Sedimentary facies delimited and compared with other modern carbonate basins. Much quantitative data.

Wilson, J. L., 1967a, Carbonate-evaporite cycles in Lower Duperow Formation of Williston Basin: Bull. Canadian Petroleum Geology, v. 15, p. 230–312, 22 pls., 14 figs., 2 tables.

Integrated stratigraphy, petrology, and sedimentation in part of an intracratonic basin. Twelve widespread cycles in lower part of formation studied in detail. Deposition was largely back-reef. Thorough description of the basic cycle and its variant as well as paleontology and environmental summary. Informative plate descriptions. Well-recommended.

Wilson, J. L., 1967b, Cyclic and reciprocal sedimentation in Virgilian strata of southern New Mexico: Geol. Soc. America Bull., v. 78, p. 805–817, 4 pls., 4 figs., 1 table.

Shelf and basin carbonate cycles documented in a narrow, elongate basin. Informative summary table.

Cassinis, Giuseppe, 1968, Stratigrafia e tettonica dei terreni Mesozoici compresi tra Brescia e Serle: Atti. Ist. Geol. Univ. Pavia, v. 19, p. 50–152, 9 pls., 30 figs.

The Italian mix of stratigraphy, sedimentary petrology, and tectonics is used to decipher the paleogeography of a small area in the southern Alps. Good illustrations. Italian, French, German, and English summaries.

Fischbuch, N. R., 1968, Stratigraphy, Devonian Swan Hills reef complexes of central Alberta: Bull. Canadian Petroleum Geology, v. 16, p. 446–556, 23 pls., 38 figs., 1 table.

Four major facies and nine stages of reef growth. Detailed reef geometry and zonation. Numerous plates of stromatoporoids.

Jenik, A. J., and Lerbekmo, J. F., 1968, Facies and geometry of Swan Hills Reef Member of Beaverhill Lake Formation (Upper Devonian), Goose River Field, Alberta, Canada: Am. Assoc. Petroleum Geologists Bull., v. 52, p. 21–56, 7 pls., 10 figs., 1 table.

Geometry, paleontology and petrography of a reef. Twenty-two rock types and eight environmental facies are distinguished and related to reef geometry.

Langton, J. R., and Chin, G. E., 1968, Rainbow Member facies and related reservoir properties, Rainbow Lake, Alberta: Bull. Canadian Petroleum Geology, v. 16, p. 104–143, 3 pls., 19 figs.

Fourteen facies related to reef geometry. Influence of diagenesis on reservoir properties. Correlation of facies, depositional environment and bathymetric zonation. Illustrates reservoir facies and corresponding capillary pressure curves.

Leavitt, E. M., 1968, Petrology, paleontology, Carson Creek North reef complex, Alberta: Bull. Canadian Petroleum Geology, v. 16, p. 298–413, 22 pls., 15 figs., 3 tables.

Geometry, petrology, and paleontology of a 24-square mile Devonian reef complex. Stromatoporoids and algae are most important reef builders. Five facies, 11 microfacies, and 32 rock types. Diagenesis has not greatly changed original sedimentary features. Informative and clear classification of reef-complex limestones. Main growth of reef complex outlined.

Lefeld, Jerzy, 1968, Stratygrafia i paleogeografia dolnej Kredy Wierchowej Tatr (Stratigraphy and palaeogeography of the High-Tatric Lower Cretaceous in the Tatra Mountains): Studia Geologica Polonica, v. 24, 116 p., 18 pls., 13 figs., 2 tables.

Cretaceous carbonates of the High-Tatric zones in the Tatra Mountains belong to five microfacies: *Calpionella*, *Saccocoma*, globigerinids, pseudooolitic, and oolitic facies as well as reef and nonreef facies. Paleoecology. English summary and subtitles.

MacQueen, R. W., and Bamber, E. W., 1968, Stratigraphy and facies relationships of the Upper Mississippian Mount Head Formation, Rocky Mountains and foothills, southwestern Alberta: Bull. Canadian Petroleum Geology, v. 16, p. 225–287, 12 pls., 11 figs., 1 table.

Unit is 500 to 1000 feet thick, widespread, and divisible into four major facies: supratidal, subtidal, shallow but open sea, and echinodermal and bryozoan banks.

Ota, Masamichi, 1968, The Akiyoshi Limestone Group: A geosynclinal organic reef complex: Akiyoshi-dai Sci. Mus. Bull. 5, 44 p., 31 pls., 17 figs., 6 tables.

Effective documentation of a Carboniferous-Permian reef complex believed to have been initiated by a volcanic rise in a geosynclinal setting. Five major facies. Several interesting flow diagrams. Stress on paleoecology and comparison with modern. Good illustrations and tables with English subtitles. Compare with Demirmen and Harbaugh (1965). Japanese with English abstract.

Pirlet, H., 1968, La sédimentation rythmique et la stratigraphie du Viséen supérieur V3b, V3c inférieur dans les synclinoriums de Namur et de Dinant: Acad. royale Belgique Mém. cl. sci., 2nd ser., v. 17, 98 p., 18 pls., 3 figs., 6 tables.

Nine short chapters with emphasis on petrography, rhythmic sedimentation, and depositional environment as well as detailed stratigraphy. Recommended for those looking for data on carbonate cycles.

Radwánski, A., 1968, Studium petrograficzne i sedymentologiczne Retyku Wierchowegs Tatr (Petrographical and sedimentological studies of the High-Tatric Rhaetic in the Tatra Mountains): Studia Geologica Polonica, v. 25, 146 p., 54 pls., 6 figs., 9 tables.

Rhaetic (Triassic) of the central Carpathians consists of terrigenous and carbonate sediments. Thorough description of skeletal debris. Vertical profiles. Environmental analysis and paleogeography. Well-referenced. English summary and subtitles.

Wilson, R. C. L., 1968, Carbonate facies variation within the Osmington oolite series in southern England: Palaeogeography, Palaeoclimatology, and Palaeoecology, v. 4, p. 89–123, 5 pls., 9 figs., 5 tables.

Five major facies within the oolite are recognized, defined, and interpreted. Facies model given in block diagram.

Wobber, F. J., 1968, Microsedimentary analysis of the Lias in South Wales: Sed. Geology, v. 2, p. 13–49, 12 figs., 6 tables.

Petrography in the Carozzi style, but with regional application, helps correlate between a sandy nearshore and a carbonate-shale offshore facies.

Friedman, G. M., ed., 1969, Depositional environments in carbonate rocks: Soc. Econ. Paleontologists and Mineralogists Spec. Pub. 14, 209 p.

An excellent volume—resulting from a symposium—with 11 papers plus an introduction by the editor and discussion by participants. Two papers treat deep water limestones, three deal with mixed and transitional ones, and four with shallow water limestones plus ecologic papers. Very well illustrated both with photomicrographs and line drawings.

Dmitrieva, E. V., *et al.*, 1969, Atlas tekstur i struktur osadochnikh gornikh porod Chast 2 Karbonatnye porody (Atlas of textures and structures of sedimentary rocks Part 2 Carbonate rocks): Moskva, Vsesoyznyi Nauchno-Issledovateliskiĭ Geologicheskiĭ Institut (VSEGEI) Ministerstva Geologiĭ SSSR, 707 p., 231 pls., 40 figs., 34 tables.

Introduction of 189 pages reviews many aspects of carbonate rock textures and structures at the mega- and microlevels; ends with 20 page glossary of Russian carbonate rock terms in which English equivalents are cited. Western literature covered through 1963, Russian literature through 1964 (but one 1966 paper). About 450 references. Plates contain over 750 figures, including thin section photomicrographs of limestones having biotic debris. However, plates not organized along biotic lines and are not indexed for biotic constituents. Major Russian contribution. Biogenic, coprolitic, and stromatolitic limestones discussed in various sections of text.

Dunham, R. J., 1969a, Early vadose silt in Townsend Mound (reef), New Mexico, *in* Friedman, G. M., ed., Depositional environments in carbonate rocks: Soc. Econ. Paleontologists and Mineralogists Spec. Pub. 14, p. 139–181, 22 figs.

Very informative qualitative observation and helpful photomicrographs. One of the first efforts to relate interframework silt and cement to paleohydrology.

— 1969b, Vadose pisolite in the Capitan reef (Permian), New Mexico and Texas, *in* Friedman, G. M., ed., Depositional environments in carbonate rocks: Soc. Econ. Paleontologists and Mineralogists Spec. Publ. 14, p. 182–191, 22 figs.

Considers the general question of why caliche is not found more commonly in ancient carbonates and then shows pisolites of the Capitan reef may in fact be caliche deposits recording emergence.

Gygi, Reinhardt, 1969, Zur Stratigraphie der Oxford-Stufe (oberes Jura-System) der Nordschweiz und des süddeutschen Grenzgebietes: Beitr. geol. Karte Schweiz, n. ser. 136, no. 11, 123 p., 6 pls., 11 figs., 9 tables.

Five parts: definition of components and rock types (31 p.), lithostratigraphy (41 p.), biostratigraphy (9 p.), chronostratigraphy (4 p.), and lithogenesis and paleogeography (5 p.). The plates contain 50 figures of hand specimens, thin sections, and electronphotomicrographs (up to 12,500×). Figure 43 shows a micrite section 1 micron thick and photographed under crossed nicols. About 180 references.

Kendall, C. G. St. C., 1969, An environmental reinterpretation of the Permian evaporite carbonate shelf sediments of the Guadalupe

Mountains: Geol. Soc. America Bull., v. 80, p. 2503–2526, 9 pls., 12 figs., 5 tables.

An outstanding, well-illustrated analysis of the primary and secondary processes that produce the limestones, dolomites, and evaporites of the Permian shelf of west Texas. Environmental model and comparison with modern is stressed. Well-illustrated. Worth careful reading.

Kendall, C. G. St. C., and Skipwith, P. A. d'E., 1969, Holocene shallow-water carbonate and evaporite sediments of Khor al Bazam, Abu Dhabi, southwest Persian Gulf: Am. Assoc. Petroleum Geologists Bull., v. 53, p. 841–869, 24 figs., 1 table.

Seven major lithofacies in a shallow-water lagoon. Skeletal alteration by blue-green algae. Detailed classification of accretionary grains plus photomicrographs and drawings of skeletal debris.

Logan, B. W., *et al.*, 1969, Carbonate sediments and reefs, Yucatán shelf, Mexico: Am. Assoc. Petroleum Geologists Mem. 11, Parts 1 and 2, p. 1-198, 10 pls., 64 figs., 13 tables.

Fourteen chapters provide comprehensive description and analysis. Chapters 4 and 5 (constituents, sediment types, and formational processes) are particularly recommended. Extended treatment of reefs. Outstanding monographic analysis of the modern that should become a classic.

Matsumoto, T., ed., 1969, Litho- and bio-facies of carbonate sedimentary rocks—a symposium: Palaeont. Soc. Japan Spec. Paper 14, 82 p., 19 pls., 24 figs., 5 tables.

Carbonate petrography and biofacies in a geosynclinal setting. Five papers. Good illustrations.

Mattavelli, L., Chilingar, G. V., and Storer, D., 1969, Petrography and diagenesis of the Taormina Formation, Gela oil field, Sicily, (Italy): Sed. Geology, v. 3, p. 59–86, 3 pls., 8 figs.

Mostly dolomitized stromatolites and breccias. Diagenesis related to petrophysics.

Purser, B. H., 1969, Syn-sedimentary marine lithification of Middle Jurassic limestones of the Paris Basin: Sedimentology, v. 12, p. 205–230, 16 figs.

Diagenetic fabrics include calcite drusy cements, echinoderm overgrowths and microcrystalline cements. Bored surfaces ("hard grounds") at tops of regressive sequences. Very well-illustrated with excellent, informative photomicrographs.

Rich, Mark, 1969, Petrographic analysis of Atokan carbonate rocks in central and southern Great Basin: Am. Assoc. Petroleum Geologists Bull., v. 53, p. 340–366, 10 figs., 3 tables.

Nine limestone types are linked to a conceptual model based on wave energy, depth, and biologic distribution.

Veevers, J. J., 1969, Sedimentology of the Upper Devonian and Lower Carboniferous platform sequence of the Bonaparte Gulf Basin: Australian Dept. Nat. Devel., Bur. Min. Res., Geology and Geophysics Bull. 109, 86 p., 43 pls., 30 figs., 12 tables.

Petrography and chemistry of reefal and nonreefal limestones classified by computer programs. Petrography and sedimentary structures of marine quartzose sandstones and conglomerates.

Zankl, Heinrich, 1969, Der Hohe Göll, Aufbau und Lebensbild eines Dachsteinkalk-Riffes im Obertrias der nördlichen Kalkalpen: Abh. Senckenbergischen Naturf. Gesell. no. 519, 96 p., 15 pls., 74 figs.

Petrography, paleocurrents, and detailed description of the components of a Triassic reef. Figure 74 gives good summary. English, French, and Russian resumés.

Logan, B. W., *et al.*, 1970, Carbonate sedimentation and environments, Shark Bay, Western Australia: Am. Assoc. Petroleum Geologists Mem. 13, 223 p.

Four papers by five authors describe and interpret modern and Quaternary sedimentary environments in a 5,000 square mile area on the west coast of Australia. Three papers focus specifically on carbonate sedimentation, including carbonate bank deposition, laminated algal mats, and a history of Quaternary carbonate sedimentation.

References Cited

Barrs, D. L., 1963, Petrology of carbonate rocks, *in* Shelf carbonates of the Paradox Basin, a symposium: Four Corners Geol. Soc., 4th Field Conf., p. 101–129, 22 figs.

Bathurst, R. G. G., 1958, Diagenetic fabrics in some British Dinantian limestones: Liverpool Manchester Geol. Jour., v. 2, p. 11–36, 1 pl., 2 figs.

— 1964, The replacement of aragonite by calcite in the molluscan shell wall, *in* Imbrie, John, and Newell, N. D., eds., Approaches to paleoecology: New York, John Wiley and Sons, Inc., p. 357–376, 4 pls., 1 fig., 2 tables.

Bissell, H. J., and Chilingar, G. V., 1967, Classification of sedimentary carbonate rocks, *in* Chilingar, G. V., *et al.*, eds., Carbonate rocks Developments in Sedimentology, v. 9A: Amsterdam, Elsevier Publishing Co., p. 87–168, 16 pls., 4 figs., 9 tables.

Bonham-Carter, G. F., 1965, A numerical method of classification using qualitative and semi-quantitative data, as applied to the facies analysis of limestones: Bull. Canadian Petroleum Geology, v. 13, p. 482–502, 3 figs., 9 tables.

Carozzi, A. V., 1960, Microscopic sedimentary petrography: New York, John Wiley and Sons, Inc., 485 p., 88 figs.

CHAVE, KEITH, 1954, Aspects of the biogeochemistry of magnesium 1. Calcareous marine organisms: Jour. Geology, v. 62, p. 266–283, 16 figs., 3 tables.

CHILINGAR, G. V., *et al.*, eds., 1967, Carbonate rocks, vols. 9A and 9B, *in* Developments in sedimentology: Amsterdam, Elsevier Publishing Co., 471 and 413 p.

CHOQUETTE, P. W., and PRAY, L. C., 1970, Geologic nomenclature and classification of porosity in sedimentary carbonates: Am. Assoc. Petroleum Geologists Bull., v. 54, p. 207–250, 13 figs., 3 tables.

COOGAN, A. H., 1969, Recent and ancient carbonate cyclic sequences, *in* ELAN, J. G., and STEWART, CHUBER, eds., Symposium on cyclic sedimentation in Permian Basin: Midland, West Texas Geological Soc., p. 5–16, 17 figs.

DOTT, R. H., JR., 1964, Wacke, graywacke and matrix—what approach to immature sandstone classification? Jour. Sed. Petrology, v. 34, p. 625–632, 3 figs., 1 table.

DUFF, P. M. D., *et al.*, 1967, Cyclic sedimentation: Amsterdam, Elsevier Publishing Co., 280 p., 91 figs. (Developments in sedimentology, v. 10).

DUNHAM, R. J., 1962, Classification of carbonate rocks according to depositional texture, *in* HAM, W. E., ed., Classification of carbonate rocks: Am. Assoc. Petroleum Geologists Mem. 1, p. 108–121, 7 pls., 1 table.

— 1969, Early vadose silt in Townsend mound (reef), New Mexico, *in* FRIEDMAN, G. M., ed., Depositional environments in carbonate rocks: Soc. Econ. Paleontologists and Mineralogists Spec. Pub. 14, p. 139–181, 22 figs.

EDIE, R. W., 1958, Mississippian sedimentation and oil fields in southeastern Saskatchewan, *in* GOODMAN, A. J., ed., Jurassic and Carboniferous of western Canada: Tulsa, Am. Assoc. Petroleum Geologists, John Andrew Allen Memorial Volume, p. 331–363, 18 figs., 7 tables.

FISCHER, A. G., *et al.*, 1967, Electron micrographs of limestones and their nanno-fossils: Princeton, Princeton Univ. Press, Mon. in Geology and Paleontology 1, 141 p., 94 figs.

FLÜGEL, ERIK, *et al.*, 1968, Review on electron microscope studies of limestones, *in* MÜLLER, GERMAN, and FRIEDMAN, G. M., eds., Recent developments in carbonate sedimentology in central Europe: New York, Springer-Verlag New York, Inc., p. 85–97, 2 pls., 5 figs., 3 tables.

FOLK, R. L., 1959, Practical petrographic classification of limestones: Am. Assoc. Petroleum Geologists Bull., v. 43, p. 1–38, 41 figs., 2 tables.

— 1962, Spectral division of limestone types, *in* HAM, W. E., ed., Classification of carbonate rocks—a symposium: Am. Assoc. Petroleum Geologists Mem. 1, p. 62–84, 7 figs., 3 tables.

— 1965, Some aspects of recrystallization in ancient limestones, *in* PRAY, L. C., and MURRAY, R. C., eds., Dolomitization and limestone diagenesis: Soc. Econ. Paleontologists and Mineralogists Spec. Pub. 13, p. 14–48, 14 figs., 7 tables.

FORCE, L. M., 1969, Calcium carbonate size distribution on the west Florida shelf and experimental studies on the microarchitextural control of skeletal breakdown: Jour. Sed. Petrology, v. 39, p. 902–934, 21 figs., 4 tables.

FRIEDMAN, G. M., 1969a, The fabric of carbonate cement and matrix and its dependence on the salinity of water, *in* MÜLLER, GERMAN, and FRIEDMAN, G. M., eds., Recent developments in carbonate sedimentology in central Europe: New York, Springer-Verlag New York, Inc., p. 11–20, 8 figs.

— 1969b, Depositional environments in carbonate rocks—an introduction, *in* FRIEDMAN, G. M., ed., Depositional environments in carbonate rocks: Soc. Econ. Paleontologists and Mineralogists Spec. Pub. 14, p. 1–3.

GOLDMAN, M. I., 1926, Proportions of detrital organic calcareous constituents and their chemical alteration in a reef sand from the Bahamas: Carnegie Inst. Washington, Marine Biology Papers, v. 23, p. 39–66, 1 fig., 13 tables.

GRIFFITH, L. S., *et al.*, 1969, Quantitative environmental analysis of a Lower Cretaceous reef complex, *in* FRIEDMAN, G. M., ed., Depositional environments, in carbonate rocks: Soc. Econ. Paleontologists and Mineralogists Spec. Pub. 14, p. 120–138, 34 figs.

HAM, W. E., ed., 1962, Classification of carbonate rocks—a symposium: Am. Assoc. Petroleum Geologists Mem. 1, 279 p.

HATCH, F. H., *et al.*, 1965, Petrology of sedimentary rocks, 4th ed. (Revised by J. T. GREENSMITH): London, Thomas Murby and Co., 408 p., 95 figs., 11 tables.

IMBRIE, JOHN, and PURDY, E. G., 1962, Classification of modern Bahamian carbonate sediments, *in* HAM, W. E., ed., Classification of carbonate rocks—a symposium: Am. Assoc. Petroleum Geologists Mem. 1, p. 253–272, 13 figs., 3 tables.

— and VAN ANDEL, T. H., 1964, Vector analysis of heavy-mineral data: Geol. Soc. America Bull., v. 75, p. 1131–1156, 12 figs., 11 tables.

IRWIN, M. L., 1965, General theory of epeiric clear water sedimentation: Am. Assoc. Petroleum Geologists Bull., v. 49, p. 445–459, 12 figs.

KAUFFMAN, E. G., 1967, Colorado macroinvertebrate assemblages, Central Western Interior, United States, *in* Paleoenvironments of the Cretaceous Seaway—a symposium: Golden, Colorado School of Mines, p. 67–143, 12 figs.

KLOVAN, J. E., 1964, Facies analysis of the Redwater reef complex, Alberta, Canada: Bull. Canadian Petroleum Geology, v. 12, p. 1–100, 9 pls., 20 figs., 8 tables.

KUMMEL, BERNHARD, and RAUP, DAVID, eds., 1964, Handbook of paleontological techniques: San Francisco, W. H. Freeman and Co., 852 p.

LAPORTE, L. F., 1967, Carbonate deposition near mean sealevel and resultant facies mosaic: Manlius formation (Lower Devonian) of New York State: Am. Assoc. Petroleum Geologists Bull., v. 51, p. 73–101, 34 figs., 3 tables.

LEIGHTON, M. W., and PENDEXTER, C., 1962, Carbonate rock types, *in* HAM, W. E., ed., Carbonate rocks—a symposium: Am. Assoc. Petroleum Geologists Mem. 1, p. 33–60, 9 pls., 3 figs., 2 tables.

MATTHEWS, R. K., 1966, Genesis of recent lime mud in southern British Honduras: Jour. Sed. Petrology, v. 36, p. 428–454, 8 figs., 5 tables.

MÜLLER, GERMAN, 1967, Methods in sedimentary petrology: Stuttgart, E. Schweizerbartische Verlagsbuchhandlung, 283 p., 91 figs., 31 tables.

MÜLLER-JUNGBLUTH, W. U., and TOSCHEK, P. H., 1969, Karbonatsedimentologische Arbeitsgrundlagen: Veröffentlichungen der Universität Innsbruck No. 8, Alpenkundliche Studien, No. 4, 32 p.

ORME, G. R., and BROWN, W. W. M., 1963, Diagenetic fabrics in the Avonian limestones of Derbyshire and North Wales: Proc. Yorkshire Geol. Soc., v. 34, p. 51–66, pls. 7–13, 1 fig.

PARKS, J. M., 1966, Cluster analysis applied to multivariate geologic problems: Jour. Geology, v. 74, p. 703–715, 5 figs., 8 tables.

PETTIJOHN, F. J., 1957, Sedimentary rocks, 2nd ed.: New York, Harper and Brothers, 718 p., 173 figs., 119 tables.

POTTER, P. E., 1968, A selective, annotated bibliography on carbonate rocks: Bull. Canadian Petroleum Geology, v. 16, p. 87–103.

— and BLAKELY, R. F., 1968, Random processes and lithologic transitions: Jour. Geology, v. 76, p. 154–170, 9 figs., 4 tables.

— and PETTIJOHN, F. J., 1963, Paleocurrents and basin analysis: Berlin, Göttingen, Heidelberg, Springer-Verlag, 296 p., 30 pls.

PURDY, E. G., 1968, Carbonate diagenesis: an environmental survey: Geologica Romana, v. 7., p. 183–228, 6 pls., 10 figs.

SHAW, A. B., 1964, Time in stratigraphy: New York, McGraw-Hill Book Co., 365 p.

STAUFFER, K. W., 1962, Quantitative petrographic study of Paleozoic carbonate rocks, Caballo Mountains, New Mexico: Jour. Sed. Petrology, v. 32, p. 357–396, 38 figs.

STOCKMAN, K. W., *et al.*, 1967, The production of lime mud by algae in south Florida: Jour. Sed. Petrology, v. 37, p. 633–648, 14 figs., 1 table.

VISHER, G. S., 1965, Use of vertical profile in environmental reconstruction: Am. Assoc. Petroleum Geologists Bull., v. 49, p. 41–61, 16 figs.

WALTHER, JOHANNES, 1894, Lithogenesis der Gegenwart: Jena, Gustav Fischer, 1055 p.

WISHART, DAVID, 1969, Fortran II programs for 8 methods of cluster analysis (Clustan I): Kansas Geol. Survey Computer Contr. 38, 112 p., 5 figs., 11 tables.

3. Identification of Biotic Constituents

Introduction

We have endeavored to provide a brief, elementary summary of the groups of fossils one is likely to encounter and recognize in thin sections at magnifications of 10 to 100 and to indicate where additional information is available. Our coverage includes 18 major fossil groups.

Very few summaries of fossils as observed in thin section exist. CAYEUX (1916, 1935), MASLOV (1937), and J. H. JOHNSON (1951) provided earlier introductions, and MAJEWSKE (1969) has provided a recent well-illustrated overview.

SCHMIDT (1924) reviewed the microscopic skeletal features of many living groups of animals as displayed in polarized light. A vast literature is available on individual fossil groups because skeletal microstructure and internal skeletal architecture have become an increasingly important aspect of paleontological studies. In order to reveal shell microstructure, as well as internal architecture, some groups (e.g., bryozoans, larger foraminifers, stromatoporoids) are studied almost exclusively from thin sections or allied techniques, such as peels and serial and polished sections. In addition, microfacies papers commonly discuss the identification of fossils encountered during individual stratigraphic studies (for example, LAPORTE, 1962).

The student of biotic debris in thin section faces several problems. First, the biotic debris appears as two dimensional cross sections of three-dimensional objects. Obviously, reconstructing the appearance of an object from a single cross section is hazardous and the best "guesstimate" will be made by those who have a good sense of geometry (how various fossils will appear when traversed in different planes) and those who have had experience with many fossil groups. Anyone not familiar with fossils and the enormous variety of sizes and shapes in which they appear should refer to standard textbooks (CAMACHO, 1966; EASTON, 1960; MOORE *et al.*, 1952; MÜLLER, 1957–1968; SHROCK and TWENHOFEL, 1953). For a more detailed overview one can consult the variously authored multi-volume Treatise on Invertebrate Paleontology, which is near completion and whose early volumes are currently undergoing revision, or comparable French (PIVETEAU, 1952–1969) and Russian (ORLOV, 1958–1964) compilations. In particular, the reader should consider some typical shapes and visualize how the shape would appear when cut by planes oriented at various angles to the shell. In large measure this is the key to identifications of major fossil groups. Many published text-figures are appropriate for this purpose; for example, MOORE *et al.* (1952, p. 282, fig. 8–4) presented an informative figure for gastropods. We discuss in a separate section below the geometry of shell architecture.

Secondly, where does one go for pictorial examples of various biotic groups? The illustrations in the microfacies or paleontological literature certainly provide a larger catalog of pictures and identifications than we can present and are indispensable when they coincide with an area or a rock sequence of particular interest to the investigator. We recommend strongly that the beginner browse among the photomicrographs of the published microfacies studies, especially those relevant to the problem at hand, for just as the best geologist is commonly the one who has seen the most geology (and thought about it) so the best identifications of fossil debris are commonly made by those with the most experience. The microfacies literature varies greatly in the level of identifications provided. This level is frequently a function of the paleontological inclinations of the investigator or the state of paleontological knowledge of the sequence studied. We have assembled an annotated list of the major microfacies monographs that have come to our attention and these are listed at the end of this chapter.

Thirdly, diagenetic solution, recrystallization or replacement alters original microstructure. Alteration of microstructure complicates identification because, without the use of shell microstructure, debris commonly cannot be definitely ascribed to a major fossil group. We have discussed this problem below in a section on post-mortem changes in shell microstructure.

Experience will provide many solutions to initial problems of identification. We repeat our admonition that the best identifier commonly will be he who has spent the most time and effort working on the problems at hand. Workers who study the microfacies of relatively restricted areas or of narrow stratigraphic intervals are commonly able to identify constituents to the smallest taxonomic levels. Such detailed identifications are related largely to fossil groups whose diagnostic characters are determined by thin-section studies.

Although we have limited our descriptions and illustrations to magnifications of approximately 10 to 100, the light microscope will magnify at least by another order of magnitude, but the depth of field rapidly decreases at higher magnifications. For the purposes of many microfacies studies, magnifications of 10 to 100 are sufficient although other paleontological purposes require higher magnifications and greater depth of field. For example, the transmission and the scanning electron microscopes recently have been applied to elucidating the shell structure of living and fossil shells. We have not considered these techniques, although we fully recognize that the scanning electron microscope may be the first sophisticated electronic instrument to have a major and revolutionary impact on traditional morphological studies in paleontology. We have cited electron microscope studies only where they appear to assist in interpreting the microstructural features observed in the optical microscope.

Geometry of Skeletal Architecture

Skeletal architecture refers to the shape or geometry of large skeletal or shell features, internal and external plates, and ornamentation. Wall microstructure is excluded and is considered separately.

Diversity of form is a measure of the successful adaptation of many organisms to a wide variety of environmental conditions. Paleontologists have used a substantial nomenclature to describe this diversity of form and, as a consequence, paleontological nomenclature has been troublesome to the nonpaleontologist who studies carbonate rocks. But as we hope to show, most shells can, in reality, be approximated by a relatively few simple geometric patterns. In discussing individual biotic groups, we have emphasized geometry over terminology and in this book formal taxonomic names are at a minimum. This is not because we regard names as unimportant; indeed, they are essential as the key to a large body of fundamental literature. However, the smallest nomenclatorial divisions of the biotic constituents in our thin sections usually are not known to us and we have generally avoided fine taxonomic discrimination in order to focus on the general features of the major biotic groups as found in skeletal debris in the sedimentary record.

For the purposes of discussion the geometry of skeletal architecture is divided into five groups: 1) straight or coiled single cones, tubes, or cylinders; 2) colonial tubes; 3) valves; 4) multi-element or multi-plated shells; and 5) reticulate or rectangular and cystose networks. Obviously other types of divisions could be prepared and certainly none of these categories is infallible with respect to designating major groups of biotic debris. However, in conjunction with other architectural elements and shell microstructure, the above shapes should permit most debris either to be identified or to indicate the most probable groups.

Table 7. *Identification of Major Fossil Constituents Based on Skeletal Architecture**

I. Tubes, cones, and cylinders

A. Straight (orthoconic) forms

1. Internal longitudinal plates perpendicular to outer wall and parallel to growth (commonly long) axis of cones

a. Corals; imperforate walls, orthogonal fibrous or trabecular microstructure. Ordovician to Recent

b. Archaeocyathids; inner and outer porous walls, very fine-grained dark wall. Lower and Middle Cambrian

2. Internal transverse plates perpendicular to long (commonly growth) axis of cones

a. Cephalopods; round or oval transverse cross sections, all microstructure commonly recrystallized (? shells originally aragonite), plates may be singly perforate, internal calcareous deposits commonly present in inner chambers. Cambrian to Cretaceous

b. Tentaculitids; ornamented usually by transverse rings or ribs, internal plates uncommon and imperforate when present. Ordovician to Devonian

3. No internal plates

a. Brachiopod spines; hollow, fibrous or lamellar wall microstructure. Most common Devonian to Permian

b. Echinoid spines; act as single crystals of calcite under crossed polarizers, best seen in transverse sections, which usually are hollow in cross section and exhibit radial pattern of pores. Ordovician to Recent

c. Sponge spicules; siliceous forms generally have hollow central canal; calcareous forms usually single calcite crystals but lack porous structure of echinoid spines. Cambrian to Recent

B. Coiled tubes

1. Internal transverse plates or walls

a. Cephalopods; recrystallized shell microstructure, single perforation of transverse plate (septum), generally large size relative to other biotic debris. Ordovician to Recent

b. Foraminifers; small size, commonly less than 1 mm in maximum dimension, numerous transverse plates, shell wall microstructure mightily varied as discussed in text. Cambrian (?), Ordovician to Recent

2. No internal structures

a. Gastropods; commonly recrystallized or shell has been leached and subsequently infilled by calcite. Cambrian to Recent

b. Worm tubes; small, generally attached on one surface. Ordovician to Recent

II. Colonial or multiple tubes

A. Bryozoans; laminated wall microstructure, apertures and tubes generally less than 0.5 mm in diameter. Ordovician to Recent

B. Corals; fibrous wall microstructure, apertures generally greater than 0.5 mm in diameter. Upper Cambrian to Recent

C. Stromatoporoids; tubular structures, present in some forms, are embedded in reticulate or cystose network (see **V** below); wall microstructure complex (see text). Cambrian to Eocene

D. Solenoporacean algae; tubes commonly less than 0.1 mm, cross partitions may be present. Cambrian to Cretaceous

E. Codiacean algae; tubes commonly less than 0.1 mm, no cross partitions, tubes commonly intertwined in interior but approach exterior at right angles. Cambrian to Recent

III. Valves

A. Ostracodes; small fibers or prisms at right angles to valve wall, commonly recurved margins of shell (duplicature), generally less than 1 mm in maximum dimension. Cambrian to Recent

B. Brachiopods; fibrous or prismatic shell microstructure; commonly exhibit architectural plates in beak area. Cambrian to Recent.

C. Pelecypods; prismatic, laminated, and cross-lamellar layers common; usually lacks internal architecture; frequently altered where shell layers were aragonitic. Cambrian to Recent

IV. Multiplates

A. Echinoderms; finely porous plates act as single crystals of calcite under crossed polarizers. Cambrian to Recent

B. Trilobites; characteristic curved and recurved ("shepherd's crook") cross sections; slender prisms at right angles to shell wall. Cambrian to Permian

C. Vertebrates; dense exterior of individual bones, vesicular interiors; phosphatic. Ordovician to Recent

D. Conodonts; laminated shell microstructure, generally less than 1.0 mm in maximum dimension; phosphatic. Middle Cambrian to Triassic

Table 7 (Continued)

V. Rectangular and cystose networks

A. Rectangular networks

1. Bryozoans; horizontal plates (diaphragms) traversing the tubes (zooecia); laminated wall microstructure, zooecia less than 0.5 mm in maximum diameter. Ordovician to Recent
2. Corals; horizontal plates (tabulae) traversing tubes (corallites); fibrous wall structure, corallites generally greater than 0.5 mm in maximum diameter. Cambrian to Recent
3. Stromatoporoids; in longitudinal section exhibit horizontal and vertical elements (laminae and pillars), wall microstructure complex (see text). Rectangular grid commonly absent in tangential sections. Cambrian to Eocene
4. Wood and calcareous algae. Rectangular cellular architecture, usually measured in microns, commonly less than 0.1 mm; tissue of calcareous algae may show much larger clear calcite areas representing sporangia (reproductive tissue). Wood: Devonian to Recent; Calcareous algae: Cambrian to Recent
5. Foraminifers: many large foraminifers greater than 1 mm exhibit very numerous chambers that present reticulate or rectangular patterns in thin section. Wall microstructure varies with type of foraminifer. Pennsylvanian to Recent
6. Sponges; rectangular pattern produced by network of rays of spicules particularly in forms where the ends of the spicules are fused to form a rigid framework. Cambrian to Recent

B. Cystose networks (curved plates)

1. Bryozoans; usually zooecial tubes in midst of cystose tissue, cysts less than 0.5 mm in maximum dimension. Ordovician to Recent
2. Corals; cysts confined to single coral or corallite, usually larger than in bryozoans, greater than 0.5 mm in maximum diameter, coral microstructure fibrous. Cambrian to Recent
3. Stromatoporoids; comparable to size in some corals, wall structure complex. Ordovician to Eocene
4. Vertebrates; no regular pattern to internal tissue, phosphatic. Ordovician to Recent

* Caution: Use with discretion, formulae variable depending on previous experience of user.

Our analysis of shell shape and form is an informal one. THOMPSON (1942) has published a formal review of the geometry of shell morphology, and RAUP (1961, 1966b, 1967) has applied mathematical analyses to the shapes of coiling in different molluscan groups.

We provide in Table 7 a cookbook approach for the zealous amateur and eager initiate into the mysteries of biotic identifications in thin section. Authorities on every group will be happy to cite numerous exceptions to this table. We still think it is a reasonable start for a beginner but please realize that exceptions do exist and that this table is not the final arbiter in difficult decisions. Finally, we wish you all the best of luck in your endeavors at identification. You will need it!

General Observations on Skeletal Microstructure

Skeletal microstructure refers to the texture or fabric of the minerals that form the shell or skeletal wall.

Different terminologies for basic units of microstructure are used in different groups and sometimes even within the same group of organisms and, as a consequence, terminology can be confusing to the beginner. For example, the prismatic structure of arthropods or the foraminifers would probably be called fibrous among the corals or some of the bryozoans. In general, however, the term prism is commonly applied when the basic units are oriented perpendicular to the shell surface and the term fiber is used frequently where the crystals are at an angle to the shell surface or wall. Of course, the secreting surface in the living animal need not be parallel to the final surface of a wall and the evidence in many groups is that it was not. Furthermore, fibrous prisms can be oriented in various radial and complex patterns that BRYAN and HILL (1940) believed were characteristic of spherulitic crystallization in the scleractinian corals. How widespread this pattern of crystallization is among shelled organisms is not clear, although it is the dominant mode of secretion in the corals and presumably is present in some bryozoans (CHEETHAM *et al.*, 1969). In addition to fibers or prisms, crystallites may be arranged in thin, apparently uniform sheets or laminae.

The thin section interpretation of prism or fiber versus laminae is influenced in part by the position of the thin section with respect to the microstructural units. For example, fibers and prisms of small size are apparently real in the arthropods and brachiopods regardless of the direction of the thin section with respect to the skeleton. This is not necessarily true in the bryozoans, and the apparent fibrous character of many bryozoan skeletons represents the cross section of continuous layers (laminae) of the shell (cf. BOARDMAN, 1960). Molluscan shell structure is even more complex as individual units may be built into larger units as shown by BØGGILD (1930) and many subsequent workers. The type of microstructure preserved is apparently dependent in large measure on the position of the thin section with respect to the shell microstructure.

Although all types of shell microstructure are not necessarily covered by the above discussion, the preceeding remarks will help introduce the student to both the variety and the over-all general plan of the microstructure of shells.

The refraction and reflection of light at crystal boundaries, particularly with fine-grained crystalline skeletal debris, is also a factor to be considered. Light is refracted and reflected at crystal boundaries, which commonly cause them to appear dark in thin section. The more boundaries the light crosses in passing through the thin section the darker the section appears because much of the light is dispersed by reflection and refraction. Consequently, the smaller the grains or crystallite sizes, the darker skeletal walls appear in thin section. In contrast, on polished surfaces, fine-grained walls appear white (porcellaneous), because the many small grains tend to reflect more light than surrounding areas. Walls with crystals between 10 and 1 or 2 microns are present in some groups of foraminifers, in the walls of coralline algae and archaeocyathids, and in the "primary" wall layer of some Paleozoic bryozoans.

Unfortunately, larger size of crystallites does not always produce clear shell microstructure as the crystallites may be encased in organic sheaths that help diffuse the light passing through the thin section. In addition, large crystals may be porous as is true of the echinoderms. The fine porosity of echinoderm plates provides many surfaces to disperse the light so that these crystals in thin section commonly have a gray cast, which is lacking in nonporous crystals of calcite cement in the same thin section. Individual crystals of echinoderm plates may be 10 centimeters or more in maximum dimension in some echinoid spines and only slightly smaller in the spinose plates of some crinoids.

Skeletal Mineralogy

LOWENSTAM (1963, p. 142–143) has summarized the occurrence of minerals within biologic groups, the vast majority of shells being composed of calcium carbonate (Table 8). Biotic constituents contribute minor amounts of silica and phosphates and make rare contributions of sulphates, oxides or fluorides. One should keep in mind, however, that the variety of minerals secreted by biologic activities illustrates that the occurrence of minerals such as magnetite or fluorite in sedimentary rocks can not be assumed always to be of inorganic origin as generally has been thought. Initial carbonate mineralogy of shells is also important because it affects the preservation of shell structure, the prime example being aragonitic shells which are more likely to be altered by leaching and infilling or recrystallization than calcitic shells. The initial carbonate mineralogy of shells is also the principal determinant of the carbonate and silica cycles in sediments.

REVELLE and FAIRBRIDGE (1957) have reviewed the effect of organisms on the planetary equilibrium of calcium and carbon dioxide. The silica cycle has been reviewed by SIEVER (1957) and the organic occurrence of silica, which is most common in plant groups, is discussed by SIEVER and SCOTT (1963). None of the higher animal

Table 8. *Skeletal Mineralogy of Fossil Groups* (Modified from LOWENSTAM, 1963, Fig. 2)

	Algae	Protozoans	Sponges	Coelenterates	Bryozoans	Brachiopods	Annelids	Mollusks	Arthropods	Echinoderms	Hemi-chordates	Vertebrates
Carbonates												
Aragonite	○	○		○	○		○	○	○		○	○
Calcite	○	○	○	○	○	○	○	○	○	○		
Aragonite + Calcite			?	○	○		○	○	○			
Vaterite								○				①
Protodolomite										②		
Monohydrocalcite												③
"Amorphous"				○			○	○				
Silicates												
"Opaline"	○	○	○					?				
Phosphates												
Dahllite								○				○
Francolite								○				
Hydroxyapatite						○						○
Undefined + Calcite								○	○			
Oxides												
Magnetite								④				
Lepidocrocite			⑤					⑥				
Goethite								○				
Magnetite + Goethite								○				
Brucite	⑦											
Amorphous (Fe) + Aragonite		○						○				
Sulfates												
Celestite		○										
Barite		?										
Gypsum				⑩								⑪
Fluorides												
Fluorite								⑧	⑧			
Oxalates												
Weddellite								⑨				

1. SUTER and WOOLEY (1968)
2. SCHROEDER *et al.* (1969)
3. CARLSTRÖM (1963); FLEISCHER (1969)
4. LOWENSTAM (1962); CAREFOOT (1965)
5. TOWE and RÜTZLER (1968)
6. LOWENSTAM (1967)
7. SCHMALZ (1965); WEBER and KAUFMAN (1965)
8. LOWENSTAM and McCONNELL (1968)
9. LOWENSTAM (1968)
10. SPANGENBERG and BECK (1968)
11. COGAN *et al.* (1958)

phyla make use of silica for only the protozoans, silicoflagellates, ebridians, and sponges utilize it.

Post-Mortem Changes in Shells and Shell Microstructure

After death a combination of physical and chemical processes commonly modifies the fossil skeleton and its internal microstructure. In general, these diagenetic changes all tend to increase the difficulty of identification of fossil debris in thin section.

Organic binders decay after death so that one commonly sees separated segments of multi-element skeletons instead of the whole. This is particularly true of the larger invertebrates. Scavengers also contribute to fragmentation in their search for food: they not only separate skeletal parts but additionally may crush them. Other modifying agents, which are easily overlooked but are very effective, are animals or plants such as algae and sponges that bore into skeletal material either before or after death. In well-washed limestones, currents separate and, if sufficiently strong, may abrade and further fragment segmented, skeletal debris. In addition to these physical processes, shell microstructure may be destroyed by recrystallization—with or without change in mineralogy—to cryptocrystalline carbonate.

Calcite, which constitutes the bulk of invertebrate skeletal material, forms as a partial solid solution series with $MgCO_3$: it may contain less than 4 mole percent (low-Mg calcite) or as much as 30 mole percent (high-Mg calcite), the latter mostly in Recent skeletal debris (CHAVE, 1954a). Factors influencing Mg content include mineralogy (aragonite seldom has more than one percent Mg), water chemistry, growth temperature, and phylogenetic level. CHAVE (1954a, 1954b; 1962), LOWENSTAM (1963) and KENNEDY *et al.* (1969) summarize at length skeletal mineralogy and some of its ecologic and phylogenetic implications.

Low-Mg calcite is more stable than either high-Mg calcite or aragonite and is, along with dolomite, the common form in ancient carbonates (e.g., FRIEDMAN, 1964; LAND, 1967). When calcite loses Mg, there commonly is no textural change of microstructure, the excess Mg being removed from the lattice by a currently little understood process. The transformation of aragonite to calcite may be accompanied by a textural transformation which obliterates original shell microstructure (Pl. 39–2). Two processes appear to be active (BATHURST, 1964): inversion of the aragonite lattice to calcite involving no change in void space so that original shell microstructure is largely preserved (at least as textural ghosts) and solution-deposition whereby a new low-Mg sparry calcite (drusy cement, see p. 7) is precipitated in the mold of the original aragonitic skeletal material. The resulting fill commonly has a distinctive fabric—a drusy surface of prismatic crystals perpendicular to the mold encloses larger, more equant anhedral crystals away from the boundary (blocky cement, see p. 7). When seen under crossed nicols this fabric of "marching men" (Pl. 43–2, 59–4) is a clear, unambiguous signal of leaching of skeletal debris for we know of very few organisms having such a primary skeletal microstructure. Groups such as gastropods, which usually are composed totally of aragonite, are characteristically identified by their drusy calcite and shape.

Ghosts outlined by either greater opaqueness or slight color contrasts may define the original skeleton in the secondary drusy fabric (Pl. 40–3). *Micritic envelopes*, very fine-grained, nearly opaque (dark in thin section) replacement borders on carbonate grains (Pls. 39–3, 40–2, 81, 99–1), may also define the original skeletal outline (BATHURST, 1964, p. 365–369 and 1966; WINLAND, 1968). Although their origin is not fully understood, boring algae are believed primarily to be responsible. SWINCHATT (1969) has suggested that algal boring may be depth dependent.

High-Mg calcite, low-Mg calcite and aragonite also can be replaced by dolomite. Ghosts may remain even if replacement is complete. No living organism is known to make its total shell with dolomite, although recently SCHROEDER *et al.* (1969) reported protodolomite in some echinoids.

Recrystallization of fine-grained carbonate mud is widespread and is of concern to the petrographer interested in fossil identification, because it can extend from the fine-grained matrix to the skeletal framework itself thus making fossil identifications more difficult. Recrystallization may involve high to low-Mg calcite, aragonite to calcite or only low-Mg calcite itself. FOLK (1965, p. 23–26) has commented at length upon its different aspects and provided a comprehensive classification. Recrystallization of skeletal debris without significant mineralogical change has been stressed by PURDY (1968). He described these transformations in subsea, subaerial, and subsurface environments and noted that much debris has a definite order of susceptibility to recrystallization. The processes causing recrystallization of both skeletal and nonskeletal (ooids, grapestone, faecal pellets, etc.) carbonate grains are incompletely understood. Some studies indicate that carbonate muds are more susceptible to recrystallization and dolomitization than sand size carbonate debris. However, few studies have been made of the selective susceptibility to dolomitization of sand-sized fossil debris. Good general sources of present knowledge about carbonate diagenesis are FRIEDMAN (1964), PRAY and MURRAY (1965), LAND (1967), GAVISH and FRIEDMAN (1969), and FÜCHTBAUER (1969). A pioneer and outstanding petrographic study of carbonate textural transformations is that of CULLIS (1904). ORME and BROWN (1963, Fig. 1) illustrate current nomenclature.

Direct solution of calcareous shells after death on the sea bottom is probably a factor only in the deep ocean where the temperature is near freezing and ocean water is undersaturated with respect to calcium carbonate. On shallow shelves, where most of the fauna that form carbonate rocks lived, chemical solution of carbonate skeletal debris at the sediment-water interface is not considered significant because ocean water is approximately saturated or even supersaturated with respect to calcium carbonate (e.g., PETERSON, 1966; WEYL, 1967).

Phosphates (Pls. 52, 53) and silica (Pls. 3, 14, 15) are two other minerals that form minor but ubiquitous shell materials. Because skeletal material composed of either phosphate or silica is uncommon, fossils composed of these two minerals are usually easily identified. Animals that utilize phosphatic materials in their shells are conodonts (Pl. 53-3, 4), vertebrates (Pls. 52, 97-2), some groups of inarticulate brachiopods (Pl. 29-6), and arthropods.

Secondary silicification of plants and animals plays an important role in their preservation and is well summarized by SIEVER and SCOTT (1963, p. 590-591). Silicification may preserve minute anatomical details or simply leave a structureless cast. There are all gradations between the two. The replacing silica may be either amorphous (chalcedony and opal, especially in the Tertiary) or crystalline in Mesozoic and older sediments. Under suitable conditions probably every type of skeletal carbonate may be silicified. In thin section silica is white or brownish in plane light and under cross nicols shows a random, radial, or fibrous arrangement of very fine-grained, low-order white crystals. Animals that utilize silica are the sponges and radiolarians. Although neither are commonly studied in thin section, diatoms and silicoflagellates, both plants, also consist of silica. Unless promptly buried, siliceous skeletal material is, however, subject to solution after death in the ocean because sea water is everywhere appreciably undersaturated with respect to silica (SIEVER, 1962, p. 137).

Although generalization is difficult, it is probably true that identification of most fossil debris smaller than 30 microns in ancient carbonates is very difficult if not impossible: smaller fragments show less original skeletal outline, show less internal microstructure, and are more likely to be affected by later recrystallization and dolomitization. As size becomes smaller, the small fragments of different organisms are more likely to be similar in petrographic appearance and thus lose their identifying characteristics (FERAY *et al.*, 1962, Fig. 2). Indeed, when working with ancient calcareous muds and silts one may be fortunate to even recognize that the particle in question is a fossil fragment. Possibly special study of crushed debris of known specimens, ones that are suspected to have made contributions may be helpful. This procedure was followed in the study of modern muds off Honduras by MATTHEWS (1966, Figs. 3, 4, and 5). Identification was facilitated because he studied loose particles with minimal diagenetic history.

Tintinnines

Skeletal architecture.—Tintinnines are free-swimming pelagic protozoans that produce a cup- or vase-shaped shell (lorica, Pls. 1, 2; Fig. 11). Some forms are ornamented with transverse or longitudinal ribs or exhibit a basal spine. Tintinnines in the fossil record are frequently studied on the basis of cross-sectional shapes revealed in thin sections. Longitudinal cross sections usually exhibit a constriction or a flange at the open end of the lorica and transverse sections are circular. Most forms have a maximum dimension of less than a millimeter (commonly 80 to 150 microns). For reviews of fossil tintinnines see COLOM (1948), DEFLANDRE (1952), and CAMPBELL (1954b). POKORNÝ (1958, p. 526–528), TAPPAN and LOEBLICH (1968), and LOEBLICH and TAPPAN (1968) contain more recent bibliographies.

Skeletal microstructure.—TAPPAN and LOEBLICH (1968) reviewed the structure of the tintinnine lorica in both recent and fossil examples and added observations of new

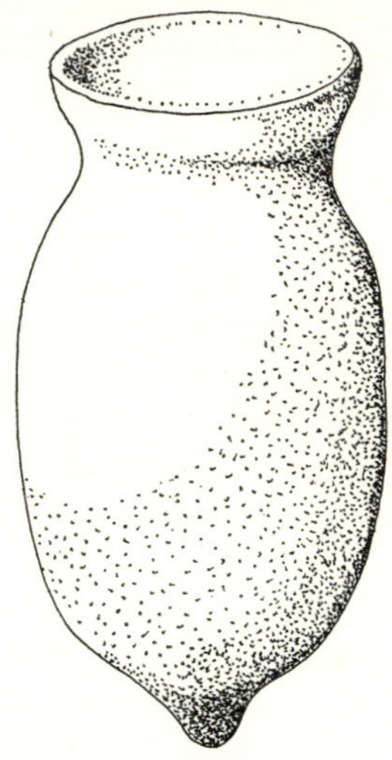

Fig. 11. Vase-shaped tintinnine shell illustrating flange or lip at open end.

materials. Tintinnine loricas have at least three types of shell composition: organic, partially calcified, or agglutinated. Calcareous agglutinated loricas commonly consist of very small coccoliths and nannocones whereas siliceous agglutinated loricas are composed of diatom shells or fine quartz grains.

REMANE (1964, p. 6–9) examined the wall structure of Jurassic and Cretaceous tintinnines and concluded that the originally aragonitic wall had been converted to calcite. TAPPAN and LOEBLICH (1968, p. 1382–1383) believed that calcite was the original mineral component of fossil tintinnine shells and cited as evidence calcite tintinnine shells in a geologic environment where aragonitic foraminifers have not been converted to calcite. All of these authors have concluded that the calcitic wall exhibits fibrous radial calcite oriented at right angles to the tintinnine wall.

REMANE (1964, p. 6) also noted that the outer boundary of the tintinnine shell wall tends to be smooth but that the inner boundary is usually irregular in coarse-grained and smooth in fine-grained sediments. Wall thickness is correlated with grain size of associated debris and tends to be thicker in coarse-grained sediments. In rocks cemented by sparry calcite, tintinnine shells appear interlocked or welded to the cement. These interlocked inner boundaries of the lorica and the rock matrix represent early diagenetic changes before sediment or subsequent cement filled the shells. REMANE indicated that the secondary wall thickening has the same crystallographic orientation as the crystals of the lorica. If true, these observations complicate the evaluation of primary and secondary microstructural characters of the tintinnine test.

REMANE (1964, p. 8) reported that a few specimens showed a thin median line of fine pyrite or limonite. The consistent position of this line suggested that an organic layer may have separated a bipartite wall. TAPPAN and LOEBLICH (1968, p. 1383) found, however, that the three-layered calcareous wall of an Eocene tintinnine consisted of 1) an inner layer of calcite rhombs, 2) a median layer of cryptocrystalline calcite, commonly of alveolar nature, and 3) a thin, smooth, and apparently structureless outer layer that was infrequently preserved. Wall details, such as those discussed above, are observed best at magnifications of several hundred times.

Distribution.—Fossil tintinnines were abundant principally in late Jurassic and Lower Cretaceous rocks of the Tethyan realm (Europe, Africa, and Asia). However, tintinnines have also been reported from sediments of Ordovician to Recent age. TAPPAN and LOEBLICH (1968) review stratigraphic occurrences. Recent tintinnines have a cosmopolitan distribution in the oceans but are most diverse in tropical waters. Only a few living species are known from fresh waters, and all fossil forms are believed to be marine.

Comparisons.—All other groups of organisms of comparable size apparently lack an imperforate calcareous vase-shaped test.

Radiolarians

Skeletal architecture.—Radiolarians are microscopic one-celled organisms that secrete a siliceous (opaline) shell, usually less than 1 mm and commonly between 0.1 and 0.2 mm in maximum diameter. The shell morphology is extremely variable but typically is a hollow perforate sphere or vase (Fig. 12) containing one or more bars or struts across the interior. Spines may extend beyond the perforate shell wall. Some forms consist only of simple tetraradiate or multiradiate branching spicules or needles. CAMPBELL (1954a) has prepared a comprehensive review, and DEFLANDRE (1952) has published a shorter general account of radiolarians. For a short bibliography see POKORNÝ (1958, p. 476–481). Articles by GOLL (1968, 1969) contain excellent drawings of some Cenozoic radiolarians in different views.

Skeletal microstructure.—Radiolarians usually are described taxonomically from isolated specimens rather than from thin sections. Consequently, taxonomic papers do not show the appearance of radiolarians in thin section, and figures illustrating radiolarians in thin sections are widely scattered in the literature, principally in microfacies reports (for example, COLOM, 1955; BONET, 1956). Petrographic information is lacking but the shell shows very low interference colors under cross nicols, which is consistent with a composition of opaline silica. Radiolarians in the Lower Jurassic limestone figures on Plate 3 exhibit low birefringence and fine-grained fibrous structure comparable to some cherts and chalcedonies. The porous structure of the shells is poorly shown in these specimens, and some additional alteration and addition of silica to the original shells may have taken place. The frequency of this type of alteration in the geologic record is not known, but opaline silica is much less stable than common quartz sand grains (SIEVER and SCOTT, 1963). DEFLANDRE (1936) and REMANE (1963, p. 27) have recorded replacement of radiolarian silica by calcite in Jurassic and Cretaceous Tethyan limestones of Europe. Examination with polarized light indicates that some calcareous replacement is present in both radiolarian figures on Plate 3.

Distribution.—Radiolarians, exclusively marine animals, have been reported from rocks of Ordovician to Recent age and are present on all continents. Concentrations of radiolarians in chert or silica beds have been interpreted as evidence of volcanic activities that provided abundant silica to initiate radiolarian blooms.

Comparisons.—Some open-lattice or spicular radiolarian skeletons are very similar to silicoflagellate and ebridian skeletons The silicoflagellates have a hollow, tubular skeleton and are usually smaller (a few dozen microns) so that they are not likely to be recognized at the magnifications of 100 or less. Silicoflagellates and ebridians usually are studied by freeing specimens from the matrix rather than by preparing thin sections. The spicular elements of ebridians and radiolarians are solid whereas many siliceous sponge spicules have a central canal along the spicule axis. Most siliceous sponge spicules recognized in petrographic work are likely to be larger than radiolarian spicules.

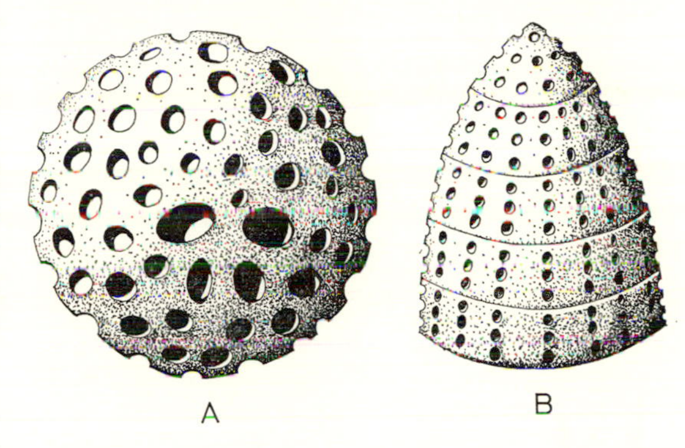

Fig. 12. Spherical (A) and conical (B) radiolarians.

Foraminifers

Skeletal architecture.—Foraminifers are one-celled animals that secrete architecturally varied agglutinated and calcareous shells. Most forms are multichambered (Pls. 4 to 13) and their chambers are arranged serially (uniserial, biserial or triserial) or are coiled (planispiral or conically spired). An indication of the wide variety of shapes can be obtained by examining figures in paleontologic textbooks (MOORE *et al.*, 1952, p. 39–72; POKORNÝ, 1958, p. 88–430).

Random thin sections through foraminifers frequently will not reveal all the chambers or the true shape from a single specimen. Biserial (Pl. 4–5) or triserial forms appear uniserial if cut parallel to one series of chambers. Planispiral (flat coiled) forms exhibit all chambers only if cut in the plane of coiling (Pl. 4–4). Sections at right angles to the plane of coiling will exhibit paired chambers (Pl. 7–2) in a manner directly analogous to shapes exhibited by coiled gastropods and cephalopods. Recognition of sections through single chambers or off axis (obliquely oriented) sections commonly depends on comparison with random sections of other specimens in the same slide. CUMMINGS (1955) has published line drawings indicating the appearance of some uniserial forms sectioned at various positions and angles across the shell.

Although foraminifers are known that are 5 centimeters or more in maximum dimension, most are smaller than 1 millimeter in maximum diameter. Large foraminifers are studied almost exclusively by thin section, and small foraminifers are studied increasingly by the same technique, especially specimens in indurated rocks.

Skeletal microstructure.—The complexity of foraminiferal wall microstructure as discussed in the literature is suggested by Table 9, which tabulates under five major headings some of the terms utilized in describing wall microstructure. NORLING (1968, p. 23–24) has used other terms to describe variations in lamellation of the shell wall, and various authors use additional terms to describe the layering in the wall and septa of lamellar foraminifers (see review by LOEBLICH and TAPPAN, 1964, p. C100). Many authors use differences in wall structure to demonstrate phylogenetic (evolutionary) relationships among foraminifers, but these interpretations are still much debated. However, nomenclatorial complexity illustrates the wide variety of wall microstructure within the foraminifers and shows that some combinations of wall characters and shell architecture are char-

Table 9

Outline of Wall Microstructure of Foraminifers
(Modified from CONKIN, 1961, p. 234—235; GLAESNER, 1963, p. 22; and NORLING, 1968, p. 24)

I. Layering
- A. Composite, consisting of several layers of different texture. For types of layers in Paleozoic foraminifers see summary of REITLINGER in LOEBLICH and TAPPAN (1964, p. C93), CONKIN (1961, p. 234–235), and text. For Mesozoic example see NORLING (1968, p. 22–24)
- B. Lamellar, consisting of layers added with each additional chamber, divided by NORLING into:
  1. Mesolamellar, consisting of newly built chamber wall covering one or more, but not all, of previous chambers
  2. Lamellar, consisting of newly built chamber wall that covers and adheres to entire previously formed test
- C. Nonlamellar, consisting of a single wall that adheres to or covers only a portion of the wall of previous chamber

II. Composition
- A. Calcite (most calcareous foraminifers)
- B. Aragonite
- C. Siliceous (primary in a few genera of one family; secondary in agglutinated arenaceous foraminifers, but see CONKIN, 1961, p. 234–235, for dissenting opinion)

III. Structure
- A. Agglutinated
  1. Arenaceous
     a. Calcareous, extraneous grains in calcareous or ferrugineous cement or both
     b. Siliceous, extraneous grains in siliceous cement
  2. Granular calcareous, equidimensional grains of calcite embedded in crystalline calcite cement
     a. Calcite granules secreted by the protoplasm, embedded in a secreted crystalline calcite cement (? pseudo-agglutinated)
- B. Fibrous
- C. Microgranular

IV. Texture
- A. Porcellaneous
- B. Granulate
- C. Radiate
- D. Single crystal
- E. Spicular

V. Perforation
- A. Imperforate (no pores)
- B. Perforate (pores)

acteristic of the foraminifers. Studies by X-ray and staining techniques indicate that less than one percent of calcareous foraminiferal shells are composed of aragonite (LOEBLICH and TAPPAN, 1964, p. C96). Most aragonitic forms belong to the families Ceratobuliminidae and Robertinidae.

Petrographic studies of arenaceous foraminifers have not been common. Consequently, most observations have been based on entire specimens, but include some rare analyses of wall cements and test mineralogy (LOEBLICH and TAPPAN, 1964; TOWE, 1967a). TOWE (1967a) discussed the arenaceous test of *Haplophragmoides canariensis* (D'ORBIGNY), which is composed of quartz and feldspar grains in an organic cement. However, the arenaceous content of shells of other arenaceous species is known to vary widely between species and within individual species that live on different substrates. TOWE (1967a, p. 149) suggested that the selection of kinds of grains by a foraminifer is a chemotaxial effect related to the kinds of organic compounds adsorbed on the surfaces of various silicate or carbonate grains. SLITER (1968, p. 83) reported that the tests of *Trochammina pacifica* CUSHMAN exhibit a progressive increase in grain size in the walls as the specimens grow larger, but the opposite effect has been reported in at least one species.

TOWE and CIFELLI (1967, p. 742) have noted that microscopic studies of the calcareous foraminiferal wall commonly emphasize one or the other of two features: (1) crystallography or (2) lamellar construction. The crystallographic approach has a long history usually cited as beginning with SORBY (1879), but WOOD (1949) listed some earlier petrographic studies of foraminiferal wall microstructure. The modern lamellar approach was initiated by SMOUT (1954) and has been pursued by a number of workers, particularly REISS (1963a, 1963b), whose interpretation of some features of lamellar growth has been questioned by HOFKER (1962, 1967).

TOWE and CIFELLI (1967) applied the electron microscope to the problems of foraminiferal wall structure. Although this technique is beyond the resolution of the optical microscope, the results are applicable to features observed by means of light microscopy. TOWE and CIFELLI inferred crystallographic orientations from observations based on the electron microscope. They recognized three types of wall structure: (1) radial, (2) granular, and (3) porcellaneous, all of which they related to optical features. Although a small number of species was studied, TOWE and CIFELLI recognized that some species could not be classified into these groups because they display intermediate structural features. Furthermore, optical characteristics also depend on the orientation of the thin section with respect to the structural units of the shell, an important point that has not always been recognized.

Radial structure (Pl. 12–1) is characterized optically by a black cross pseudofigure under crossed nicols when the shell wall is cut perpendicular to the shell surface. The pseudofigure results from the orientation of the c-axes of the crystallites perpendicular to the shell wall. Sections cut parallel to the shell wall exhibit an isotropic extinction under cross nicols as the c-axes of the crystallites are oriented perpendicular to the plane of the stage of the microscope. Optically indistinct radial walls are present when the crystallographic c-axes are not uniformly oriented perpendicular to the shell wall as viewed under crossed nicols

Granular structure is characterized by an apparent lack of preferred orientation of crystallites as viewed under crossed nicols. TOWE and CIFELLI (1967, p. 748) indicated that the crystallites have a preferred orientation (parallel to the rhombic cleavages of calcite). However, the rhombic cleavages do not provide a mechanism for a preferred (unidirectional) orientation of the c-axis with respect to the shell wall. This results in an apparent lack of orientation as viewed optically.

The porcellaneous wall (Pl. 5–5) contains a "true" random orientation of crystallites and is imperforate, whereas the radial and granular walls have perforations. Further, the crystallites in the radial and granular walls may be arranged into layers, as discussed at length by SMOUT (1954) and REISS (1963a). Layering is not apparent in the porcellaneous wall.

The studies by TOWE and CIFELLI (1967) were based on Recent foraminifers but clearly apply to Mesozoic and Cenozoic species. The applicability of this study to Paleozoic foraminiferal wall structure is less apparent probably due to problems of preservation and to differences in terminology among workers as well as differences in the types of Paleozoic foraminiferal wall.

REITLINGER (1950; as summarized by LOEBLICH and TAPPAN, 1964, p. C93) recognized six types of wall structure among Paleozoic smaller foraminifers: (1) dark micrograined walls with very fine calcite grains and fine perforations, (2) two-layered walls of gray or yellow color, consisting of small elongate, light-colored grains giving a fibrous appearance, and including some larger angular grains, (3) coarse-grained and agglutinated walls with much calcareous cement, (4) coarsely perforate walls with simple to dendritic pores and possibly with agglutinated material, (5) hyaline-radial perforate walls which may have an interior darker layer, and (6) micrograined walls of grayish, yellowish or brownish color.

REITLINGER (1958, p. 62) recognized five types of wall structure in the Paleozoic endothyrids and made the following freely translated comments:

"In the endothyrids the wall of the shell is calcareous, secreted not uncommonly with agglutinated particles, sometimes even in large numbers. In the wall, layers are partly differentiated by having different granularity and color. Sometimes they develop an additional interior vitreous layer. In several instances pores are observed, simple or complex, of alveolar character. It is possible to distinguish the following types of walls:

I. Wall, single layered, heterogeneous, small or fine-grained, usually with separate aggregates of larger grains or separate large grains (majority of Tournaisian plectogyrids).

II. Wall, homogeneous, fine-grained (majority of Visean plectogyrids and endothyrids), rarely with inner vitreous wall (*Quasiendothyra*). Within the limits of these groups in the wall of some endothyrids a thin dark integumentary layer (tectum) is isolated and simple pores are developed (group of *Plectogyra omphalata*).

III. Wall, multilayered, up to 4 layers: 1) thin dark integumentary layer (tectum), 2) light gray heterogeneous-granular layer, 3) dark fine-grained layer, 4) additional vitreous layer (*Globoendothyra*).

IV. Wall, rough-grained with agglutinated material, mainly grains of calcite, rarely with small shells of foraminifers, quartz, and similar materials. Usually an additional vitreous layer is observed.

V. Wall, rough-grained with agglutinated material (grains of calcite, small foraminifers and small fragments of skeletons of other organisms), with large pores, often of alveolar character (*Bradyina*)."

Although individual wall layers have similarities to wall microstructure observed in post-Paleozoic foraminifers, the organization of the wall layers of many Paleozoic smaller foraminifers apparently differs from that of most later foraminifers.

As reviewed by THOMPSON (1964, p. C364), the number of layers in the wall of Paleozoic fusulinids varies from two to four. Primitive early Pennsylvanian fusulinids have a three-layered wall consisting of a thin dark very fine-grained primary layer (tectum) over and underlain by lighter colored layers (upper and lower tectoria). In four-layered forms a lighter vitreous layer (diaphanotheca) is added below the tectum (Fig. 13 C). The advanced fusulinid wall (schwagerinid type, Pl. 6-1) consists of two layers (Fig. 13 D), a thin dark tectum and a distinctive honeycomb-layer (keriotheca).

In summary, the foraminifers illustrate well the complexity of wall microstructure that is possible among organisms secreting a monomineralic shell (Fig. 13). The number of textural and other descriptive terms used for describing the foraminiferal wall reflects this diversity (Table 9). The calcite of a porcellaneous wall, which possibly is analogous to the tectum and related layers in Paleozoic calcareous foraminifers, is very fine-grained, and the numerous grain boundaries internally reflect so much light that the wall appears dark. The larger the crystal size the more light penetrates and the layers appear lighter colored. In some cases crystals are so aligned that a calcite layer appears vitreous (clear or glassy). Preferred alignment of crystal units and crystalline unit boundaries also may produce a finely prismatic shell layer, usually seen in multiple layers as in lamellar wall structure. Variations in prismatic layers are probably responsible for references to fibrous and radiate types of wall structure in the foraminifers. The wall may be further modified by pores, which may be aligned to produce alveolar (for example, schwagerinid wall structure among the fusulinids) or inflated and less regularly distributed pores that form vesicular walls as in *Bradyina*. The walls are modified in some forms by admixture of calcareous or siliceous cement with grains of varying composition obtained from the substrate. The grains may be arranged in rough or smooth exterior patterns or in a graded or size arrangement within the shell wall. These types of wall structure are combined in various ways to form multilayered foraminiferal walls. As an additional complexity, the calcareous microstructure commonly is obcured by diagenesis, particularly among calcareous Paleozoic smaller foraminifers. Consequently, only the fellowship of foraminiferal ferrets can distinguish the individual sheep among the phlox.

Distribution.—In their stratigraphic review of foraminifers LOEBLICH and TAPPAN (1964, p. C135–136) reported that 1) arenaceous foraminifers first appeared in the Cambrian, 2) calcareous forms appeared first in the Ordovician, 3) porcellaneous walled foraminifers first appeared in the Carboniferous, and 4) perforate hyaline forms appeared first in the Permian. Foraminifers have been a common constituent of limestones since the middle Paleozoic.

Foraminifers have a worldwide distribution in all marine environments from the Ordovician to the Recent. A very few Tertiary and Recent forms have been reported from fresh water. The large foraminifers of the late Mesozoic to Recent have been interpreted as living in shallow warm marine waters. By analogy, the same interpretations have been applied to the large foraminifers (fusulinids) of the late Paleozoic.

Foraminifers are used increasingly for environmental (paleoecologic) interpretations as well as for chronologic correlations. Extrapolations from environmental studies of Recent foraminifers are apparently applicable as far back as the Cretaceous. Students of foraminifers frequently make the distinction in Mesozoic and Cenozoic examples between pelagic) free floating) and bottom dwelling (benthic) foraminifers. The former are esteemed as widespread chronologic markers and the later as paleoecologic indicators.

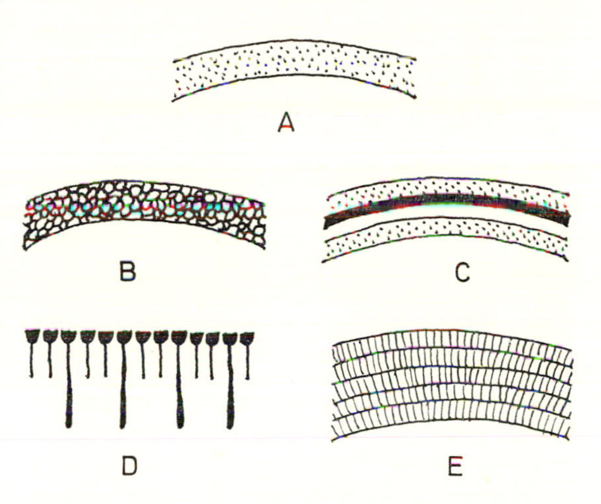

Fig. 13. Foraminiferal wall microstructure. A, porcellaneous microstructure having uniform texture; B, granular microstructure showing no preferred orientation of crystallites; C, multilayered microstructure exhibiting dense fine-grained and clear inner layers and lighter fine-grained outer layers; D, schwagerinid wall microstructure showing porous and characteristic comb-like wall; E, lamellar wall microstructure consisting of layers of small crystalline units oriented perpendicular to layers.

Comparisons.—Most foraminifers are characterized by their small size and multichambered test. Coiled gastropods are not chambered and usually exhibit recrystallized or leached and infilled wall microtexture. Cephalopods are usually larger than foraminifers, have much larger chambers, and their aragonitic walls are commonly recrystallized to sparry calcite. Large coiled foraminifers are characterized by numerous chambers and usually complex internal structure.

Sponges (Porifera)

Skeletal architecture.—Sponges take almost any imaginable shape. However, many entire sponges are characterized by globular, vase, or cup shapes. The skeleton is formed of units of mineral matter called spicules that either exist independently in the body wall or fuse or interlock to form various types of networks. The spicules are either calcareous or siliceous, although one small group has "spiculoids" composed only of organic substances. In addition, nonspicular calcium carbonate may form a major part of the skeleton of some calcareous sponges and organic fibers (spongin) form a variable part of the skeleton of most sponges. One group of sponges, including the common bath sponge, has a skeleton made solely of organic fibers. Sponges may also incorporate sand and other foreign particles into their skeletons. Sponges containing only organic fibers are rare in the fossil record.

The size range of sponge skeletons commonly varies from a few centimeters to a meter or more. Spicules range in size from less than a millimeter to a few centimeters in maximum length but transverse diameters are usually less than a millimeter.

In thin section, articulated skeletons commonly exhibit a definite organization of the spicules (Pl. 14–6) which are arranged to form open areas (canals) in the midst of the spicular network (Pl. 14–4, 14–5). Some fused spicular nets resemble the models used in chemistry classes to illustrate the atomic structure of molecules (Pl. 14–6). Individual disarticulated spicules show a great variety of rayed shapes both ornamented and unornamented (Pl. 14–1, 14–2, 14–3). De Laubenfels (1955) gives illustrations of spicules.

Skeletal microstructure.—The microstructure of sponge spicules exhibits initially a rarely preserved concentric layering about the axis of each ray. Opaline silica is relatively unstable, and siliceous sponges are not uncommonly replaced by calcite. The replacement is analogous to the alteration of opaline radiolarian tests by calcite. Equally common is the replacement of calcareous sponge skeletons by silica.

Most originally siliceous sponge spicules are characterized by a central canal or hollow center which may be obliterated during diagenesis (Pls. 14–3, 15–1). Spicules of calcareous sponges were originally single crystals of high-magnesian calcite. Siliceous spicules commonly exhibit a fine structure analogous to many fine-grained cherts.

Distribution.—Sponges are known from the Cambrian to the Recent. The existence of Precambrian sponges is considered questionable.

Comparisons.—The reticulate nets of sponge spicules are comparable to the internal features of some corals, stromatoporoids, and bryozoans, but these groups always exhibit calcareous skeletons and have various kinds of lamellar and fibrous wall structure, lacking in most sponges. In addition, the skeletal network of sponges is commonly more irregular, and the structural elements shorter and less continuous than in corals or bryozoans. The vesicular interior of some vertebrate bones lacks the regular network exhibited by many sponges and vertebrate bone is phosphatic. Individual large spicules may resemble spines of brachiopods but lack the fibrous wall structure characteristic of brachiopod spines. Some plates (sclerites) of holothurians (echinoderms) resemble the triradiate spicules of calcareous sponges, but most holothurian spicules have a different repertory of shapes from those of sponges. Alcyonarian corals have calcareous spicules, many of which resemble the siliceous spicules of some lithistid sponges, but they differ in mineralogic composition as well as in details of form.

Archaeocyathids

Skeletal architecture.—Archaeocyathids possess cuplike, vase-shaped or cylindrical calcareous skeletons (Fig. 14), commonly 1—10 cm in maximum dimensions, frequently consisting of porous inner and outer walls (Pl. 16–4), which are joined by vertical septa (parieties). Additional structures include porous flat and curved plates (tabulae and dissepiments) analogous to the internal structures of corals (Pl. 16–2). Some walls exhibit very complex porous vesicular or spicular architecture (Vologdin, 1962, Figs. 8–21; Hill, 1965, Figs. 4, 5). Less commonly the skeleton is colonial and exhibits chainlike (catenoid) or branching (dendroid) growth forms.

Archaeocyathids have been reviewed by Okulitch (1955), Zhuravleva (1960), Vologdin (1962) and Hill (1965). Okulitch (1955) presented line drawings of archaeocyathids, and Vologdin (1962) and Hill (1965) have published line drawings and photographs. Archaeocyathids are studied commonly by peels or thin sections so that published cross sections, although oriented, usually exhibit features comparable to those observed in randomly oriented sections.

Skeletal microstructure. —Okulitch (1955, p. E8) characterized archaeocyathid wall microstructure as granular-laminar. Hill (1965, p. 25) reported that the microstructure of archaeocyathids "consists of a very fine granular mosaic of calcite, the crystals being about 0.02 mm in diameter and uniform in color." Laminar textures also have been observed, especially in secondary deposition of carbonate at the base of the skeleton. The limited number of sections we have examined show a dark, fine-grained wall of equidimensional grains of calcite.

Distribution.—According to Hill (1965, p. 30), archaeocyathids are known from all continents except South America and are characteristic of Lower Cambrian calcareous facies. Most reports of pre- and post-Lower Cambrian archaeocyathids are questioned by Hill (1965, p. 30 ff).

The paleoecology of archaeocyathids as reviewed by Hill indicates that they preferred warm waters (greater than 25° C) at depths of 20 to 50 meters. Bioherms of archaeocyathids developed at these preferred depths but individuals lived less commonly at depths interpreted to be as shallow as 10 meters or as deep as 100 meters.

Comparisons.—The conical or cylindrical shape, internal septa, and internal flat and curved plates give the archaeocyathids a general resemblance to solitary cup corals. Perforate walls differentiate archaeocyathids from cup corals, which are not known from the Cambrian. Archaeocyathids have a different spicular architecture than sponges.

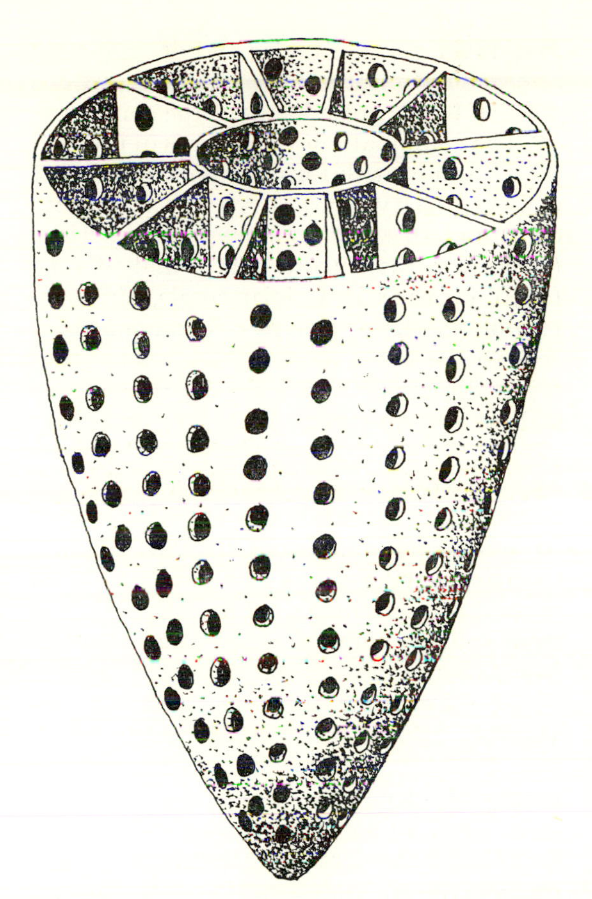

Fig. 14. Archaeocyathid skeleton displaying septa and inner and outer porous walls.

Corals (Anthozoan Coelenterates)

Skeletal architecture.—Although the different groups of living corals vary significantly in the nature of their soft parts, the major living and fossil groups have produced similar types of solitary or colonial skeletons. The skeleton, which is exoskeletal in origin, is characterized commonly by an outer wall and internally by a series of vertical plates (septa) arranged radially perpendicular to the outer wall and which may or may not meet in the center of the skeleton (Pls. 17–4, 18–2). Horizontal plates (tabulae) and convex upwardly or outwardly curved plates (dissepiments) are common in many corals and may be restricted to areas between septa and near the outer wall or extend throughout the coral skeleton (Pl. 19–5). Septa are observed most easily in transverse sections perpendicular to the growth axis of the coral. Tabulae and dissepiments appear most clearly in longitudinal sections parallel to the growth axis of the coral.

The combinations of patterns of internal plates and growth forms provide an almost endless variety of coral types. The nomenclature of these types is reviewed by HILL (1956) for the rugose corals and WELLS (1956) for the scleractinian corals, which are two of the major groups of coral contributors to the fossil record. Both groups contain cylindrical or conical solitary corals as well as loosely joined (Pl. 19–2) or tightly packed (Pl. 19–5) colonial individuals (corallites). Colonial corals exhibit a variety of shapes ranging from large encrusting masses to branching and hemispherical colonies. Colonies vary from a few millimeters to several meters in maximum dimension, but most solitary corals are 1 to 10 centimeters in maximum dimension.

The Paleozoic tabulate corals are invariably colonial and have numerous small round or polygonal corallites. The corallites commonly lack true septa and septalike structures. In some forms the walls of the corallites are pierced by numerous small openings (mural pores) and in others the wall is solid or less commonly tubules connect adjoining corallites. Horizontal or curved partitions (tabulae) are almost always present. Colonies may consist of loosely joined or tightly packed corallites. One group of tabulate corals (heliolitids) consist of corallites, usually having twelve short septa, which commonly are poorly developed or absent in thin sections. Individual corallites are separated by tubular, vesicular or tabulate skeletal tissue.

Alcyonarian corals do not produce massive calcareous skeletons but contain numerous small calcareous elements (spicules) within the organic tissue of the colony. The spicules vary in size from a few tenths of a millimeter or less to five millimeters in maximum dimension and exhibit a variety of shapes, most commonly very ornamented, spinose rods. These spicules are significant contributors to modern sediments in tropical carbonate environments.

Skeletal microstructure.—The microstructure of corals is the subject of a large literature which is difficult to review because workers have introduced a large number of terms to describe both altered and unaltered microstructural characters. KATO (1963) has presented an introduction to the literature, and additional reviews include those of ALLOITEAU (1952; 1957), HILL (1956), WELLS (1956), SOKOLOV *et al.* (1962), and SCHOUPPÉ and STACUL (1966). Workers frequently illustrate microstructural features by means of line drawings, presumably because the finely fibrous character of the coral wall does not yield good photomicrographs. The scanning electron microscope should provide assistance in this problem.

Most Paleozoic corals apparently secreted a calcitic skeleton, and many post-Paleozoic corals deposited an aragonitic skeleton. However, the same basic microstructural features are present in skeletons of both mineralogic types (KATO, 1968). The basic feature of many well-preserved coral skeletal microstructure is the deposition of fine calcareous fibers perpendicular to the surface of the secreting tissue. Because of topographic variations and shifts in the secreting surface, the fibers can be arranged in more than one pattern. However, the secreting surface does not have to be parallel to all the final skeletal architectural surfaces. FLOWER (1961) and KATO (1963) have published line drawings illustrating the major patterns of fibers that form distinctive categories of skeletal microstructure.

FLOWER (1961) recognized three types of microstructural walls in some North American Ordovician corals (Fig. 15): 1) fibrous, 2) rugosan, and 3) trabecular. The fibrous wall exhibits a relatively simple continuous wall of parallel fibers arranged obliquely to the wall surfaces so that they slope up and in from the outer corallite wall. A discontinuity of microstructure (axial plane) is present at the junction of closely appressed corallite walls (Fig. 15A). Rugosan wall microstructure (Fig. 15B) is characterized by a uniform dark fine-grained layer (axial plate), which lies between corallites at the position of the axial plane or represents the outer wall of loosely packed corallites.

In the trabecular wall (Fig. 15C), the axial plane or plate of the rugosan type of wall has broken into numerous small planes or points from which the fibers radiate upward and outward to form chevron (herringbone) or cone-in-cone patterns when viewed respectively in longitudinal or transverse cross section. Recrystallization or alteration of the trabecular wall commonly produces clear rods, which some authors call monacanthine wall microstructure. FLOWER (1961, p. 26) regarded the coral wall as primitively composed of parallel fibers (fibrous wall) that subsequently evolved into walls containing groups of radiating fibrous bundles (sclerodermites), which were arranged into columns of varied complexity (trabeculae). In post-Ordovician corals the organization of trabeculae frequently displays considerable complexity because multiple and irregularly arranged trabeculae appear in the wall of the skeleton (KATO, 1963, p. 587).

Electron microscopy of aragonitic scleractinian corals indicates that skeletal crystals grow in the form of laths, blades, or needles arranged as individuals, in spherulitic arrays, or in clusters called fasciculi (WISE, 1970, and references cited therein).

KATO (1963) has prepared a series of instructive line drawings illustrating variation in wall structure of Paleozoic rugose corals caused by 1) topographic variations and discontinuities in the secreting surface, 2) position of thin sections with respect to structural and microstructural units of the skeleton, and 3) alteration of original structures. KATO (1963, p. 604) also recognized lamellar microstructure consisting of lamellae (layers) parallel to the median septal dark line or added as an overgrowth to other types of wall microstructure. He distinguished between a "true" lamellar (layered) wall and a lamellarlike wall marked by growth increments in the fibrous wall microstructure. KATO (1963) included summaries of terminology and line drawings of the wall microstructure of numerous Paleozoic coral genera. ALLOITEAU (1952, 1957) has pro-

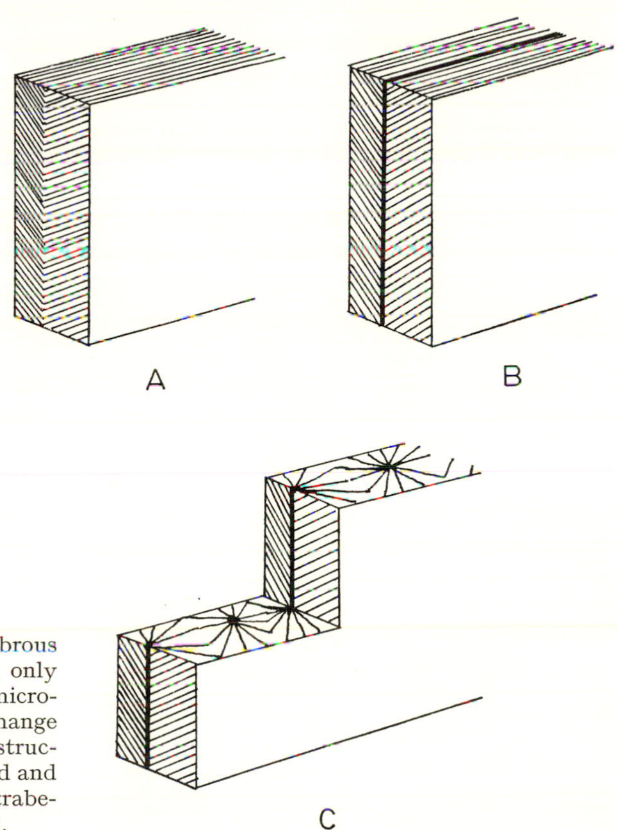

Fig. 15. Coral wall microstructure. A, fibrous microstructure having axial plane marked only by change in direction of fibers; B, rugosan microstructure having axial plate at junction of change of orientation of fibers; C. trabecular microstructure characterized by fibers radiating upward and outward in cone-in-cone fashion from solid trabeculae (modified from FLOWER, 1961, Fig. 3).

vided numerous illustrations of microstructure of post-Paleozoic corals.

Distribution.—Rugose corals range from the Ordovician through the Permian. Scleractinian corals are of Triassic to Recent age. Tabulate corals are of Ordovician through Permian age, but a few post-Paleozoic tabulates are reported from rocks as late as Eocene.

Fossil corals are interpreted as marine animals and were most abundant in clear shallow carbonate seas where they were major contributors to reefs and banks from the Ordovician to the Recent. Corals have a worldwide distribution throughout their stratigraphic range.

Comparisons.—Solitary corals are similar to archaeocyathids but commonly lack porous walls. The wall structure of solitary corals is fibrous, whereas that of the archaeocyathids is uniformly fine-grained. Many post-Paleozoic corals (including all scleractinians) were composed originally of aragonite but usually exhibit recrystallized or infilled sparry calcite walls in thin section.

Colonial corals are separated most readily from bryozoans or stromatoporoids by the presence of septa and by differences in wall structure. Bryozoans usually have lamellar walls, and stromatoporoids have rather complex microstructure readily subject to alteration. The layering produced by the laminae of stromatoporoids is usually absent in bryozoans and corals. The diameter of corallites is usually much larger (greater than a millimeter) in corals than are the zooecial apertures of bryozoans (less than a millimeter, commonly 0.15 to 0.35 mm).

Calcareous alcyonarian spicules have shapes similar to those of lithistid sponge spicules, which are, however, siliceous. Holothurian (echinoderm) plates (sclerites) generally have a different set of shapes than alcyonarian spicules.

Stromatoporoids

Skeletal architecture.—The early Paleozoic stromatoporoids are colonial organisms generally assigned to the coelenterates although their position within the coelenterates is not clearly known. Many of the early Paleozoic stromatoporoids, especially those in the Ordovician, are composed of outwardly or upwardly convexly curved, overlapping plates called cysts (Pl. 22-1). An area of smaller cystose plates commonly surrounds a central axis containing very large cystose plates in cylindrically shaped skeletons. However, stromatoporoid skeletons most commonly resemble a rectangular network in sections parallel to the growth direction as seen in longitudinal or vertical sections (Pl. 21-2). The horizontal plates, and in some cases parallel layers of rods, are called laminae and the vertical rodlike elements are designated pillars. The spacing and arrangement of the pillars and laminae vary widely and the enclosed spaces are called galleries. Other structures include upturned laminae (Pl. 21-1, 21-4) that form wartlike protuberances (mamelons) on the exterior surfaces and radiating canals (astrorhizae) associated commonly with the mamelons. The rectangular network does not appear in tangential sections perpendicular to the direction of growth.

Most stromatoporoids are massive forms a few decimeters in maximum dimension although larger forms are known. The shape of massive colonies varies with environmental conditions but hemispherical and tabular forms are common. Individual thin sections obviously will encounter only a fraction of a large colony. Some Devonian ramose colonies have branch diameters of a few millimeters. Modern monographic studies, based almost exclusively on thin sections, and reviews of lower Paleozoic stromatoporoids have been published by LECOMPTE (1951, 1952, 1956), GALLOWAY (1957), and YAVORSKY (1955, 1957, 1961, 1963, 1967). FLÜGEL and FLÜGEL-KAHLER (1968) have published a comprehensive synonomy and bibliography of the Paleozoic stromatoporoids. DEHORNE (1920) reviewed the Mesozoic stromatoporoids.

As HUDSON (1960, p. 182) has noted, the systematic position of Mesozoic and Cenozoic "stromatoporoids" is the subject of argument among various workers. On structural and stratigraphic grounds HUDSON (1960, p. 182-183) indicated that Mesozoic stromatoporoids represent a different lineage than either the Paleozoic stromatoporoids or the Tertiary hydroids, but he preferred to refer Mesozoic forms to the Stromatoporoidea (*sensu lato*). Architecturally the Mesozoic stromatoporoids exhibit vertical (pillars) and horizontal elements analogous to Paleozoic forms. However, some Mesozoic forms lack a rectangular network because of the irregularity of the horizontal elements. HUDSON (1958, p. 90) recognized two groups of Mesozoic stromatoporoids: 1) the milleporidiids, lacking astrorhizae and mamelons (HUDSON, 1956, p. 716) and 2) the remaining (?) non-hydroidean families which exhibit astrorhizae and related structures. The astrorhizal system (astrocorridors; astrorhizal canals) appears to be more complex in Mesozoic stromatoporoids than in Paleozoic forms.

Skeletal microstructure.—The skeleton of lower Paleozoic stromatoporoids apparently was highly porous and readily subject to alteration. STEARN (1966, p. 78) reported no less than 14 types of wall microstructure described by different students of stromatoporoids (Table 10 and Fig. 16). Workers are not in complete agreement on which types of microstructure are primary and which secondary. Nevertheless, thin section studies of wall microstructure are considered essential for differentiation of taxa at all levels within the Paleozoic stromatoporoids. STEARN (1966, p. 83) and ST. JEAN (1967, p. 420) have proposed somewhat different mechanisms of altering original skeletal tissue to obtain some of the microstructural features presently observed (Figs. 17 and 18). Microstructure is generally observed best at magnifications of 50 to 100 times. The reviews of shell microstructure by STEARN (1966) and ST. JEAN (1967) contain diagrammatic text-figures as well as unretouched photographs.

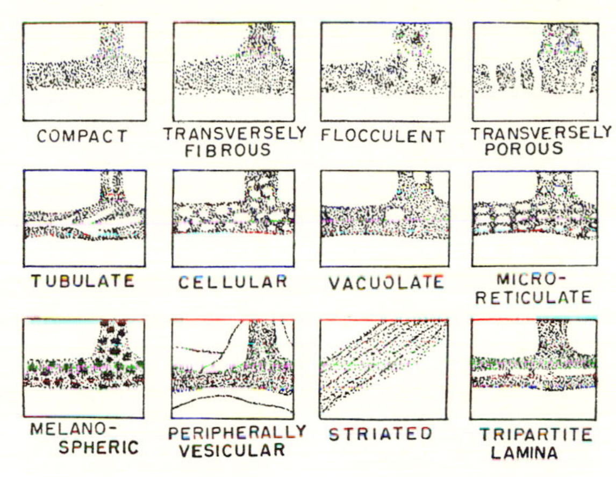

Fig. 16. Stromatoporoid wall microstructure (STEARN, 1966, Fig. 1, by permission of the Palaeontological Association and the author).

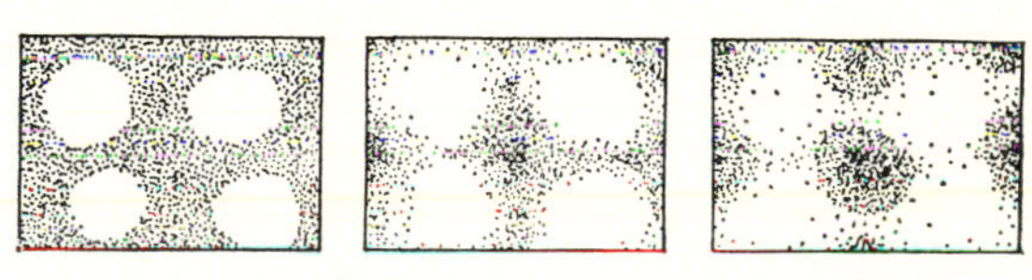

Fig. 17. Transition from cellular to melanospheric microstructure, left to right (STEARN, 1966, Fig. 2, by permission of the Palaeontological Association and the author).

Table 10. *Skeletal Microstructure in Lower Paleozoic Stromatoporoids*
(Modified from STEARN, 1966, p. 78)

Cellular (maculate of some authors). Tissue filled with closely spaced subspherical voids

Compact. Tissue composed of evenly distributed specks or evenly colored calcite

Flocculent. Tissue composed of unevenly distributed specks or coloring

Maculate. See cellular

Melanospheric. Tissue filled with dark subspherical groups of specks

Meshed fiber. Lamina longitudinally fibrous, composed of intermeshed fibers and dissepiments

Microeticulate. Tissue contains cellules arranged in vertical and horizontal series

Peripherally vesicular. Structural elements bordered by a layer of vesicles or a continuous membrane

Striated. Tissue filled with dark or light, thin, rod-like bodies

Transversely porous. Tissue transversed by pores which open into galleries

Tripartite laminae. Lamina has a central light zone that may break up into a line of cellules

Tubulate. Tissue contains curved and branching tubes, commonly horizontal

Vacuolate. Tissue contains subspherical voids, larger and more distinctly spaced than cellules

Water jet. Tissue marked by fibers that spray outwards and upwards from a central zone

As discussed by HUDSON (1960, p. 182; 1959, p. 29), the microstructure of Jurassic stromatoporoids consists of two layers. A central fine, dark, granular, less commonly maculate (spotty) layer and a fibrous layer. The fibrous layer consists either of uniformly arranged fibers perpendicular to the wall surface (orthogonal fibers) or bundles of fibers (fascicles, trabeculae or clinogonal fibers). DEHORNE (1920) contains an earlier detailed discussion of the wall microstructure of Mesozoic stromatoporoids.

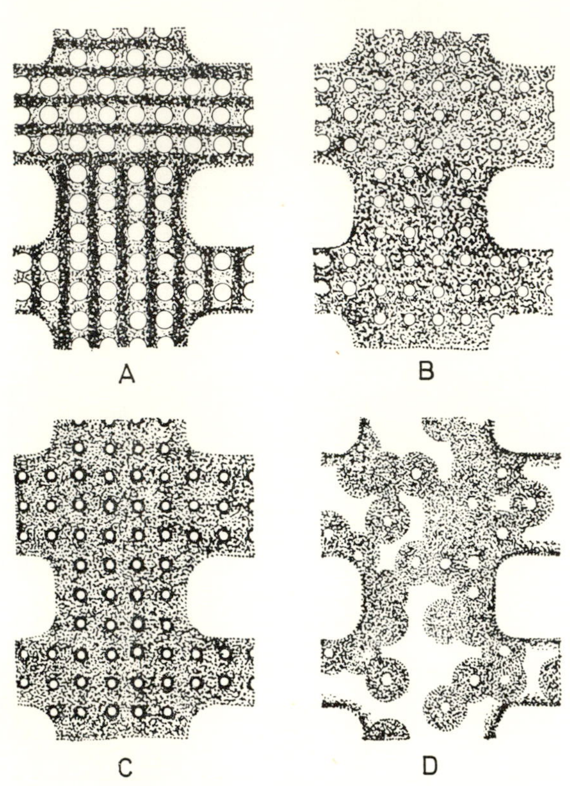

Fig. 18. Sketches illustrating proposed relations between maculate and microreticular stromatoporoid microstructure. A, vertical or horizontal microlaminae indicated by arrangement of dark and light tissue; B, tissue interpreted either as microlaminate or as aligned hollow centers of maculae; C, as B but hollow centers having dark outlines that accentuate maculate appearance of tissue; D, fused and partially destroyed maculae (ST. JEAN, 1967, Fig. 1).

Distribution.—Early Paleozoic stromatoporoids had a world-wide distribution in the Ordovician, Silurian, and Devonian. They are associated principally with carbonate rocks, presumably because they preferred clear shallow waters. They are an important constituent of Silurian and Devonian biohermal and/or biostromal communities.

Mesozoic stromatoporoids apparently preferred conditions comparable to their lower Paleozoic analogues, but some forms apparently were adapted better to muddy environments. Mesozoic stromatoporoids are distributed most widely in the Tethys region (Europe—Asia—North Africa) and are best known from Jurassic and Cretaceous rocks.

Comparisons.—Stromatoporoids of the lower Paleozoic lack the clearly double-layered wall, well-developed fibrous layer, and complex astrocorridors of the Mesozoic stromatoporoids. The rectangular patterns in some longitudinal sections of stromatoporoids might be confused with some tabulate corals, but the anthozoan corallites usually are much better delimited than the individual chambers or galleries in stromatoporoids. The rectangular patterns of the calcareous cellular algae or of wood are much smaller than the chambers in stromatoporoids. Bryozoans differ in wall structure from the stromatoporoids, and contain, in contrast to the stromatoporoids, well delimitated zooecial chambers. A few stromatoporoids produce skeletons resembling sponge spicule networks, but sponges lack the types of wall structure of the stromatoporoids.

Bryozoans

Skeletal architecture.—Bryozoans are colonial animals whose calcareous skeletons exhibit a wide range of shapes. The most common shapes (Pls. 23 to 28) are encrusting (flat, hemispherical, or irregular), branching (hollow or solid; flattened or circular), and fenestrate (resembling a window screen). These categories of shapes are convenient for discussion purposes, but the categories overlap as some fenestrate bryozoans and probably all hollow ramose colonies were encrusters. Colonial sizes vary from a few millimeters to half a meter but most colonies (zoaria) or fragments of zoaria exhibit a range of 1 to 10 cm in maximum dimension.

Internally the individual animals lived in round or polygonal elongate tubes or polygonal boxes (zooecia) that may or may not exhibit a number of different kinds of transverse (diaphragms, Pl. 28–2), curved (cystiphragms, Pl. 25–3) or "incomplete" plates (hemiphragms). In addition, smaller tubes such as acanthopores (usually containing no internal structures) and mesopores (commonly containing numerous closely spaced horizontal plates, Pl. 23–2) are also present. Acanthopores containing diaphragms are distinguished from mesopores by their laminated "cone-in-cone" structure pointing outward within the zooecial wall laminae. In general, mesopores in lower Paleozoic bryozoans contain numerous horizontal plates that usually are lacking in upper Paleozoic mesopores. CUFFEY (1967, Figs. 14, 16, 17) has illustrated diagrammatically a number of internal structures in a single Paleozoic bryozoan species, but many of these internal structures are widely distributed in Paleozoic bryozoans (Fig. 19).

BOARDMAN and CHEETHAM (1969) assigned bryozoans to two broad informal categories based on the geometry and mode of growth of the zooecia within colonies: 1) tubular bryozoans and 2) box bryozoans. The tubular bryozoans include the trepostome, cryptostome, cyclostome, and crystoporate bryozoans, which have terminal apertures at the surface of the colony and grow by addition to the surficial end of the zooecial tubes. In many Paleozoic trepostome and cryptostome bryozoans, the zooecial walls are thin in an inner zone (endozone, immature zone) and thick in an outer zone (exozone, mature zone) of the colony (Pls. 23–6, 28–2). The Paleozoic cyclostome bryozoans are characterized by simple tubes (zooecia) commonly with wall of constant thickness and generally without internal structures. Most cyclostomes are post-Paleozoic and exhibit minutely porous walls and thick laminated wall tissue in the exozone. The Paleozoic cystoporate bryozoans generally have a vesicular to cystose skeletal framework between zooecia and simple zooecial tubes (Pl. 24–2), which commonly are modified at their outer ends by a hood (lunarium) partially covering the aperture. The lunarium usually is reflected internally in tangential thin sections by a crescent-shaped notch in the zooecial wall and a thickening of wall tissue behind the notch.

The boxlike bryozoans consist principally of the Mesozoic to Recent cheilostome bryozoans that commonly have differentiated zooecia that serve specific protective and reproductive functions. However, the Paleozoic fenestrate bryozoans, although presently placed in the cryptostome bryozoans, exhibit boxlike zooecia, that is, the zooecial chamber usually did not enlarge significantly during growth of the colony as

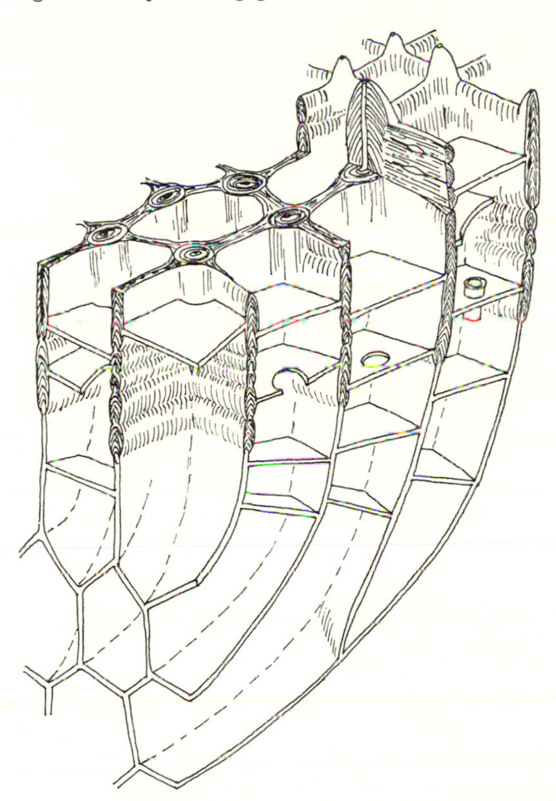

Fig. 19. Bryozoan architecture showing interiors of zooecial tubes displaying internal flat or curved plates (diaphragms, cystiphragms), laminated walls, and cone-in-cone microstructure of acanthopores present within the zooecial walls and projecting above the surface of the colony at the upper right (modified from CUFFEY, 1967, Fig. 16).

in most tubular bryozoans. The characteristic window screen growth habit of Paleozoic fenestrate bryozoans easily differentiates them from most boxlike cheilostome bryozoans.

Tangential sections (parallel to and directly below the colony surfaces) of bryozoans reveal the cross sectional shapes of zooecia (Fig. 19); differences in zooecial diameters that may represent differentiated zooids (individual animals) or accessory tubes such as acanthopores or mesopores;

and the presence or absence of vesicular or cystose tissue between zooecia. Longitudinal sections (perpendicular to the colony surface) show in profile the orientation and length of zooecial and accessory tubes within the colony, progressive variations in wall thickness or wall character, internal plates, and interzooecial vesicular or cystose tissue along the zooecial length.

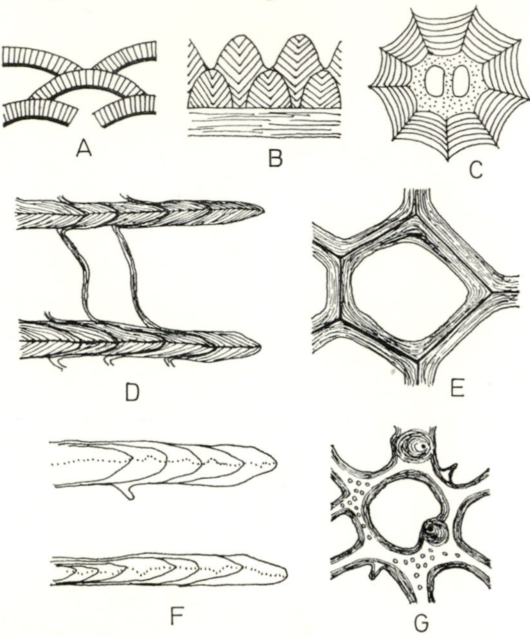

Fig. 20. Bryozoan wall microstructure. A, thin fine-grained primary layer and thick fibrous or blocky secondary layer; B, lamellar (below) and fibrous (above) layers in the cheilostome bryozoan *Metrarhabdotos* (modified from CHEETHAM *et al.*, 1969, Fig. 1C); C, clear granular inner layer (stippled) surrounding zooecia and laminated outer layers pierced by thin rods extending from the inner granular layer (modified from TAVENER-SMITH, 1969, Fig. 1A); D, E, thinly laminated wall microstructure in longitudinal and tangential sections; F, G, thickly laminated wall microstructure in longitudinal and tangential sections; acanthopores present in wall in tangential section G (D–G, modified from BOARDMAN, 1960, Figs. 12, 13).

Skeletal microstructure.—Most Paleozoic bryozoans presumably had calcitic skeletons. RUCKER (1969) has determined the mineralogy of a number of Recent species of the box-shaped cheilostome bryozoans, a post-Paleozoic group, and found calcite (38 species), aragonite (8 species), and mixed calcite-aragonite (15 species) mineralogy in individual colonies. CHEETHAM *et al.* (1969) have interpreted the aragonitic skeleton in one genus of cheilostome bryozoans as a recent evolutionary development, which is in accord with an earlier statement by RUCKER (1967) that a calcite skeleton is the more primitive mineralogic condition in cheilostome bryozoans.

On the basis of reviews and reports of bryozoan microstructure by BOARDMAN (1960), BOARDMAN and CHEETHAM (1969), CHEETHAM *et al.* (1969), TAVENER-SMITH (1969), and UTGAARD (1968a, 1968b), bryozoans exhibit no less than three types of wall microstructure distributed in various combinations in different bryozoan groups (Fig. 20). The types of bryozoan wall microstructure are: 1) laminated, 2) granular, and 3) fibrous.

Laminated wall tissue is present in some members of all the major tubular and box-like bryozoan groups and is one of the distinctive features of much bryozoan debris as revealed in thin section. The closely spaced parallel lineations of the laminae of the wall tissue (Fig. 20) appear fibrous in cross section or clear and structureless in sections parallel to the laminae (BOARDMAN, 1960, p. 27). Laminated tissue is most common in the walls of the exozone. Granular tissue is present in two forms: 1) thin dark uniform very fine-grained layer and 2) thick or thin clear layer. The first form is common in the endozone of many bryozoans or as the initial wall layer in many Paleozoic cystoporate genera. The second form is most common as the primary layer in fenestellid bryozoans (TAVENER-SMITH, 1969, p. 285) or possibly in the special zooecial (lunarial) deposits of some cystoporate genera. One might suggest that either grain size or uniform crystal orientation are the cause of these differences in granular walls, but the electron microscope work of TAVENER-SMITH (1969) on the fenestellids does not appear to support either hypothesis. Fibrous tissue is present also in two forms: 1) as blocky, crystalline or prismatic aggregates perpendicular to the zooecial walls (Fig. 20) in some cystoporate genera (UTGAARD, 1968a, p. 1033) and as 2) radially fibrous ("water jet") structure at high angles to the zooecial wall surface in the aragonitic superficial layer of the frontal wall (Fig. 20) of at least one cheilostome genus (CHEETHAM *et al.*, 1969, p. 131).

In many tubular bryozoans having laminated walls in the exozone, the laminae of the zooecial walls are arched either broadly or sharply convexly outward and produce a cone-in-cone or chevron structure (Fig. 20) in cross section. BOARDMAN and TOWE (1966) and NYE (1969) have reported that some Tertiary cyclostome bryozoans exhibit a reverse cone-in-cone wall structure, that is, the laminae are arched inward with respect to the exterior of the colony.

Distribution.—Modern calcareous bryozoans are marine and are present at all latitudes and depths although they are most abundant in shallow continental seas (SCHOPF, 1969b). By analogy, abundant fossil calcareous bryozoans, which are distributed worldwide from the Ordovician to the Recent, are assumed to have lived in exclusively marine shallow continental seas. Bryozoans are not very common nor very well known in Triassic or Jurassic rocks. As is typical of many colonial animals, the shape of bryozoan colonies (zoaria) is commonly variable and reflects environmental conditions. DUNCAN (1957), SCHOPF (1969a), and CUFFEY (1970) provide discussions and an introduction to the literature on bryozoan ecology and paleoecology.

Comparisons.—Finely comminuted fragments of the laminated layer of the bryozoan wall could not be differentiated in thin section from similar appearing brachiopod debris. However, most bryozoan fragments are large enough to show some zooecia in cross section and are unlikely to be mistaken for brachiopods. The laminated structure in the mature wall (exozone) of many bryozoan groups distinguishes bryozoans from superficially similar forms such as tabulate corals or stromatoporoids as viewed in longitudinal sections. In addition, the diameter of corallites is usually greater than 0.5 mm, whereas bryozoan zooecia are usually less than 0.5 mm in diameter. The cellular structure of wood or the coralline algae, which superficially resemble heavily tabulated exozones of some bryozoans as viewed in longitudinal sections, is an order of magnitude smaller than the zooecial diameters of bryozoans.

Brachiopods

Skeletal architecture.—The brachiopod shell consists of two unequal valves or curved plates, which usually are hinged along one margin. External outlines of valves usually are round, ovate, or subtriangular. The curvature of individual valves is variable

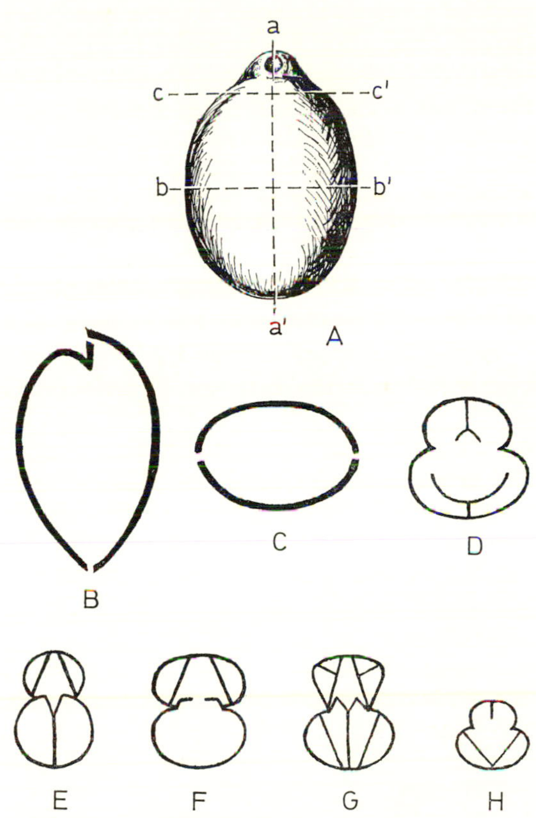

Fig. 21. Brachiopod architecture. A, top (brachial, ventral) view of valves of brachiopod; B, cross section through line aa′ of A; C, cross section through line bb′ of A; D–H, cross sections through line cc′ illustrate variations in number and distribution of plates in the beaks of some brachiopods.

and includes flat, convex, concave, and flexed forms that commonly are elongated either parallel or perpendicular to the hingeline. The shell surface may be smooth, plicated, corrugated, or spinose. Brachiopods are commonly 1 to 10 cm in maximum dimension, but a few forms are 30 cm or more in maximum dimension.

Internally, several patterns of vertical and horizontal plates may be present principally in the beaks or along the hingeline within one or both valves (Fig. 21). The involved terminology of these plates is defined and illustrated by MOORE *et al.* (1952) and WILLIAMS and ROWELL (1965). The vertical plates appear singly (median septum) or doubly (dental or crural lamellae or double septa) in either valve. The double plates may converge, diverge, or remain parallel to one another but usually extend for only a portion of the distance from the hingeline or beaks toward the front of the shell. The vertical plates may support variable flat or spoon-shaped horizontal plates (hinge plate, spondylium, curalium). Less commonly the spiral internal calcareous supports (spiralia) of the feeding organ (lophophore) are preserved. In cross section the spiralia would appear as a series of paired structures as shown in Plate 99–1.

Brachiopods are divided into two major groups: 1) inarticulates and 2) articulates. Inarticulates usually lack the complex internal plates and calcified spiralia found in the articulates and have either phosphatic or calcareous shells. Articulate brachiopods usually have internal plates, less commonly calcified spiralia, and calcareous shells. The plane of symmetry in brachiopods is generally at right angles to the hingeline and perpendicular to the plane of juncture of the two valves (commissure). In contrast, the plane of symmetry and the plane of commissure are usually coincident in the pelecypods.

For the purposes of discussion we consider three types of cross sections of articulated brachiopod shells (Fig. 21): 1) parallel to the hingeline and perpendicular to the plane of juncture (commissure) of the two valves (transverse vertical section), 2) perpendicular to the hingeline and perpendicular to the commissure (longitudinal vertical section, Pl. 29–2, includes plane of bilateral symmetry of GAURI and BOUCOT, 1968, p. 81), and 3) parallel to the commissure (Pl. 30–1). The first type of cross section, when cut through the beaks near the hingeline, reveals internal plates. At midshell the first type of cross section would show bilateral symmetry of the valves and a cross section of the spiralia, which would appear as discontinuous arcs where preserved. The second type of section reveals the symmetry or asymmetry of the valves and, if cut medially near the longitudinal shell axis, may reveal one or more internal plates in the beaks. The third section cut above or below the commissure would show a single closed shell and might traverse the spiralia if they are preserved.

Skeletal microstructure.—The shell microstructure of inarticulate brachiopods varies, partly because they include both phosphatic and calcareous shells. The calcareous shells of the craniids have three types of shell structure arranged in two layers (as discussed by WILLIAMS and ROWELL, 1965, p. H74–H75): 1) a thin, locally absent, outer layer, commonly pale brown, yellow, or even clear and apparently structureless, which is traversed by finely branching holes (puncta), and 2) an inner layer, which is composed of a lamellar layer whose laminae are festooned between the shell punctae and whose fibers are oblique to the shell surfaces and an innermost layer composed of fibers generally perpendicular to the shell surface, exhibiting fewer puncta, and representing the site of attachment of muscles. The shell structure of other calcarous inarticulates is poorly known but apparently they are impunctate and exhibit a crude layering in which the lamellae make an oblique angle with the shell surface.

Among the phosphatic inarticulate brachiopods the phosphatic lingulaecans have a thinly laminated shell of alternating phosphatic and organic layers which are finely punctate (Pl. 29–6). The puncta are very small but are coarser in phosphatic than in organic layers. The laminae are oblique to the shell surface. GRAHAM (1970, p. 165 to 167) has discussed and figured Scottish Carboniferous lingulacid brachiopods that exhibit these features. The discinid shell contains a large number of lamellae disposed crudely into two layers: 1) an outer layer of oblique lamellae stacked one above the other and resembling imbricating tiles on a roof and 2) an inner layer of lamellae oblique to the inner surface of the valve. Both layers contain fine puncta that are probably difficult to see in fossil shells at low magnifications. Among the acrotretids the lamellae are distributed as in the discinaceans but are composed of two types of lamellae that alternate with one another: thicker lamellae (the bulk of the shell) have a strongly fibrous structure in which the fibers are arranged perpendicular to the lamellae and thinner bands that lack conspicuous fibrous structure. The latter may represent organic bands phosphatized since the deposition of the shells within the sediments.

More recently, JAANUSSON (1966) has inferred that some members of the inarticulate Ordovician and Silurian trimerellid brachiopods were aragonitic because they are preserved consistently as molds or as casts filled with sparry calcite in rocks bearing other groups of unaltered calcareous brachiopods.

WILLIAMS and WRIGHT (1970) have reviewed the electron microscopy of the inarticulate calcareous brachiopods. They recognize in these forms a primary layer of acicular needles apparently disposed at some angle to the shell surface and a secondary laminated layer of platy crystallites. The shell beneath the muscles (myotest) is composed of platelets disposed at high angles to the shell surface or to shell laminae.

According to WILLIAMS (1968) the shell of most articulate brachiopods consists of two layers, namely, a primary outer layer and a secondary inner layer (Fig. 22). The normally very thin primary layer frequently not preserved in fossils, appears finely granular in thin section and is composed of minute polygons normal to the shell surface as viewed with an electron microscope. The secondary layer consists of fibers usually oblique to the shell surface (Pls. 29–5, 30–3). WILLIAMS (p. 33) believed that a reduction in the organic sheaths around the individual fibers and an orientation of the fibers perpendicular to the shell surface produced a prismatic layer (Pl. 29–3), which he regarded as a modification of the secondary layer.

WESTBROEK (1967) recognized three types of shell microstructure within the secondary layer of articulate brachiopods (Fig. 22): 1) fibrous structure, 2) prismatic structure, and 3) the myotest. The first two types are described above. The myotest was originally described by KRANS (1965, p. 95) as consisting of finely granular nonfibrous calcite. WESTBROEK (1967, explanation to Plate 14) characterized the microstructure of the myotest as "chaotic" and indicated that in rhynchonellid brachiopods the myotest is not granular but more finely fibrous and, hence, more blurred than is observed in the normal fibrous secondary layer (WESTBROEK, 1967, p. 28). The myotest apparently is associated with areas of muscle attachment in the interior of the valves. WESTBROEK (1967, p. 28–29) also recorded stratification in the fibrous layer of some uncinulid brachiopods. He regarded the stratification as a result of systematic change in

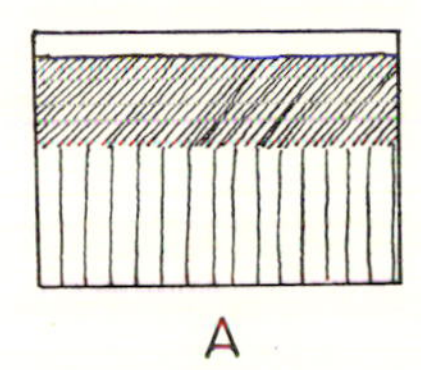

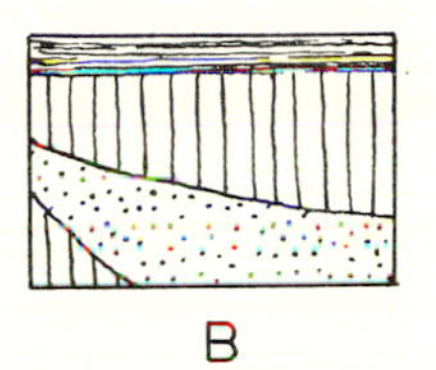

Fig. 22. Brachiopod shell microstructure. A, thin clear layer at top, fibrous median layer, and prismatic layer; B, laminated upper layer, prismatic layer interrupted by granular myostracal layer.

the orientation of the fibers, which appear to follow complicated trajectories that he compared with spirals.

In their discussion of the structure of pentamerid brachiopods, GAURI and BOUCOT (1968, p. 87–88) reported that the shell microstructure consisted of an outer lamellar layer, commonly thin and absent because of spalling and weathering in many free specimens, and a usually thick, inner prismatic layer (Pl. 29–3). However, the proportions of these layers are reversed in some species and additional lamellar and prismatic layers are present in other species. GAURI and BOUCOT (1968, p. 86) also recognized a granular myotest in pentamerid brachiopods; they subsequently (1970, p. 130 reported fibrous, prismatic, and myotest layers in a spiriferid brachiopod.

ARMSTRONG (1969, p. 311, 319) used the term crossbladed to refer to series of elongate tabular units of rectangular cross section that formed sheets (laminae) in the shells of strophomenid and pholidostrophid brachiopods. TOWE and HARPER (1966, p. 153) had used the term crossed-lamellar for the same fabric in pholidostrophid brachiopods. Cross-bladed microstructure produces nacreous (pearly) luster observed on the exterior of some brachiopod shells.

On the basis of published studies, articulate brachiopods may be placed in three groups based on shell fabric. The first group contains brachiopods (mostly those forms not cited in the second and third groups) characterized by a primary (commonly porous) layer of calcite crystallites and a secondary fibrous layer. The second category consists of brachiopods (some pentamerids, spiriferids, and koninckaceans) that have primary and secondary layers as in the first group as well as a tertiary layer (WILLIAMS, 1968, p. 34), which is questionably equivalent to the prismatic layer of GAURI and BOUCOT (1968, p. 87) as developed in some pentamerids. The third group consists principally of the strophomenids and pholidostrophids and is characterized by a cross-bladed fabric. Because brachiopod microstructure is an active field of investigation, terminology and interpretations are not stabilized, and revisions of present concepts undoubtedly will occur as more information becomes available.

WESTBROEK (1967) and GAURI and BOUCOT (1968) have provided especially good photographs of the shell microstructure they discuss. WILLIAMS (1968) has presented an excellent series of electron photomicrographs of the shell microstructures he recognizes among articulate brachiopods.

The shells of articulate brachiopods also are characterized as endopunctate, impunctate, and pseudopunctate. Endopunctate shells have in their secondary layer holes (puncta) which continue as fine branches in the primary layer, but the branches do not reach the shell surface. However, when the thin primary layer is partly eroded or absent, the shell surface appears punctate, that is, the holes pierce the entire thickness of the secondary layer as observed in transverse sections (Pl. 29–4). Impunctate shells (Pl. 30–1) exhibit no puncta in their secondary layer, and pseudopunctate shells (Pl. 29–5) display clear rods usually in an inner shell layer distributed much as puncta might be in endopunctate shells. Puncta are not always resolved by the optical microscope and some punctae are revealed clearly only by use of the electron microscope (SASS, 1967). Cross sections of spinose shells in which the hollow spines have been broken also produce a pseudopunctate shell.

Distribution.—Brachiopods are distributed worldwide from the Cambrian to the Recent but were most abundant in the Paleozoic where they are a significant contributor to skeletal limestones. They probably occupied all marine ecologic niches but were most abundant in shallow continental seas. None are known from fresh water.

Comparisons.—The fibrous shell microstructure of brachiopods could be confused with bryozoan wall microstructure in finely comminuted debris. However, the shell architecture of bryozoans, which consists of numerous small tubes or boxes, and the shell architecture of brachiopods consisting of two valves is unlikely to cause confusion. In the absence of a well-defined fibrous layer, shell fragments of brachiopods and pelecypods are difficult to distinguish with an optical microscope. Future studies utilizing the electron microscope may provide a means of differentiating brachiopod from molluscan shell microstructure in very small fragments. Pelecypods generally lack internal structures and may exhibit a crossed-lamellar layer, which is absent in brachiopods. Altered or recrystallized fragments of brachiopods and pelecypods are impossible to differentiate in thin section. The valves of ostracodes are usually much smaller and thinner walled than brachiopods, show a very fine prismatic microstructure, and may display recurved shell edges (duplicature).

Worms

Skeletal architecture.—Calcareous worm shells are typically straight, slightly curved, or spirally coiled (Pl. 53–5) unpartitioned tubes, usually exhibiting a round or ovate cross section. Calcareous shells are commonly a few millimeters in maximum diameter and a few centimeters to 10 centimeters or more in maximum length.

Worm tubes are also constructed of phosphate, sand and silt grains, and organic materials. They exhibit the same morphologic shapes as calcareous tubes although straight or slightly curved unpartitioned tubes are more common than spirally coiled forms. Most of the phosphatic tubes considered worms by HOWELL (1962, p. W165) are designated a different animal group (hyolithelminthids) by FISHER (1962, p. W132).

The disarticulated parts of the jaw apparatus of some worms (scolecodonts) are generally less than 1 mm and are comparable in size and shape to the cones or serrate bars and blades of conodonts but consist of acid insoluble organic compounds. HOWELL (1962) has prepared a taxonomic review and a short bibliography of fossil worms.

Skeletal microstructure.—The microscopic character of pre-Tertiary worm shells is poorly known. SCHMIDT (1951, 1955) has described the microscopic shell structure of Tertiary calcareous serpulid worm tubes and concluded that they exhibit two layers (Fig. 23). In longitudinal section the outer layer shows chevron (cone-in-cone) microstructure and a clear or laminated inner layer. In tangential section the shell layers appear as concentric rings or laminae. BATHER (1923) indicated that the shell of *Cornulites* consists of a series of laminae composed of minute prisms of calcite and of vesicles. Alternating zones of laminae and vesicles are characteristic of the species BATHER studied.

Systematic study of the petrography of noncalcareous worm tubes is not very advanced. FISHER (1962, p. W130) reported that phosphatic shells (hyolithelminthids) usually exhibit thick, laminated shells and have a smooth interior surface. In thin section, sand or silt tubes would be recognized only by contrast in grain size or mineralogy with the surrounding matrix. Because the sand or silt grains are held together by organic cements, they frequently disaggregate before they can be preserved in the fossil record.

Scolecodonts are opaque in thin section, but some opaque shapes (cones, serrate bars) might be diagnostic. SCHWAB (1966, p. 416) bleached some scolecodonts and reported that in transmitted light they display two layers: 1) a dense, lamellar, dark-brown to black outer cuticle which is rich in organic matter and 2) a less dense, light-brown inner lining that contains numerous minute "tu-

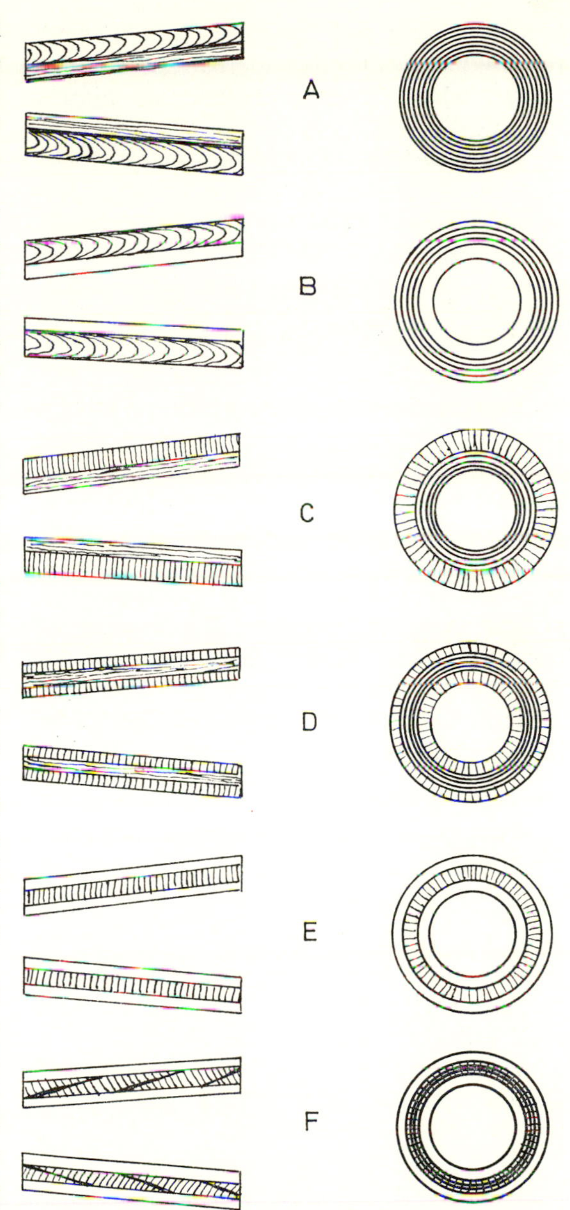

Fig. 23. Comparison of serpulid worm, scaphopod, and vermetid gastropod shell microstructure. A, longitudinal and transverse sections of serpulid shell displaying a parallel laminated inner layer and cone-in-cone laminated outer layer; B, longitudinal and transverse sections of serpulid shell exhibiting clear inner layer and laminated cone-in-cone outer layer; C, longitudinal and transverse sections of vermetid gastropod having prismatic outer layer and laminated inner layer; D, longitudinal and transverse sections of a vermetid gastropod show prismatic inner and outer layers and laminated inner layer; E, longitudinal and transverse sections of scaphopod illustrate prismatic median layer and clear inner and outer layers; F, as E but prismatic layer is itself layered (modified from SCHMIDT, 1955, Pl. 1).

bules". TASCH and SHAFFER (1961, p. 370) reported as fibrous microstructure the feature which SCHWAB regarded as minute "tubules".

Distribution.—Calcareous worm tubes have a worldwide distribution from Cambrian to Recent, principally in marine deposits. Phosphatic hyolithelminthids are restricted to the Cambrian and Ordovician. Cornulitids range from the Ordovician to the Carboniferous, and scolecodonts range from the Ordovician to the Recent.

Comparisons.—Comparable Tertiary tubular shells among the vermetid gastropods and the scaphopods (Fig. 23 E, F) exhibit at least one prismatic shell layer, which is absent in the serpulid worm shell. Pteropods have either one layer of typically molluscan crossed-lamellar structure or homogeneous molluscan microstructure. The vesicles of cornulitids are smaller than those in the stromatoporoids, but fragments would be difficult to distinguish from some Paleozoic bryozoans.

Mollusks

The mollusks, a large and diversified group, have been studied extensively because of their economic value for food and for jewelry. We review the shell architecture and shell microstructure of the mollusks as a whole as well as providing discussions of three major molluscan groups: gastropods, pelecypods and cephalopods.

Molluscan Skeletal Architecture

The mollusks exhibit a wide variety of shell shapes including the following common ones: 1) two valves (pelecypods), 2) straight tubes containing partitions (some cephalopods) or 3) open-ended tubes without partitions (scaphopods), 4) tubes closed at one end and lacking partitions (peteropods), 5) coiled tubes containing partitions (cephalopods), and 6) coiled tubes lacking partitions (gastropods). Exceptions to these generalizations are easily found; for example, some gastropods are cap-shaped, and a few have internal partitions or uncoiled tubes. Chitons, not cited above, have a shell or carapace of a linear series of overlapping calcareous plates. KNIGHT *et al.*, (1960) present information on the skeletal architecture of scaphopods and chitons, which are not discussed further by us.

Internal plates and internal calcite deposits are utilized in the classification of most cephalopods, but, except for shallow muscle scars, interior skeletal architecture usually is lacking in the shells of other molluscan groups. Exceptions are present principally among the rudistids and comparable pelecypods that were adapted for life in reef environments and in the internal ribbing in some gastropods.

Molluscan Skeletal Microstructure

BØGGILD (1930, p. 245–256) recognized eight types of shell microstructure in his monograph and summary of the petrography of molluscan shells. Most of these structures and their corresponding mineralogy are listed in Table 11. Homogeneous structure shows a uniform character in ordinary light and under crossed nicols commonly displays crystal units with c-axes perpendicular to the shell surface. BØGGILD recognized that this type of structure grades into prismatic structure (Pl. 34), which is characterized by shell elements that commonly have polygonal cross sections and a unit extinction. The prisms may be simple or branched and show straight or irregular boundaries and usually are elongated perpendicular to the valve surface. Optic axis orientations are variable. OBERLING (1964, p. 16) and MACCLINTOCK (1967, p. 14) have called very elongated narrow prisms that are 1 or 2 microns in diameter and as much as 220 microns long fibrillar structure. Foliated structure (Pl. 35–2) is characterized by regular parallel leaves or laminae composed of calcite. According to MACCLINTOCK (1967, p. 16), the laminae (sheets, folia) intersect the depositional shell surface at an angle (Pl. 35–2) and the optic axes are parallel to the sheets. BØGGILD (1930, p. 250) considered nacreous microstructure as the aragonitic equivalent of foliated structure, but MACCLINTOCK (1967, p. 16) noted that, in contrast to foliated

Table 11. *Microstructure of Molluscan Shells* (Based on BØGGILD, 1930, and SCHENCK, 1934)

Micro-structure	Mineralogical Composition	Ordinary Light	Crossed Nicols
Homogeneous	"Typically developed among calcite shells." Sometimes aragonite	No visible structure	Extinction in one direction; main axes parallel; usually normal to surface of shell
Prismatic	Aragonite (rare) and calcite (common)	Prismatic, with prisms generally normal to surface of the shell	Normally, each prism is a single crystallographic individual
Foliated	Always calcite	Layer built up of more or less parallel leaves	May resemble crossbedding in sandstone
Nacreous	Always aragonite	Consists of thin leaves (less than 0.001 mm) of equal thickness separated by equally thin leaves of some organic substance	Optic axes always normal to the leaves
Grained	Calcite and aragonite	Irregularly formed grains	Optic orientation irregular
Crossed-lamellar	Generally aragonite	Layer built up of larger "lamels", each rectangular; short axis generally vertical. Length of the single "lamel" of the first order may be several mm. This structure an aggregate, as in serpentine	Although each large "lamel" is built up of smaller "lamels", each one is a single crystal individual. Acute bisectrix forms an angle of 75° with edge of "lamel"
Complex	Always aragonite	Layer consists of sub-layers of two kinds: one finely prismatic and the other complex crossed lamellar	Prismatic layers very thin; irregular extinctions

microstructure, the aragonitic nacreous sheets are parallel to the depositional shell surface and the optic c-axes are perpendicular to the laminae or sheets. Grained microstructure consists of irregularly formed grains that also exhibit irregular optic axes orientations. Grained structure may represent recrystallization of previous shell structure. BØGGILD (1930, p. 251) reported a pelecypod species and a belemnite genus that exhibited shells composed in whole or in part of a single large crystal of calcite, certainly a rare type of structure in mollusks.

Crossed-lamellar structure (Pl. 37–6) is apparently confined to the mollusks and is found in all major groups except the cephalopods. Crossed-lamellar structure is characterized by lamellae or laths (first-order lamellae) that are composed of smaller lamellae (second-order lamellae) that lie at fixed angles within each first order lamella and display opposite inclinations in adjacent first order lamellae (Fig. 24). This is an obvious adaptation for adding strength to the

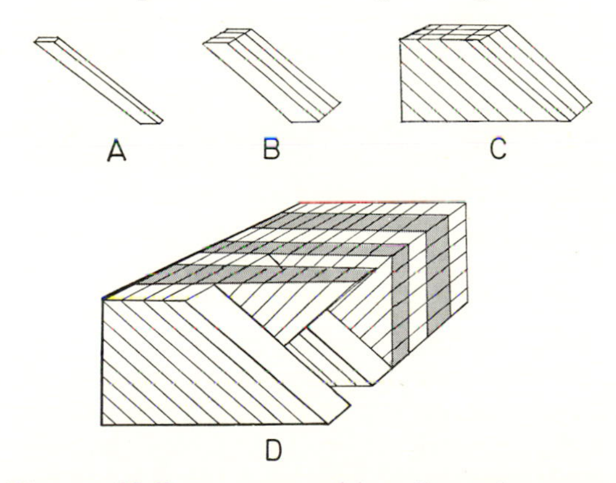

Fig. 24. Molluscan crossed-lamellar microstructure. Third order lamels (A) are combined into second order lamels (B) that compose third order lamels (C). The "tops" of third order lamels, arranged as shown in D, are the lamels most easily seen microscopically (modified from MACCLINTOCK, 1967, Fig. 19).

shell. The first-order lamellae have different orientations in different shell layers. The second order lamellae within each first order lamella are very small (less than 1 micron according to BØGGILD, 1930, p. 253). Complex microstructure is subdivided by BØGGILD (1930, p. 254) into crossed-lamellar and prismatic. Subsequent workers have interpreted complex microstructure as the result of a complicated geometric arrangement of individual lamellae or prisms.

Pelecypods

Skeletal architecture.—Pelecypods have two valves hinged along one margin and their appearance in thin section depends on the shape, ornamentation, and orientation of the thin section with respect to the entire shell. Many pelecypods are approximately equivalved. When both valves are attached and closed, cross sections perpendicular to the hingeline and to the plane of junction of the valves (commissure) commonly display valves that are mirror images of one another (Pl. 33–6). Heart-shaped cross sections are formed when a well developed groove exists along the hingeline in the position of the ligament that holds the valves together (Pl. 33–6). Parallel to the hingeline and perpendicular to the commissure the valves show rather simple arcs thinning at the shell margins (Pl. 33–5). Pelecypods commonly exhibit greatly thickened hinge areas where large articulating teeth and sockets are present in many forms (Pl. 33–1). These features are observable most easily in sections across the hinge area perpendicular to the commissure. In contrast to the brachiopods, plates in the hinge area are uncommon. COX *et al.* (1969) provide a comprehensive review of the morphology of the pelecypod shell.

Some pelecypods have markedly unequal valves, and some groups have especially heavy shells probably adapted for turbulent waters; one valve may be developed into a large cup or bowl shape and the other valve acts as a cover over the bowl. Comparable features exist in some Permian reef-building brachiopods which apparently never grew as large as the Jurassic and Cretaceous pelecypods that adapted this mode of shell shape.

Many pelecypods are too large for the entire shell outline to appear within the confines of a section of a few square centimeters. Entire shapes are more likely to be apparent in peels and polished surfaces of hand specimens. Most adult pelecypods probably fall in the range of 1 to 15 centimeters in maximum dimension, but exceptional forms are known up to 2 meters in largest dimension.

Skeletal microstructure.—BØGGILD (1930) reported eight types of molluscan microstructure in the pelecypods. However, TAYLOR *et al.* (1969) recognized only six (Table 12) of BØGGILD's eight microstructural types in the pelecypods and added another type (myostracal prisms). Myostracal prisms are irregularly elongate and range in size from a few to 50 microns (0.05 mm). The prisms generally increase in size away from their originating surface.

OBERLING (1964) reviewed the shell layering of pelecypods and used the terms given in Table 13. He found that the inner or outer layer commonly merges with the middle layer and introduced the terms mesectostracum and mesendostracum. On the basis of the combination of shell microstructure and shell layering, OBERLING recognized three major groups of pelecypods: 1) nacro-prismatic group, consisting of primitive pelecypods typically with a nacreous inner layer (mesendostracum) and a prismatic outer layer (ectostracum), 2) foliated group, consisting typically of one or more foliated layers, and 3) complex-lamellar group, typically with a complex inner layer (endostracum) and a crossed-lamellar outer layer (mesectostracum). The nacro-prismatic type of layering is present also in cephalopods and primitive gastropods. In contrast to OBERLING, TAYLOR *et al.* (1969) simply referred to pelecypod shell layers as inner, middle, and outer. They provided the most recent review of pelecypod microstructure for various taxonomic categories and their taxonomy follows the usage of COX *et al.* (1969).

The rudist pelecypods commonly display a "cellular-prismatic" microstructure characterized by striking rectangular or radial patterns (Pls. 36–1,2,3; 37–1; 38–2). DECHASEAUX (1952, p. 356–357, figs. 214, 215)

Table 12. *Pelecypod Microstructure*
(Modified from TAYLOR *et al.*, 1969)

Nacreous structure
Rounded or euhedral tablets arranged either in thin sheets, separated by sheets of organic matrix (sheet nacre), or in columns (lenticular nacre). Invariably aragonitic

Foliated structure
A dendritic aggregate of folia built of lath-shaped elements with euhedral terminations, joined in side-to-side contact to form sheets. Pl. 35–2

Prismatic structure
Columnar, usually polygonal blocks of carbonate, which may be simple or composite. Simple prisms are oriented normal to the shell surface, with each prism separated by a thick conchiolin wall. Simple prisms may be aragonitic or calcitic. Composite prisms lie parallel to the shell surface and are built up of fine acicular elements, which usually lack a thick conchiolin wall. Composite prisms are invariably aragonitic. Pl. 34

Crossed-lamellar structure
Elongate, interdigitating sheets oriented with the major and minor axes in the plane of the shell surface. These sheets are the first order lamels. They are built of inclined sheets (second order lamels) which in turn are built up of fine laths joined in side-to-side contact. Crystal orientation is uniform within a lamel but different in adjacent lamels. Invariably aragonitic. Pl. 37–6

Complex crossed-lamellar structure
Blocks built up of laths arranged into second order lamels, as in crossed-lamellar structure, but with several different attitudes of lamels in the various blocks. Invariably aragonitic

Homogeneous structure
A very fine aggregate of granules with uniform optical orientation over wide areas of shell. Invariably aragonitic

Myostracal prisms
A "prismatic" aggregate laid down under areas of muscle attachment. Invariably aragonitic

has provided two very fine photographs of this type of microstructure. Subsequent alteration of all or part of these patterns suggest that they were variably aragonitic (KENNEDY and TAYLOR, 1968).

Because aragonitic shells are usually recrystallized or leached and infilled by sparry calcite, altered pelecypod shells can be recognized only by general skeletal shape and broken or fragmental pelecypod debris can not be identified with certainty.

Table 13. *Molluscan Layering*
(After OBERLING, 1964)

Palliostracum—secreted by mantle
Periostracum
Organic, not preserved in fossils
Ectostracum
Outer shell layer
Mesostracum
Middle shell layer, includes hinge and shell outside of pallial line
Endostracum
Inner shell layer within pallial line

(Ectostracum + Mesostracum = *Mesectostracum*; Mesostracum + Endostracum = *Mesendostracum*)

Myostracum—secreted at the muscle attachment areas
Pallial myostracum
Secreted along the pallial line
Adductor myostracum
Secreted at area of attachment of adductor muscles
Other myostracal layers
Secreted in shell beneath other muscle attachment areas

Distribution.—Pelecypods are distributed worldwide from the Cambrian to Recent and are present in marine and fresh waters. The earliest freshwater forms are of Pennsylvanian age. Pelecypods are very important contributors to post-Paleozoic limestones. In the Paleozoic they apparently were much less common in carbonate rocks than brachiopods.

Comparisons.—The general shape of pelecypods is also characteristic of brachiopods and ostracodes. Brachiopods are differentiated from pelecypods on the basis of their fibrous shell microstructure but in cases of alteration or recrystallization the two groups are difficult or impossible to differentiate from one another. Ostracodes are generally smaller and have a finely prismatic microstructure. Ostracodes are commonly found with both valves together and are characterized by an overlap of the margins of one valve around the other valve. In addition many ostracodes have a recurved edge of the shell (duplicature) along the free (unhinged) margin, which is commonly observed in thin sections. Internal calcified structures are usually lacking in pelecypods and ostracodes but are common in many brachiopods.

Gastropods

Skeletal architecture.—Gastropod shells are usually coiled tubes which, because of geometry and ornamentation, can form a number of different shapes. Cox (1960) has reviewed the morphology of gastropods. As indicated by the investigations of many workers, for example Raup (1966b), the coiled tubes geometrically are logarithmic spirals. Cross sectional views of gastropods vary considerably depending on whether they are parallel, perpendicular, or at some angle to the axis of coiling. Cross sections parallel to and passing through the axis of coiling commonly show paired openings arranged in linear or pyramidal patterns depending on whether the gastropod was coiled in a plane (planispiral) or wound successively higher (high-spired) around the axis of coiling. Linear patterns are smaller paired openings bracketed by larger ones (Pl. 41–3) and pyramidal ones consist of smaller paired openings above larger ones (Pl. 40–2). Many forms exhibit open coiling about an axis, but others are tightly coiled and leave no open area (umbilicus) about their axis of coiling. Internal structures such as cross partitions (septa) and subsequent chamber fillings (cameral deposits) are rare in the gastropods. However, some groups, such as Tertiary land snails and advanced marine gastropods, produce additional shell deposits along the shell aperture (teeth) or around the pillar (columnellar plications, Pl. 40–1) surrounding the axis of a coiled shell (columnella). In cross section these additional deposits fill part of the internal cross section of the shell and impart characteristic shapes to the internal cross sections of the coiled shell. These cross sections are probably diagnostic chronologically and taxonomically, but internal studies of these gastropods have not been extensive. Cross sections also show external ornamentation of ridges and nodes.

Cross sections perpendicular to the axis of coiling may exhibit two concentric circles with a single bar of shell between them. This is present principally in forms having an open umbilicus. The outer circle is the outer shell wall, and the bar is where the section crosses the spirally ascending or descending coil of the shell tube. Where an open umbilicus is lacking, these cross sections display a single "septum" extending into the circle of shell (Pls. 39–1, 39–2, 40–3). An open spiral shell outline represents sections perpendicular to the axis of coiling at the level of the final body whorl or spiral.

Sections neither parallel nor perpendicular to the axis of coiling may be difficult to interpret. However, such cross sections commonly show sufficient coiling or curvature to suggest gastropods. Sections slightly oblique to the axis of coiling will probably show both paired openings and single openings as will sections parallel to but not passing through the axis of coiling.

One group of specialized gastropods (pteropods) have shells resembling very narrow straight-sided cones and are usually 2 centimeters or less in maximum dimension. These shells are commonly very thin and easily crushed.

Skeletal microstructure.—As a group, gastropods display all the types of skeletal microstructure described by Bøggild (1930), except single crystal structure. Prismatic-nacreous microstructure is dominant in primitive gastropods and is common in Paleozoic snails wherever microstructure is preserved. Crossed-lamellar microstructure is usually encountered in Mesozoic to Recent gastropods (Pls. 39–1, 40–3) and is less frequent in Paleozoic representatives. MacClintock (1967) has presented a modified arrangement and division of molluscan shell structures based on his study of some cap shaped gastropods (patelloids). Four of the major types of shell microstructure recognized by Bøggild are further subdivided by MacClintock as shown in Table 14. Within the patelloid gastropods these structures form four to six shell layers and MacClintock (1967) recognized 17 associations (shell structure groups) or sequences of layers of shell microstructure. MacClintock's exhaustive and well illustrated discussion of patelloid shell structure is recommended for passionate pursuers of molluscan microstructure and mediative malacologists who wish to mine molluscan monographs.

Bøggild (1930, p. 319) found pteropod shells to consist of aragonite and to have homogeneous microstructure. Cayeux (1916, p. 490) reported three shell layers in

Table 14
Shell Microstructure of Patelloid Gastropods
(MacClintock, 1967)

Prismatic. Major and minor crystals oriented at angle greater than 10 degrees to growth surface

Simple. Large blade-shaped prisms which have their long axes oriented radially and their intermediate axes normal to growth surfaces

Fibrillar. Fibrils 2 microns in diameter, reclined at angles of 48 to 53 degrees with growth surfaces and at smaller angles to shell surface

Complex. Prisms composed of aggregates of fibrils having same extinction position under crossed polarizers; grades into fibrillar microstructure

Dependent. Prisms or bundles of prisms whose optical and structural orientation is controlled by overlying shell layer

Foliated. Composed of thin flat sheets or folia of a uniform thickness of about 1 micron and which intersect growth surfaces at a low angle (four to seven degrees)

Regular foliated. Sheets built of long parallel-sided blades

Irregularly tabulate foliated. Sheets built of irregularly shaped tabulae

Crossed. Dips of second order lamellae disposed in opposite directions in adjacent first order lamellae

Crossed-lamellar. Always aragonite according to Bøggild. Second order lamellae arranged at angles of 16 to 44 degrees to growth surface

Crossed-foliated. Always calcite according to Bøggild. Second order lamellae exhibit low dip angles of 3 to 27 degrees; first order lamellae about six times as wide as those in crossed-lamellar first order lamellae

Complex-crossed. Cone-in-cone arrangement of major or first order prisms, composed of conical second order lamellae and radial third order lamellae

Complex crossed-lamellar. Lamellae have dips of about 45 degrees to growth surface

Complex crossed-foliated. Lamellae have dips of about 5 degrees to growth surfaces

a Tertiary pteropod shell; the upper and lower layers displayed calcitic homogeneous microstructure and a middle layer showed crossed-lamellar microstructure.

Distribution.—Gastropods are a very successful group that have adapted to marine, fresh water, and terrestrial environments. They are distributed worldwide from the Cambrian to the Recent in marine environments. The earliest fresh-water gastropods are probably Carboniferous, and the earliest undoubted land snails are Cretaceous (Knight *et al.*, 1960). Pteropods are rarely found in the fossil record and apparently are of Cretaceous to Recent age. They are pelagic swimmers whose remains are common in some deep oceanic sediments.

Comparisons.—Gastropods usually lack the numerous internal partitions (septa) of cephalopods, but small specimens could be confused with small coiled calcareous worm tubes. Coiled worm tubes, however, are usually attached and would display one flattened side. *Spirorbis* and some other calcareous worm tubes have walls of large sheets of laminae or folia and contain a very high percentage of organic matter. The questionable coiled worm tube shown in Plate 53–5 exhibits a finely laminar shell microstructure and possibly a thin clear median layer of homogeneous very fine-grained calcite. Uncoiled vermetid gastropods are differentiated from worm tubes by the presence of inner and outer prismatic layers as shown in Figure 23.

Cephalopods

Skeletal architecture.—Cephalopods have coiled (Pl.44) or straight (orthoconic, Pl.45–1) chambered shells. The coiled forms commonly are coiled in a plane (planispiral), although high-spired and uncoiled examples are known. Planispiral coiled forms yield thin section patterns comparable to those in planispirally coiled gastropods. However, cephalopods commonly have involute (last spiral covers all previous ones) coiling. Cephalopods range in maximum dimension from a few millimeters to nearly 10 meters but most forms probably fall in the range of 2 to 10 cm. Only the smaller forms will be found in recognizible form in thin sections. External ornamentation consists of nodes and longitudinal and transverse ribs of varying sizes and shapes. The coiled or straight tubes have round or ovate to subrectangular cross sections.

Internally, sections may reveal an initial chamber (protoconch) and partitions (septa) that produce a multichambered shell. The septa are straight to slightly curved away from the mouth or may be highly contorted toward the junction of the septa with the outer wall of the shell. Each septum is pierced by a single hole (position of the siphuncle in the living animal), which connects the chambers to one another. The chambers and the area around the siphuncle may be filled by additional calcareous deposits (cameral deposits).

The belemnites, in addition to having a chambered shell (phragmocone), had a thick calcite sheath (guard, rostrum) that enveloped the phragmocone. The guard is the most common part of the shell found in the fossil record (Pl. 46). The chambered phragmocone may exhibit a siphuncular septal opening and cameral deposits as in the other groups of fossil cephalopods.

Reviews of the morphology of cephalopods are presented by TEICHERT *et al.* (1964) for the nautiloid cephalopods, by ARKELL *et al.* (1957) for the ammonoid cephalopods, and by JELETZKY (1966) for the belemnites.

Skeletal microstructure.—Most cephalopods had an aragonitic shell. Consequently, their shells frequently are recrystallized or leached and subsequently infilled by calcite cement. Available information on wall microstructure of Recent and fossil cephalopods suggests the shell wall contained three layers: 1) an outer prismatic layer, 2) an inner nacreous layer, and 3) an inner prismatic layer (MUTVEI, 1967, p. 157; ERBEN *et al.*, 1969, p. 6). However, authors vary somewhat in their use of terms for cephalopod microstructure (ERBEN *et al.*, 1969, p. 6, Table 1). The initial chamber of the shell (protoconch) and the calcareous deposits around the end of the siphuncle (caecum) apparently had a rather complex series of shell layers. Internal cameral deposits, including those surrounding the siphuncle, have been discussed by FISCHER and TEICHERT (1969).

The belemnite guard consists of radiating prisms of calcite as illustrated in transverse sections on Plate 46. JELETZKY (1966) has discussed and illustrated the microstructure of the belemnite phragmocone.

Distribution.—Cephalopods are distributed worldwide in marine environments from the Ordovician to the Recent. However, they are significant contributors of biotic debris only in Paleozoic and Mesozoic rocks. Belemnites range from Carboniferous to Eocene but are most abundant in the Mesozoic.

Comparisons.—Cephalopods are commonly characterized in both coiled and uncoiled (orthoconic) forms by altered shell walls, internal septa, and cameral deposits. Foraminifers are commonly smaller, may exhibit perforate walls that are not present in cephalopods or gastropods, and display a different wall microstructure, generally calcitic and less frequently altered. Orthoconic cephalopods are compared with other orthoconic shells in Table 7.

Arthropods

The arthropods are a large and diverse group characterized by an outer skeleton commonly not calcified, and numerous jointed appendages (legs, antennae, various mouth parts). We discuss only trilobites and ostracodes, which are the principal calcified arthropod groups contributing skeletal debris to sedimentary rocks. RICHARDS (1951, Ch. 13) reviewed calcification in arthropods and DENNELL (1960) and TRAVIS (1960) have provided additional information.

Trilobites

Skeletal architecture.—Trilobite shells (exoskeletons, carapaces) are generally less than 1 millimeter thick and are composed of a number of individual pieces, commonly corresponding to single segments or fused segments of the animal. Most calcified segments appear variably arched (Pl. 48–2) in cross section (see MOORE *et al.*, 1952, p. 482, 484, figs. 13–4, 13–5). The lateral border of the head shield (cephalon), the tail shield (pygidium), or the lateral tips of each body (thoracic segment) may be recurved in which case the cross sections may resemble a shepherd's crook (Pls. 47–1, 48–1). Although trilobites nearly a meter in length are known, most species are only a few centimeters long, and individual parts of the carapace can be measured in tens of millimeters or less.

Skeletal microstructure.—SORBY (1879, p. 67) noted that the shell microstructure of trilobites and ostracodes was comparable to that of modern crustacean arthropods and consists of poorly defined layers or laminae roughly parallel to the shell surface. The layers in turn consisted of very fine calcite prisms perpendicular to the shell surface. Under crossed nicols the fine prisms extinct when their axes coincide with the polarizing nicol. In trilobite sections exhibiting recurved shells, a black extinction cross forms under cross nicols and moves across the shell as the stage is rotated. CAYEUX (1916, p. 443 ff), who also examined petrographically the shells of some modern arthropods in order to draw analogies to the trilobite shell, cited only two references other than SORBY in his review of trilobite structure.

CAYEUX (1916, p. 446–448) indicated that the following petrographic features are common among trilobites: 1) many fine pores or canals are perpendicular to the shell surface, 2) fine calcite prisms are perpendicular to the shell surface and are also visible in cross sections parallel to the shell surface, and 3) fine growth lines are parallel to the shell surface. The growth lines do not interrupt the prisms of the shell but only mark their growth. CAYEUX also reported coarse and fine spongy tissue which has not been cited by subsequent workers.

STØRMER (1930) published the first systematic work on the petrography of the trilobite skeleton of individual genera and species. KIELAN (1954) subsequently recorded the shell structure of an additional trilobite species. JOHNSON (1951) and HARRINGTON (1959) have reviewed the work of STØRMER.

The microstructure of trilobites appears to be an open field for investigation. The lack of work in this area is curious because trilobites represent approximately 50 percent of the Cambrian faunas and perhaps 10 percent of Ordovician ones. Some Cambrian limestones are trilobite coquinas and most Ordovician bioclastic limestones contain some trilobite fragments. Especially needed are studies of the pore structure of trilobite carapaces (Pl. 92). Pores are not conspicuous in the skeletal microstructure of previously studied Cambrian trilobites but are common in some Ordovician species. PALMER (1964, p. F4) described pores in a Lower Cambrian trilobite in which the internal position of pores is displayed externally as a strong reticulate meshwork of raised ridges.

Distribution.—Trilobites are exclusively marine arthropods distributed throughout the world in all sedimentary environments from the Cambrian through the Permian. At the specific and generic level trilobites commonly are environmentally controlled.

Comparisons.—The finely prismatic shell structure of trilobites is present in ostracodes and foraminifers, which are not likely to be confused with trilobites because they are both smaller and contain different shell architecture; for example, ostracodes are bivalved and foraminifers are chambered.

Ostracodes

Skeletal architecture.—Ostracodes consist of two valves that can be smooth or ornamented. The valves are joined along a hinge (dorsum), represented commonly by a groove or channel, and overlap along the free margins. The overlap gives a slightly asymmetrical appearance to the articulated valves (Pl. 49-1). Internally the valves frequently show a recurved shell or shelf (duplicature) along the free margins (Fig. 25). Many forms are less than 1 millimeter in maximum diameter but forms up to 3 centimeters in maximum size are known.

Skeletal microstructure.—Fossil ostracodes are little studied in thin section, although LEVINSON (1961) has indicated that the shell structure varies and may be important in taxonomic and stratigraphic (chronologic) studies. According to LEVINSON (1961), a few genera (*Eridoconcha, Cryptophyllus*) exhibit a multiple-layered shell formed by incomplete moulting of valves. Older shells are cemented to newly formed ones. Other genera exhibit 2 to 9 shell layers, and the number of layers frequently is diagnostic of a taxonomic group. In some two layered forms (Kirkbyidae) the inner layer contains knob-shaped pore canals that open to the interior of the shell but not to the shell surface. LEVINSON also reported that the inner shell layer may be laminated and the outer one prismatic and that the proportional thicknesses may vary even to the exclusion of one or the other type of shell layers.

MÜLLER (1964, p. 6) reported a phosphatic shell layer, surrounded by uncalcified (chitinous) inner and outer lamellae, in some Upper Cambrian ostracodes. ANDRES (1969, p. 166) described a three-layered phosphatic shell in two genera of Middle Cambrian ostracodes. A thick light-colored middle layer is over- and underlain by thin darker layers. Other optical features are not indicated, but WILLGALLIS (1969, p. 182, Fig. 1a) published a figure revealing laminate microstructure. WILLGALLIS (1969, p. 183) also has determined that these shells contain 70 to 80 percent calcium phosphate.

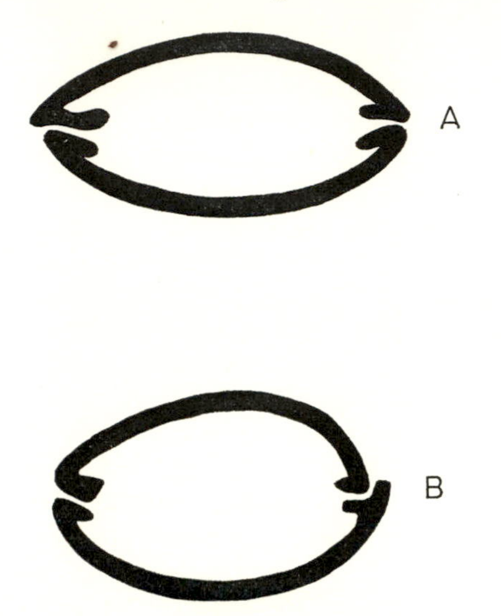

Fig. 25. Ostracode cross sections illustrating recurved edges of shell forming the duplicature.

Many of the ornamented features of the exterior shell surface are also visible in thin section. A central depression in each valve (sulcus) is commonly represented by a marked thickening of the shell as observed in thin section. Some smooth-shelled genera are characterized by inner extensions of the shell (duplicature; calcified portion of the inner lamella) that are observable in most cross sectional views.

Although many ostracode valves are rather simple smooth curves in thin sections, others are highly ornamented and spinose and develop characteristic brood pouches that should present some rather characteristic shapes (cross sections). HENNINGSMOEN (1965) and MELIK (1966) have illustrated cross sections and hinge structures of some Paleozoic ornamented ostracodes.

Pores are not conspicuous in thin sections of ostracode shells, perhaps because they have been filled by calcite and appear as calcite prisms or because they are too fine to be observed readily at magnifications of less than a hundred.

Distribution.—Ostracodes are known from the Cambrian to the Recent in most aquatic environments, both marine and fresh water (beginning in the late Paleozoic), from all over the world.

Comparisons.—The combination of fine prismatic shell structure, small size, overlap of valves, and internal recurvature (duplicature) should distinguish most ostracodes from other bivalved shells.

Echinoderms

Skeletal architecture.—Echinoderms are composed of individual plates of calcite which commonly number in the hundreds or thousands. Illustrations of entire animals may be found in standard textbooks (MOORE *et al.*, 1952) or reference works (SHIMER and SHROCK, 1944; BEAVER *et al.*, 1967; DURHAM *et al.*, 1966). Usually echinoderm skeletons rapidly disarticulate, and the skeletal plates are scattered very soon after the death of the animal. Consequently, thin sections are more likely to reveal cross sections of individual calcareous plates than of entire animals, many of which are larger than the size of a thin section. Calcareous internal structures within the body cavity are uncommon.

Entire echinoderms vary greatly in size. Although some stems of attached echinoderms are many meters long, the bodies of most free or attached echinoderms are commonly a few centimeters to 10 or 20 centimeters in maximum dimension. Individual plates are commonly a millimeter to a few centimeters in maximum dimension.

Some echinoderm plates reveal distinguishing shapes or cross sections in thin sections. The columnal or stem plates of attached forms, such as crinoids, blastoids, and cystoids, appear round (Pl. 51–5) or less commonly pentagonal when cut perpendicular to the axis of the stem. The stem plates are centrally perforated by a hole (lumen), which is commonly round or pentapetaloid. Sections parallel to the stem exhibit rectangular cross sections. If the stem plates are attached in series, then patterns such as those shown in Figure 26 are present. MOORE and JEFFORDS (1968, p. 27) have presented line drawings of actual and theoretical cross sections of crinoid columnals as well as many fine photographs.

Body plates will yield more complex and less readily interpretable patterns. The pore plates of echinoids show characteristic pore patterns, if cut parallel to the plate surface (see DURHAM *et al.*, 1966 for numerous figures of pore patterns). Some crinoid arm plates reveal characteristic lunate patterns when cut perpendicular to the arm axis and show only a rectangular cross section or two parallel plates when intersected parallel to the arm axis (Fig. 26).

Cross sections through some crinoid cups may be diagnostic. KNAPP (1969, p. 351) has proposed that a major group of late Paleozoic crinoids are characterized by downward flaring infrabasal plates, which are the first circlet of body plates in this group (Fig. 26). Properly oriented cross sections would reveal this feature in thin sections.

Blastoids have within their body cavity internal structures called hydrospires, which are infolds of calcite below the feeding grooves (ambulacral areas). These infolds are characteristic of this group of echinoderms (BEAVER *et al.*, 1967, especially Figs. 180 and 192).

Skeletal microstructure.—Each individual plate of the echinoderm skeleton acts optically as a single crystal of calcite. Consequently, each plate will extinct optically at a single position under crossed nicols in a polarizing microscope. This feature is rare in other fossil groups.

The body plates of many echinoderms exhibit an open meshwork (porous) structure within the plates and a finer more dense meshwork towards the plate surfaces. In thin section, the fine meshwork imparts a characteristic gray color to the plates as the light is refracted along the numerous boundaries between the plate meshwork and infilling calcite cement. The position of

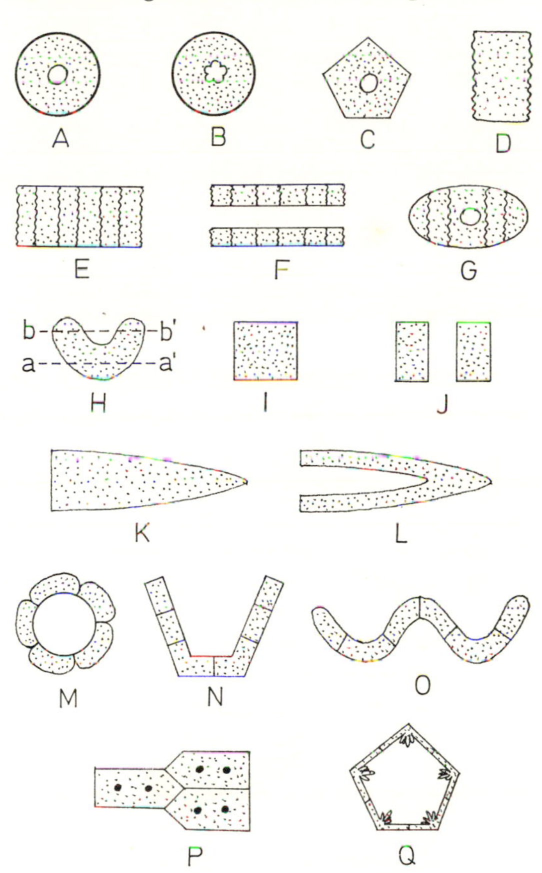

Fig. 26. Echinodermal cross sections. A, transverse section of circular stem plate having round lumen; B, transverse section of circular stem plate showing pentapetaloid lumen; C, transverse section of pentagonal stem plate exhibiting round lumen; D, perpendicular section of stem displaying serrated outline of articulating surfaces; E, series of stem plates showing serrated outlines of articulating surfaces; F, series of stem plates parallel to and passing along the lumen; G, stem plates cut obliquely and displaying lumen in central plate; H, lunate crinoidal arm plate; I, lunate crinoidal arm plate sectioned along aa′; J, lunate crinoidal arm plate sectioned along bb′; K, echinodermal spine; L, hollow echinodermal spine; M, transverse view of crinoid calyx exhibiting five plates and hollow interior; N, crinoid calyx having flat base and body plates of wall; O, crinoid calyx showing invaginated base of calyx; P, echinoid pore plates parallel to surface of plates and exhibiting pores; Q, blastoid cross section perpendicular to axis of growth in ambulacral area; interior displays folded internal hydrospires.

the optic axis in individual plates of echinoids bears specific relations to the entire skeleton and may perform a functional role in adaptations to lighting conditions in the environment (RAUP, 1966a, p. 391).

Echinoid spines (Pl. 51-1, 51-2) exhibit characteristic internal radial patterns (MOORE *et al.*, 1952, p. 710, Fig. 21-31). These patterns may be useful taxonomically and for stratigraphic zonation, although very little study has been made of them. Many echinoid spines are hollow along at least part of their length so that cross sections vary as the orientation changes (Fig. 26). Published reports (RAUP, 1966a, p. 385) indicate that the calcite c-axis is aligned parallel to the long direction of the spine. Consequently, cross sections perpendicular to the spine length will exhibit either low birefringent colors or extinction under cross nicols.

Because individual echinodermal plates act as single crystals, they commonly serve as nucleation sites for calcite cement in carbonate rocks. Calcite is added in optical continuity to the echinoderm plates. Many crystalline limestones are cemented accumulations of echinodermal plates that have broken along the calcite cleavages. LUCIA (1962) and CAIN (1968) have discussed and illustrated the petrography of some crinoidal sediments. RAUP (1966a) reviewed the optical microscopy of echinodermal shell microstructure, and TOWE (1967b), NISSEN (1969), and DONNAY and PAWSON (1969) presented information on the electron microscopy of echinodermal plates.

Distribution.—Echinoderms have a worldwide distribution from the Cambrian to the Recent. Echinoids are not found in the Cambrian. Echinoderms are exclusively marine, usually open marine, animals.

Comparisons.—Echinoderms are the only group of invertebrates whose shell is composed of numerous porous plates each of which acts as an individual calcite crystal.

Plates (sclerites) of holothurians exhibit some shapes similar to sponge spicules or alcyonarian spicules. However, the analogous sponge spicules are siliceous rather than calcareous and the alcyonarian spicules are more highly ornamented and spinose.

Conodonts

Skeletal architecture.—Conodonts are small toothlike microfossils exhibiting the shape of cones, blades, serrated bars and ornamented platforms. They are a fraction of a millimeter to about 3 mm in length and consist principally of calcium phosphate.

Skeletal microstructure.—From external and internal (thin section) observation recent workers recognize three types of conodont material in thin section: 1) the lamellar part or laminated layer, 2) white matter or nonlaminated layer, and 3) basal filling and basal cavity filling. In the lamellar layer thin sections of the microstructure of conodonts exhibit very thin lamellae (Pl. 53-4), initially deposited about a basal cavity. Very fine hollow interlamellar areas separate the lamellae from one another. The lamellae terminate along a free edge, observable in specimens freed of matrix as faint parallel lines. The lamellae are generally concentric and parallel to one another, but variations in lamellar thickness provide the means for producing the large array of nodes, pustules, ridges, and additional external morphologic features of many conodonts.

The white matter or nonlaminated layer is not laminated as viewed in an optical microscope and contains very fine pores or subcircular depressions. The basal filling is more finely laminated than the laminated layer.

Conodonts are characteristically brown or gray in thin section. These colors are common in phosphatic fossil debris as viewed in thin section. LINDSTRÖM (1964, p. 15-30) summarized the microstructure of conodonts and provided many excellent line drawings. Important subsequent work includes papers by SCHWAB (1965) and PIETZNER *et al.* (1968).

Distribution.—Conodonts have a worldwide distribution in marine rocks from the Upper Cambrian through Triassic. MÜLLER and MOSHER (1969) suggest that post-Triassic reports of conodonts represent reworking of Triassic faunas. Petrographers rarely report conodonts, although conodonts are persistent minor elements in some limestone sequences.

Comparisons.—Fragments of phosphatic inarticulate brachiopods or fish scales would probably be indistinguishable from conodonts at least in some views. Phosphatic inarticulate brachiopods and fish scales apparently have smoother non-crenulate laminae and lack hollow interlamellar layers.

Vertebrates

Skeletal architecture.—The phosphatic bones of vertebrates exhibit a wide variety of shapes and of sizes, ranging from a few millimeters to more than 2 meters. However, most of the bones encountered in petrographic studies probably will be a few centimeters or less in maximum dimension and will display ovate, round, or subrectangular shapes. Some Paleozoic fish fin spines exhibit lunate cross sections. Teeth show as many variations in cross section as bones. Teleost fish scales are thin, flat, and elongate in transverse cross sections (Pl. 52–2, 52–4).

Skeletal microstructure.—Fossil bones and teeth of vertebrates display clear or brownish colors in thin sections. Under cross nicols they may exhibit very low birefringence in the first order grays or whites or they show highly birefringent lamellar structure. The reasons for these differences are probably due to the amount of crystallization of the mineralogic components of teeth and bone. Generally the wall of bone is dense, but the interior (marrow) is vesicular and reveals an open meshwork in fossils (Pl. 52–1). The dense, compact microstructure of teeth usually does not exhibit coarse vesicular nets except in trabecular dentine. Teleost fish scales display fine laminae parallel to the scale surfaces.

Histology textbooks, such as BLOOM and FAWCETT (1968) or HAM (1969), contain chapters on the microstructure of vertebrate bone and provide an introduction to the literature on this subject. As reviewed by BLOOM and FAWCETT (1968, p. 224–227), vertebrate bone is composed of lamellae 3 to 7 microns thick. The lamellae are arranged concentrically around canals called Haversian systems or form thin sheets around the exterior of the bone. In addition to the Haversian systems, the bone lamellae are pierced by fine tubular passages called canaliculi which radiate from bone cell spaces located within the bone structure. COOK *et al.* (1962) have described Haversian systems, lamellae, and related features in five fossil vertebrate genera.

PEYER (1968) discussed the teeth of all vertebrate groups, and most of the following is summarized from his work. The bulk of a vertebrate tooth is composed of dentine and is covered externally by a thin layer of enamel, usually where the tooth surfaces are exposed, and, principally in the mammals, by tooth cementum, usually where the tooth is embedded in a tooth socket. Because dentine can form through much of the life of the animal, it may show growth layering (PEYER, 1968, p. 22). The boundary between enamel and dentine in fish teeth is not clear and some controversy exists concerning the presence of both substances in fish teeth or in other lower vertebrate groups (PEYER, 1968, p. 24).

PEYER (1968, p. 99–100) also reviewed variations in dentine (orthodentine, trabecular dentine, modified dentine, etc.). Orthodentine is generally dense material containing minute tubuli and displays low birefringent grays in polarized light. Modified dentine contains fewer tubuli (i.e. is denser) and generally is found on the caps of teeth. Trabecular dentine always occurs in association with orthodentine at the base of a tooth and exhibits a vesicular or open network as viewed in thin section. All the dentines exhibit low birefringence in polarized light and commonly reveal brownish colors in plane light. Enamel displays a characteristic negative birefringence (PEYER, 1968, p. 310) in polarized light. The differences in polarized light between dentine and enamel presumably result from differences in the degree of crystallization of the phosphate, i.e. dentine is better crystallized and has a prismatic structure according to PEYER (1968, p. 204).

GROSS (1967) has discussed the microstructure (histology) of the dermal denticles, of some middle Paleozoic vertebrates and provided several drawings illustrating the internal canal systems and growth patterns within these plates. GROSS also cited other works on the microstructure of vertebrate remains of middle Paleozoic age. JENSEN (1966) and SOCHAVA (1969) review the petrography of fossil vertebrate eggs.

Distribution.—Vertebrates are distributed worldwide in marine environments since the Ordovician and in nonmarine environments since the Devonian.

Comparisons.—The phosphatic vesicular character of bones will distinguish them from other fossil groups. Fish scales are difficult to differentiate from other groups forming laminated phosphatic shells such as inarticulate brachiopods and conodonts. The fish scales figured on Plate 52 do not reveal crenulated laminae which are characteristic of some conodonts.

Faecal Pellets

Skeletal architecture.—Pellets are common in many limestones, and most are believed to be faecal materials of small organisms. Individual pellets commonly are characterized in thin section by 1) round or ovoid shapes, 2) internally homogeneous, equigrained micrite, 3) dark color, and 4) small size, commonly less than 0.15 mm. Aggregates of pellets are usually well sorted. Because the pellets are composed of minute grains, the edges of the pellets may be lighter than the centers, which are thicker and disperse more light. If packing density is low and the pellets are embedded in sparry calcite cement, individual pellets are readily distinguished (Pl. 100). However, if packing density is high, the individual pellets commonly tend to lose their identity and a clotted texture is preserved (Pl. 54–4).

Some Mesozoic and Cenozoic pellets have a distinctive internal pattern of holes (Pl. 54–1) that indicate they were extruded from the gut of known groups of anomuran crustaceans (arthropods). These pellets (ELLIOTT, 1962) are ovoid in transverse section and subrectangular in longitudinal section and have been given taxonomic names. See HÄNTZSCHEL *et al.*, 1968, for an annotated bibliography.

Aggregates of fine calcite grains might be produced by inorganic processes. However, we favor the organic origin for these pellets, because it seems a simpler explanation and Occam's razor is a suitably succinct procedure at this juncture. Some pellets are altered so that no minute grains are visible in the dense almost opaque interiors, and only the overall outline suggests pellets (Pl. 54–3). Identifications are complicated because many small grains of skeletal or other debris have micrite rims and round shapes so that they resemble pellets or pellet-sized material (Pl. 54–6). These forms may represent rounded, comminuted debris with micrite rims produced by algae or some other organic or inorganic process. Obviously, much work remains to be done on pellet limestones, and their abundance and diversity suggests they are a fertile field for paleoscatological and other investigations.

Skeletal microstructure.—Other than the internal size and pattern of holes cited above, faecal pellets do not display any internal features in thin section, although elongate pellets might be expected to show orientation of elongate particles parallel to their surface as a result of extrusion. The crustacean pellets cited above have round or lunate holes as viewed in sections perpendicular to the axis of extrusion and long narrow canals in longitudinal sections (Pl. 54–1, 2)

Distribution.—Worldwide from the Ordovician to Recent, principally from marine limestones. HÄNTZSCHEL *et al.* (1968) did not report any pre-Ordovician examples, but there is no compelling reason why faecal pellets should not be found in Cambrian and late Pre-Cambrian rocks.

Comparisons.—Pellets consisting of very fine-grained small dense masses with or without perforations are unlikely to be confused with other skeletal groups, although confusion with intraclasts and "lumps" certainly is possible at first.

Wood

Skeletal architecture.—Most plant tissue decays after death, and only the most resistent tissues are preserved in identifiable form. Primary mineralized plant tissue is common only among the diatoms and calcareous algae as discussed in the next section. Secondary mineralized tissue can preserve almost any plant part but is observed most commonly in the woody tissue of higher plants.

Entire cross sections of woody roots or stems of higher plants usually are round, ovate, triangular, or rarely stellate. Trunks of woody plants generally are as much as 1 meter or more in diameter and a few tens of meters tall. However, the fragments encountered most commonly in thin sections of sedimentary rocks will be a few millimeters in maximum dimension and will have subrectangular, round, or ovate cross sections.

Skeletal microstructure.—The woody parts of plants usually are preserved by infiltration of various minerals such as calcite, quartz, or pyrite, which also may enclose and preserve carbonized remnants of organic material of the cells. In many examples cell shapes in wood (Pls. 55–1, 56–1) as well as some of the characteristic cell features, such as the pattern of secondary wall thickening, are preserved. The cell patterns usually are irregularly disposed in primary wood and are radially arranged in transverse sections of secondary wood, which also may display growth rings shown by cyclic changes in cell diameters. In longitudinal section the cells of secondary wood are very elongate (fibrous) in gymnosperm woods of which conifers such as pine are typical examples. ANDREWS (1961, p. 23–28) provides an introductory account of wood features and terminology as well as illustrations and diagrams of wood. GREGUSS (1967) published many excellent pictures of Permian to Pliocene woods from Hungary.

Distribution.—Woods are encountered worldwide from the Devonian to the Recent. The plants probably lived in terrestrial or coastal swamp environments. However, rivers often carried woods to marine environments where they were entombed.

Comparisons.—The cellular structure of wood is comparable in size and distribution principally to the coralline red algae, but is distinguishable, where preserved, by secondary thickening of the cell walls, lack of fruiting bodies within the woody tissue, and secondary mineralization. The organic material of the cell wall commonly is preserved in woods, sometimes carbonized, but is almost always lacking in the coralline algae in which the cell walls are defined by primary carbonate deposition.

Calcareous Algae

Calcareous algae are marine and nonmarine aquatic plants that display internal and/or external calcification. Except for the small planktonic coccolithophores, calcareous algae are attached bottom dwellers (sessile benthos). JOHNSON (1961, and references cited therein) has presented numerous reviews of fossil calcareous algae as well as three bibliographies (1943, 1957, 1967). MASLOV (1956), MASLOV *et al.* (1963), NĚMEJC (1959), and PIA (1926) reviewed and illustrated all aspects of fossil calcareous algae. Both fossil and Recent calcareous algae are studied almost exclusively in thin section. Table 15 indicates the division of algal groups discussed below.

Table 15. *Divisions of Calcareous Algae*

Red Algae (Rhodophyta)
- Corallinaceae
- Solenoporaceae
- Gymnocodiaceae
- Other red algae

Green Algae (Chlorophyta)
- Codiaceae
- Dasycladaceae
- Characeae

Blue-Green Algae (Cyanophyta)
- Porostromata
- Spongiostromata

A Problematic Alga

Calcispheres

Red Algae (Rhodophyta)

Corallinaceae

Skeletal architecture.—Coralline algae display a wide variety of shapes, including crusts, nodules, rigid and articulated (jointed) branches. Individual plants are commonly a few centimeters in maximum dimension, and internally, cell dimensions usually are measured in microns. However, fragments of fossil red algae generally have a maximum diameter of a few millimeters. Articulated branches frequently break at their joints so that fossils usually are represented by individual segments of an originally much larger plant. The encrusting forms range in thickness from a single layer of cells to many hundred (Pl. 58–1) or thousands of cell layers.

Internally coralline algae exhibit two basic cell types: large basal cells (hypothallus) and small cells (Pl. 57–2) forming the bulk of the plant skeleton (perithallus), divided into regular cells and somewhat larger megacells. The basal cells (hypothallus) of encrusting coralline algae are relatively large cells that may appear in three arrangements: 1) simple basal cells arranged curvilinearly, the cell rows bending upward toward the main portion of the plant (Pl. 55–2), 2) co-axial basal cells (Pl. 58–2) consisting of arched (arcuate) or curved layers of cells as viewed in perpendicular sections, and 3) plumose basal cells composed of curved layers of cells that may curve upwards or downwards from a central hypothetical zone as viewed in longitudinal sections parallel to the central zone.

The perithallus, which commonly forms the bulk of the calcareous skeleton, is composed of smaller cells (commonly by a factor of five or more) than the basal cells of the hypothallus. The layers of small perithallic cells tend to be arranged in linear or broadly curvilinear horizontal rows or vertical columns. Megacells, two or three times the size of regular perithallic cells, are also present in some encrusting coralline algae (Pl. 57–3). Megacells are usually arranged either vertically or horizontally in single columns or rows. Cell types are commonly rectangular and are measured in tens of microns. Consequently, cells usually are not readily visible at magnifications of less than 50, and the smallest cells are studied at magnifications of 100 and above.

The perithallus contains empty areas, usually filled with clear calcite cement in fossils, that represent the fruiting organs (sporangia, conceptacles), which are an order of magnitude larger than the cells. The conceptacles (sporangia) may display multiple ovate chambers (Pl. 58–1), which in some forms exhibit one or more openings or apertures from which the spores were dispersed.

Articulating coralline algae typically exhibit central areas of arched rows of larger hypothallic cells and lateral exterior layers of smaller perithallic cells (Pl. 56–2). The sporangia, where preserved, are on the side of the branches enclosed in the perithallic tissue.

Skeletal microstructure.—Insofar as is known, the calcareous skeleton of coralline algae is deposited principally within the cell walls and consists of very fine equigranular grains of calcite. The microstructure of the cell walls is poorly known because of the very small size of the wall components.

Distribution.—Coralline algae are distributed worldwide from Late Jurassic to Recent and are important constituents of Cenozoic and Recent reef environments. As with most algae, they are essentially restricted to the zone of penetration of light (photic zone). Most living species are found in depths of less than 75 feet (about 23 meters), although they are known from waters as deep as 600 feet (180 meters). Recent species are common in arctic as well as tropical waters. Both fossil and Recent genera and species may have restricted stratigraphic, or ecologic ranges.

Comparisons.—Coralline algae contain much smaller rectangular cell patterns than the stromatoporoids or any tabulate corals, which they might resemble superficially. They are likely to be confused only with the cell patterns in wood from which they differ by the arrangement of the cells and the presence of sporangia. Fossil wood is likely to contain more carbonized organic tissue than displayed by the coralline algae. The arrangement of large cells in some primitive coralline algae resembles the chamber patterns in large benthonic foraminifers. Fragmental debris of either would be difficult to differentiate in thin section.

Solenoporaceae

Skeletal architecture.—The solenoporoid algae are related to, and are presumably ancestral to, the coralline algae and differ in lacking calcified sporangia, which presumably were external and not embedded in calcified tissue. Solenoporoid algae consist principally of encrusting or nodular masses a few millimeters to several centimeters in maximum dimension and having small internal tubes with or without cross partitions. Differentiation into two types of cellular tissue as in the coralline algae is rare or absent. Cross sections of tubes are round or polygonal. Where cross partitions are developed, the cells may display horizontal layers rather than the multitubular vertical pattern of specimens lacking cross partitions.

Skeletal microstructure.—The calcareous cell walls contain equigranular grains of calcite comparable to those found in the coralline algae. The fine-grained cell wall appears dark in thin section, and no definite arrangement of the grains in the wall is observed.

Distribution.—Solenoporoid algae have a worldwide distribution in the Paleozoic and Mesozoic seas. Their environmental requirements were apparently similar to the coralline algae. Presumably they were abundant only within the zone of light penetration in shallow marine waters.

Comparisons.—The solenoporoid algae differ from the coralline algae in their generally larger cell size, lack of calcified reproductive organs, poorly developed or frequently absent transverse cell partitions, and poorly calcified undifferentiated cells.

Gymnocodiaceae

Skeletal architecture.—The gymnocodiacean algae are weakly calcified in comparison to the coralline and solenoporoid algae. Gymnocodiaceans are commonly composed of segments, a few millimeters in maximum dimension, which rarely bifurcate and are circular or ovate in cross section. Calcification is developed mostly in the outer portions of the plant wall, and internal structures are poorly calcified. Internally the gymnodiacean skeleton exhibits pores or tubes arranged obliquely to the surface of the segments and, at the ends of some segments, conspicuously larger round or ovoid fruiting bodies (sporangia). The porous or tubular internal gymnocodiacean skeleton constrasts strikingly with the cellular tissue exhibited by coralline and solenoporoid algae.

Skeletal microstructure.—Gymnocodiacean calcification commonly produces clear sparry calcite similar to that found in the dasycladacean algae.

Distribution.—Most forms in this group have been described either from the Permian or Cretaceous when they enjoyed worldwide distribution. Their paleoecology is unknown, but by analogy with other red algae they were restricted to, and abundant in, the zone of light penetration in the oceans.

Comparisons.—Gymnocodiacean algae display internal structure and pores oblique to the branch surface in contrast to the dasycladacean algae which have a hollow main stem and pores perpendicular to the branch surface. The gymnocodiacean pores are much smaller than the zooecia of bryozoans or the corallites of colonial corals. In addition, the algal branches lack bryozoan or coral wall microstructure, and the algal pores never exhibit internal plates or other structures so common in the bryozoans and corals. Codiacean algae have branching or interiorly intertwined tubes, features which are lacking in the gymnocodiaceans.

Other Red Algae

Skeletal architecture.—A few types of red algae that are important contributors to late Paleozoic limestones cannot be placed in any of the previous categories. These include the genera *Archaeolithophyllum*, *Cuneiphycus*, *Komia*, and *Ungdarella*, which form crusts or rigid branches a millimeter or more in thickness or diameter but usually extending for several centimeters. Some forms have a cellular tissue closely resembling modern coralline red algae and may represent an ancestral stock of the coralline algae. *Komia*, which has been assigned by many authors to the algae, has been placed in the stromatoporoids by WILSON (1969). The characters of these forms are tabulated in Table 16.

Table 16. *Characters of Some Nonconforming Red Algae*

Name	Growth Habit	Cell Arrangement	Sporangia
Archaeolithophyllum	Mainly small irregular platy or crustose unattached masses; crusts commonly superimposed and may extend for 15–20 cm; slender cylindrical branches probably protuberances of crustose forms	Hypothallus, large co-axial (internal) polygonal cells; perithallus, small exterior layers of rectangular cells	Embedded in cell tissue; circular or arched, single aperture
Cuneiphycus	Probably segmented, cylindrical or wedge-shaped segments or branches	Cell sizes not differentiated; large elongated rectangular wedge-shaped cells	Unknown
Komia	Strongly branched, cylindrical	Hypothallus, small cluster of elongate filaments; parallel to and running along branch axis; perithallus, dense mass of dichotomously branching filaments perpendicular to branch surface	Unknown
Ungdarella	Branched, cylindrical	Hypothallus single row of axial cells; perithallus, simple or branching filaments, sometimes in contorted arrangement, usually parallel to branch surface	Unknown

Skeletal microstructure.—Cell or filament walls are composed of fine-grained, equi granular calcite, which appears dark in thin section.

Distribution.—*Ungdarella* is reported only from the U.S.S.R., but the other genera are distributed throughout the northern hemisphere, principally in the Pennsylvanian (Upper Carboniferous) and Lower Permian.

Comparisons.—The cellular wedge-shaped cross sections of *Cuneiphycus* are distinctive. The other forms are the only red algae in the Paleozoic to have such well-developed cell structure or differentiated tissue (perithallus and hypothallus). Wood lacks the differentiated cellular patterns displayed by these red algae.

Green Algae (Chlorophyta)

Codiaceae

Skeletal architecture..—Codiacean algae exhibit two forms: 1) crustose or nodular growths, and 2) erect plants, commonly consisting of segmented branches. Internal round or subpolygonal branching filaments exhibit no cross partitions. The character of the filament branching is used to distinguish different types of crustose or nodular forms (Johnson, 1961, Table 10, p. 95), and the shape and segmentation of the branches and the internal organization and branching (Pl. 59–1) of the filaments differentiates the branching forms. Figure 27 illustrates some types of filament branching found in crustose and nodular codiacean algae.

Skeletal microstructure.—Presumably the calcified walls are of fine equigranular crystals of calcite or aragonite which display a dark appearance in thin sections.

Distribution.—Living codiacean algae are distributed worldwide in marine waters principally in warm shallow seas. Fossil representatives presumably occupied similar habitats. Codiaceans are rare in the Cambrian but are important algal contributors to Ordovician and later rocks.

Comparisons.—The crustose codiaceans resemble solenoporoid algae, but differ in 1) having a less compact arrangement of filaments, 2) lacking cross partitions, and 3) displaying characteristic branching filaments as shown in Figure 27. The branching codiaceans resemble the gymnocodiacean and dasycladacean algae. Codiaceans typically show internal filaments in the central branch axes, which are not calcified in the dasycladaceans. Gymnocodiaceans either lack calcified central branch axes or display straight rather than intertwined filaments or pores. Tangential sections of some codiaceans cannot be differentiated from dasycladacean algae.

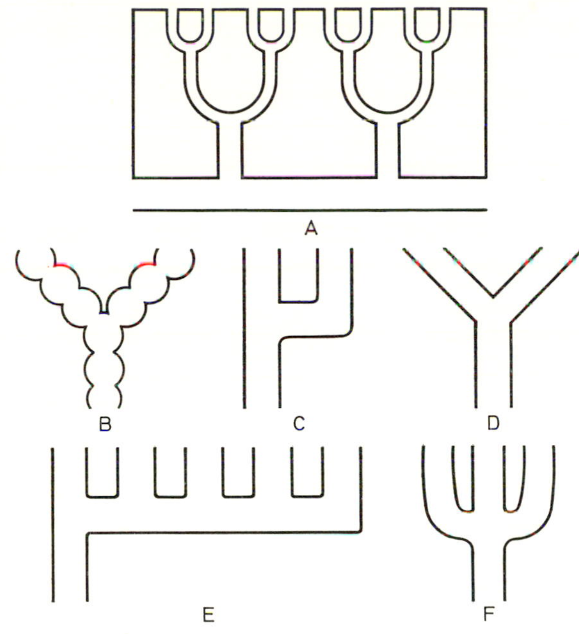

Fig. 27. Types of branching filaments in codiacean algae. A, *Halimeda*; B, *Bevocastria*; C, *Cayeuxia*; D, *Garwoodia*; E, *Hedstroemia*; F, *Ortonella* (B–F, modified from JOHNSON, 1961, Table 10).

Dasycladaceae

Skeletal architecture.—Dasycladaceans are erect plants but typically are preserved as calcified cylindrical segments of branches or stems, which are hollow in the center (Pl. 60–2) and are perforated laterally at right angles to the main stem (Pl. 60–2) by numerous "pores" representing primary or secondary branches. Calcification is limited to the external surfaces of branches so that only the branching pattern is preserved.

Skeletal microstructure.—The calcite formed on the exteriors of the branches is usually clear and sparry when observed in fossils.

Distribution.—Dasycladacean algae are found worldwide from the Cambrian to the Recent in marine environments. Living species are restricted to warm, tropical waters 12 meters (40 feet) or less in depth and are most abundant immediately below low tide level.

Comparisons.—Dasycladacean algae lack the central filaments of the main branches that characterize the codiacean algae. However, tangential sections of these two groups would be difficult to differentiate on this basis. The dasycladacean skeleton typically has a more sparry appearance than the codiacean skeleton. Gymnocodiacean algae display faint internal structure along their branch axes and have pores oblique to the branch surface. The dasycladacean algae lack the wall microstructure and internal plates characteristic of many colonial bryozoans and corals.

Characeae

Skeletal architecture.—Most characeans calcify only the reproductive parts (oogonia or gyrogonites). Calcified oogonia, usually 0.5 to 1 mm in maximum dimension, commonly are hollow and are composed of spiral tubes which have a characteristic external spiral ornamentation as protuberances on the exterior (Pl. 60–5, 6). Internally the wall may be smooth or scalloped. In some specimens the interior is calcified, and thin sections show cross sections of the spiral chamber that contained the eggs.

Skeletal microstructure.—According to reviews by PECK (1934, p. 97; 1957, p. 6), the calcareous wall of oogonia probably was composed of calcite deposited in thin layers. However, recrystallization has produced a radial fibrous pattern in many forms. The junction between the spiral tubes is a dark line in cross sections of oogonia.

Distribution.—Living representatives of *Chara* are distributed worldwide in fresh or brackish waters at depths usually less than 10 meters (33 feet). Fossil forms are known since the early Devonian and occupied similar habitats. However, characeans also are found in marine limestones in the geologic record; whether they lived in marine environments or were washed in from nearby brackish or fresh-water sources is not clear.

Comparisons.—Cross sections of characeans resemble the problematic organisms called calcispheres, which differ in their smaller size (less than 500 microns) and lack of internal structure. Characean oogonia having internal structures resemble some dasycladacean algae, which, however, exhibit branching tubes lacking in the characean oogonia.

Blue-Green Algae (Cyanophyta)

Porostromata

Skeletal architecture.—These algae have simple or branched unsegmented filaments that apparently assisted in the accumulation of calcareous nodules (Pl. 62). The best known form is *Girvanella* (Pl. 62–2) which forms nodules and encrusting masses consisting of individual round sinuous tubes 10 to 50 microns in diameter and without cross partitions. These forms were presumably green or blue-green algae but their evolutionary position is still conjectural.

Skeletal microstructure.—The walls of the filaments are composed of fine grains of calcite and appear dark in thin section.

Distribution.—These forms occupied marine environments throughout the world from Cambrian to Cretaceous and commonly are interpreted as inhabiting very shallow, even intertidal, waters.

Comparisons.—The small irregularly arranged tubes that characterize these algae are unlikely to be confused with other fossils.

Spongiostromata

Skeletal architecture.—These forms are characterized by irregular layers of sediment that formed principally by the trapping and binding of fine calcareous sediment by mucilaginous sheaths surrounding noncalcareous filamentous blue-green algae, which grew as algal mats. The mats die when too deeply covered with sediment, and the algae grow through the sediment and spread out on top to form a new mat and entrap a new layer of sediment. Attached laminated sequences, commonly displaying columnar shapes, are called stromatolites (Pl. 61–3). Oncolites typically are flattened hemispheroids, which are unattached and can roll about on the bottom (Pl. 61–2). They usually display internal layering that surrounds a nucleus, commonly consisting of a fossil shell or shell fragment. These forms are in sharp contrast to previously discussed algae, including the Porostromata, which secreted calcite that mimicked in some form the shape of the structure of the plant or some of its parts.

Skeletal microstructure.—Usually no microstructure is preserved except the general layering which results from entrapped sediment.

Distribution.—Oncolites and stromatolites are distributed worldwide from the Precambrian to the Recent, principally in marine environments. In recent years geologists consistently have interpreted stromatolites as very shallow water or intertidal deposits. PLAYFAIR and COCKBAIN (1969) have challenged this assumption by presenting evidence for deepwater algal stromatolites (45 meters) in some Australian Devonian reefs. On the basis of observations of Recent algae, GARRETT (1970) has suggested that the post-Precambium stromatolite record is poor because under normal marine conditions grazing and burrowing organisms destroy the layering produced by algal mats.

Comparisons.—Lack of any consistent defined structure, other than a textural layering, differentiates these algae from other fossils. However, some altered or dolomitized stromatoporoids can display a similar type of layering.

A Problematic Alga

Skeletal architecture.—Some fossils (*Tubiphytes*) have a thick, dark, fine-grained calcareous wall pierced by a few relatively thin internal tubes filled by clear calcite cement. MASLOV (1956) and most subsequent workers (TOOMEY, 1969) have assigned these forms to algae of unknown affinities. However, RIGBY (1958) referred North American specimens (Pl. 63–1, 2) which displayed two sizes of tubes, to the hydrozoan coelenterates. Specimens are a few millimeters to a few centimeters in maximum dimension but commonly are present as fragments.

Skeletal microstructure.—Variations in density (darkness) or the calcareous walls are believed to represent growth zones in the Paleozoic specimens. The tissue usually has a vague indistinct tubular or cellular appearance (Pl. 62–2) not unlike that of some *Girvanella* nodules.

Distribution.—These forms were distributed throughout the northern hemisphere in the late Paleozoic and were important contributors to some Permian reefs. No post-Paleozoic specimens are reported but see the discussion of Plate 63–3.

Comparisons.—The dense tissue pierced by a few tubes is not comparable to any of the major groups of invertebrates.

Calcispheres

Skeletal architecture.—Calcispheres are problematic organisms commonly less than 500 microns in maximum dimension and displaying a round central cavity, lacking apertures to the exterior, and surrounded by a variably preserved wall.

Skeletal microstructure.—In the radiosphaerid calcispheres (STANTON, 1967, p. 465)

Table 17. *References to Some Fossil Groups Not Discussed in Text*

Major Groups	Age	References
Coccoliths, discoasters and nannocones	Jurassic to Recent	FISCHER *et al.* (1967); LOEBLICH and TAPPAN (1966); TREJO H. (1960)
Diatoms	Jurassic to Recent	MASLOV *et al.* (1963)
Dinoflagellates (calcareous forms)	Jurassic to Recent	WALL and DALE (1968)
Oligosteginids	Jurassic and Cretaceous	ADAMS *et al.* (1967); POKORNÝ (1958)
Microproblematica		
Aeolisaccus (? pteropod)	Permian to Jurassic	ELLIOTT (1958)
Ancientia (? cornulitid)	Upper Ordovician	ROSS (1967)
Ampullites	Ordovician	FLOWER (1961)
Ancestrulites (? bryozoan)	Ordovician	FLOWER (1961)
Cheneyella	Ordovician	FLOWER (1961)
Cystosphaera	Ordovician	FLOWER (1961)
Distichopax (? alga)	Paleocene	ELLIOTT (1962)
Eliasites	Ordovician	FLOWER (1961)
Fentonites (? foraminifer)	Ordovician	FLOWER (1961)
Goldringella (? gastropod)	Ordovician	FLOWER (1961)
Harjesia	Ordovician	FLOWER (1961)
Hensonella (? alga)	Lower Cretaceous	ELLIOTT (1960; 1962)
Ivesella	Ordovician	FLOWER (1961)
Kruschevia	Ordovician	FLOWER (1961)
Lacrymorphus (? alga)	Triassic	ELLIOTT (1958)
Mooreopis	Ordovician	FLOWER (1961)
Moundia	Ordovician	FLOWER (1961)
Niccumites	Ordovician	FLOWER (1961)
Nuia (? alga)	Ordovician	TOOMEY (1967)
Pedicillaria	Ordovician	FLOWER (1961)
Pseudovermiporella (? alga)	Permian	ELLIOTT (1958)
Slocomia	Ordovician	FLOWER (1961)
Tholella	Ordovician	FLOWER (1961)
Warthinites (? gastropod)	Ordovician	FLOWER (1961)
Wellerites (? bryozoan)	Ordovician	FLOWER (1961)

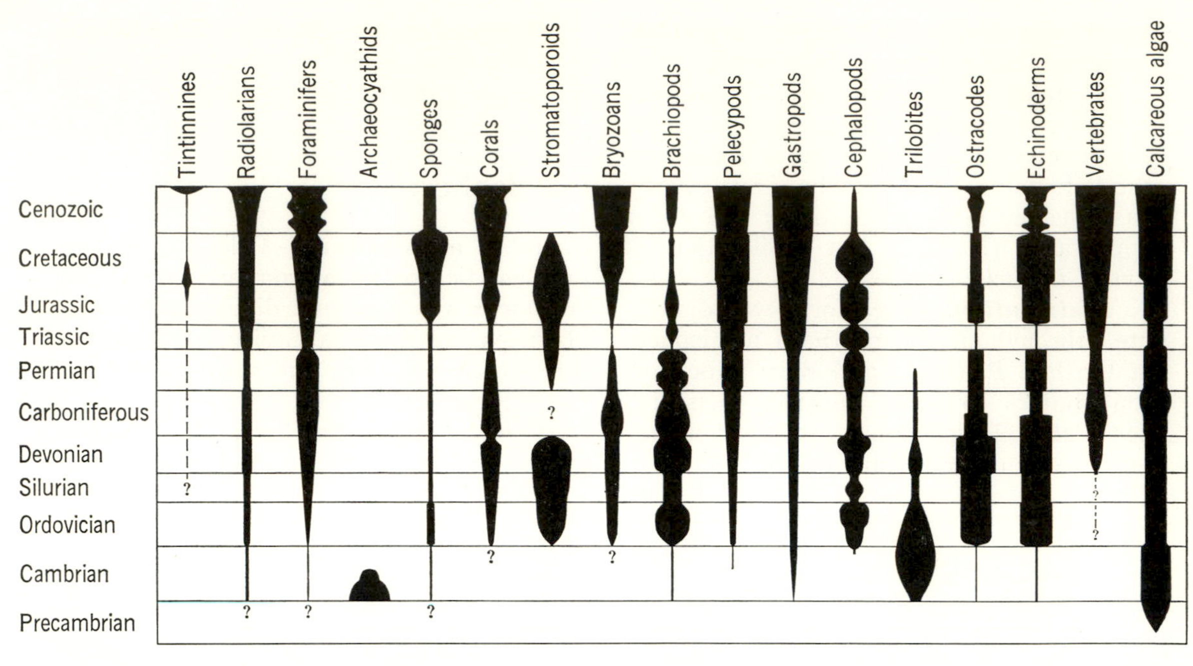

Fig. 28. Age range and taxonomic diversity of some fossil groups. Scales are not the same for the different groups. Taxonomic diversity is not necessarily equal to relative abundance although these graphs are interpreted usually in this manner. Compiled from many sources.

the wall consists of a thin fine-grained layer surrounding the inner cavity and a radially prismatic outer layer. The individual prisms of the outer layer are pierced by a very small canal about 6 microns in diameter.

Distribution.—Calcispheres are observed chiefly in Devonian and Mississippian rocks where they are conspicuous elements of some limestones. These forms are assumed to be marine plankton.

Comparisons.—Calcispheres are similar to some single chambered foraminifers (Pl. 61–5) or characean oogonia but lack the aperture(s) found in these other groups. When preservation is poor these forms are difficult to differentiate from one another.

Some Additional Fossil Groups Not Discussed

Table 17 cites some references to fossil groups that we have not discussed. These groups either require magnifications greater than 100 or they are not abundant. The references of Table 17 represent recent papers that provide an introduction to the literature.

Microfacies References

AGIP Mineraria, 1959, Microfacies Italiana: S. Donato Milanese, AGIP Mineraria, 35 p., 145 pls.

Beautifully illustrated, generally two or three photomicrographs per plate; little text. Plates arranged by age (Carboniferous to Miocene). Biotic constituents, principally foraminifers and algae, commonly are identified to generic and specific levels and names are indexed. Pertinent bibliography.

Bonet, Frederico, 1952, La facies Urgoniana del Cretácico Medio en la región de Tampico: Asoc. Mexicana Geólogos Petroleros Bol., v. 4, p. 153–262, 50 figs.

Fifty photomicrographs (1 to 3 per page) illustrate the major facies and biotic constituents of this middle Cretaceous sequence.

— 1956, Zonificación microfaunística de las calizas Cretácicas del este de Mexico: Asoc. Mexicana Geólogos Petroleros Bol., v. 8, p. 389–488, 31 pls., 4 figs., 3 tables (reprinted XX Congreso Geol. Internatl., 102 p.).

Principally taxonomic discussion and description of foraminifers, tintinnines, and problematic fossils that were used to zone the late Jurassic and Cretaceous rocks of eastern Mexico. Mostly two figures per plate; excellent photomicrographs for such high magnification (commonly 100 to 500 power). Identifications largely to specific level. English abstract.

BORZA, KAROL, 1969, Die Mikrofazies und Mikrofossilien des Oberjuras und der Unterkreide der Klippenzone der Westkarpaten: Bratislava, Slovenskej Akademie Vied, 124 p., 88 pls., 12 figs.

Not seen.

BOZORGNIA, FATHOLLAH, and BANAFTI, SALEH, 1964, Microfacies and microorganisms of the Paleozoic through Tertiary sediments of some parts of Iran: Teheran, National Iranian Oil Company, 22 p., 158 pls.

Brief stratigraphic text and appropriate references introduce excellent photomicrographs arranged by age (Precambrian to Oligocene); biotic debris of Cambrian to Oligocene age illustrated. Identifications, largely of foraminifers, are to generic and specific level.

CAROZZI, A. V., and TEXTORIS, D. A., 1967, Paleozoic carbonate microfacies of the Eastern Stable Interior (U.S.A.): Leiden, E. J. Brill, 41 p., 100 pls., 13 figs. (Internationsl Sedimentary Petrographical Series 11).

General stratigraphic introduction and short bibliographies for individual Paleozoic systems (Ordovician through Permian). Two photomicrographs per plate. Full range of biotic debris illustrated; identifications to major fossil group. Brief environmental interpretations.

CITA, M. B., 1965, Jurassic, Cretaceous and Tertiary microfacies from the southern Alps (northern Italy): Leiden, E. J. Brill, 99 p., 117 pls., 17 figs. (International Sedimentary Petrographical Series 8).

Stratigraphic introduction, bibliography, and stratigraphically arranged plates. Biotics, principally foraminifers, algae and tintinnines, identified to genus and species. Two figures per plates; biotics indexed.

CUVILLIER, JEAN, and SACAL, VINCENT, 1956, Stratigraphic correlations by microfacies in Western Aquitaine, 2nd ed.: Leiden, E. J. Brill, 33 p., 100 pls. (International Sedimentary Petrographical Series 2).

Brief introduction and stratigraphically arranged plates mostly of Jurassic to Oligocene age. Emphasis on foraminifers but many other biotic groups illustrated. Identifications of foraminifers commonly to genus and species.

DERIN, B., and REISS, ZEEV, 1966, Jurassic microfacies in Israel: Tel Aviv, Israel Inst. Petroleum Spec. Pub., 43 p., 320 photos, 2 tables.

Brief stratigraphic introduction and list of pertinent references, including citations to comparable microfacies in other areas. Photomicrographs arranged twelve to a page; captions contain formation, age, locality, and brief comments on rock type and fossil debris present. Many identifications of foraminifers and algae to genus and species.

FABRICIUS, F. H., 1966, Beckensedimentation und Riffbildung an der Wende Trias/Jura in den bayerisch-tiroler Kalkalpen: Leiden, E. J. Brill, 143 p., 27 pls., 24 figs., 7 tables, 1 map. (International Sedimentary Petrographical Series 9).

Petrography integrated with regional stratigraphic setting. Plates illustrate fossil types, especially from reef facies; identifications commonly to genus and species. Long English abstract; plate captions in German and English.

FORD, A. B., and HOUBOLT, J. J. H. C., 1963, The microfacies of the Cretaceous of western Venezuela (Las microfacies del Cretaceo de Venezuela Occidental): Leiden, E. J. Brill, 55 p., 55 pls., 8 figs., 1 chart. (International Sedimentary Petrographical Series 6).

Full range of illustrated microfacies includes nonfossiliferous rock types. Foraminifers identified to genus and species, other biotics identified only to major group. Stratigraphic introduction and bibliography.

GLINTZBOECKEL, CHARLES, and RABATE, J., 1964, Microfaunes et microfaciès du Permo-Carbonifère du Sud Tunisien: Leiden, E. J. Brill, 46 p., 108 pls. (International Sedimentary Petrographical Series 7).

Brief stratigraphic introduction and bibliography. Usually two figures per plate. Short plate descriptions; fossils, principally foraminifers and algae, are identified to generic level.

GRUNAU, H. R., 1959, Mikrofazies und Schichtung ausgewählter, jungmesozoischer, radiolaritführender Sedimentserien der Zentral-Alpen. Mit Berücksichtigung elektronenmikroskopischer und chemischer Untersuchungsmethoden: Leiden, E. J. Brill, 179 p., 90 figs. (International Sedimentary Petrographical Series 4).

Upper Jurassic and Cretaceous radiolarian chert sequence in the central Alps described in detail. A few photomicrographs of radiolarians, tintinnines, nannocones, and some problematic microfossils.

HAGN, HERBERT, 1955, Fazies und Mikrofauna der Gesteine der bayerischen Alpen: Leiden, E. J. Brill, 29 p., 71 pls., 8 tables.

Brief introduction and bibliography. Plate explanations in German and English; wide variety of biotic and nonbiotic constituents illustrated. Foraminifers commonly identified to species.

HANZAWA, SHŌSHIRŌ, 1961, Facies and microorganisms of the Paleozoic, Mesozoic and Cenozoic sediments of Japan and her adjacent islands: Leiden, E. J. Brill, 117 p., 148 pls., 6 figs., 2 tables, 4 stratigraphic charts (International Sedimentary Petrographical Series 5).

Excellent integrated geologic introduction. Bibliography. Stratigraphically arranged plates, descriptions short, 2 figures to plate. Wide range of biotics illustrated with many identifications to genus and species.

KHVOROVA, I. V., 1958, Atlas karbonatnykh porod srednego i verkhnego karbona russkoĭ platformy (Atlas of carbonate rocks of Middle and Upper Carboniferous of the Russian Platform): Moscow, Acad. Nauk S.S.S.R., Geol. Inst., 170 p., 67 pls., 4 figs., 5 tables.

Extensive introduction containing sections on methods of preparation and study, classification and nomenclature, general characters of studied sequence, and general descriptions of types of limestone and dolomite encountered. 397 figures arranged on 67 plates. Diverse biotic constituents figured and identified to major fossil group. Bibliography contains most of the important Russian monographs on carbonate rocks.

LEFELD, JERZY, 1968, Stratygrafia i paleogeografia dolnej Kredy Wierchowej Tatr (Stratigraphy and palaeogeography of the high-tatric Lower Cretaceous in the Tatra Mountains): Studia Geologica Polonica, v. 24, 116 p., 18 pls., 13 figs., 2 tables.

Short section on systematic description of fossils. Eleven of plates illustrate principally tintinnines, foraminifers, algae, and some problematic fossils. Identifications to species. Polish with long English resumé; plate and figure captions in Polish and English.

LEHMANN, E. P., *et al.*, 1967, Microfacies of Libya: Tripoli, Petroleum Exploration Society of Libya, 80 p., 37 pls., 1 fig.

Illustrations of Cambrian to Miocene subsurface rocks accompanied by description of mineralogy, texture, constituents, and interpretation. Identifications commonly to generic and specific level.

MIŠÍK, MILAN, 1966, Microfacies of the Mesozoic and Tertiary limestones of the west Carpathians (Mikrofácie vápencov mezozoika a terciéru západných Karpát): Bratislava, Slovak Acad. Sci., 278 p., 101 pls., 2 maps.

Brief introductory text outlines structure and sedimentary environments of west Carpathians in Czechoslovakia. Plate descriptions and text in English and Czech. Full range of biotic debris illustrated with identifications commonly to genus and species. Figure captions contain environmental interpretations and references for biotic identifications. Long bibliography. Constituents indexed.

OTA, MASAMICHI, 1968, The Akiyoshi Limestone Group: A geosynclinal organic reef complex: Akiyoshi-dai Sci. Mus. Bull. 5, 44 p., 31 pls., 17 figs., 6 tables.

Two photomicrographs per plate illustrate the full range of biotic debris (foraminifers, corals, bryozoans, brachiopods, algae) in a Permo-Carboniferous reef complex. Uses FOLK's classification of limestones. Japanese; plate, figure, and table captions in English.

PERCONIG, E., 1968, Microfacies of the Triassic and Jurassic sediments of Spain: Leiden, E. J. Brill, 63 p., 123 pls., 11 figs. (International Sedimentary Petrographical Series 10).

Compact stratigraphic summary, bibliography, and stratigraphically arranged plates. Full range of biotic debris illustrated and indexed.

RADIOČIĆ, RAJKA, 1960, Microfaciés du Crétacé et du Paléogène des Dinarides externes de Yougoslavie: Titograd, Int. Rech. Geol. R. P. Crna Gora, Paleont. Dinarides Yougoslaves, ser. A, Micropaleont., v. 4, no. 1, 172 p., 67 pls., 1 fig., 1 table.

Short stratigraphic introduction, pertinent bibliography of microfacies and microfaunal papers. Identifications to genus and species for a number of microfossil groups including foraminifers, algae, tintinnines, and faecal pellets. Text and plate descriptions in Slavic and French.

— 1966, Microfaciés du Jurassique des Dinarides externes de la Yougoslavie: Geologija Razprave in Poročila, v. 9, p. 5–377, 165 pls., 11 tables.

Brief introduction lists facies, stratigraphy, fossil ranges, and references. Two figures per plate each with short descriptions. Most photomicrographs illustrate algae and foraminifers usually to generic or specific level.

RADWÁNSKI, ANDRZEJ, 1968, Studium petrograficzne i sedymentologiczne Retyku Wierchowego Tatr (Petrographical and sedimentological studies of the high-tatric Rhaetic in the Tatra Mountains): Studia Geologica Polonica, v. 25, 146 p., 54 pls., 6 figs., 9 tables.

Petrographic study of Upper Triassic limestones. Commonly 2 to 6 photomicrographs per plate. Full range of biotic debris illustrated; identifications of algae and foraminifers commonly to genus and species. Polish with long English resumé; captions of plates and figures in Polish and English.

REISS, ZEEV, 1960, Lower Cretaceous microfacies and microfossils from Galilee: Bull. Res. Council Israel, Sec. G., Geo-Sciences, v. 10G, p. 223–246.

Short stratigraphic and environmental discussion, references, and brief descriptions of 107 photographs. Many generic identifications of foraminifers and algae.

REY, MARCEL, and NOUET, G., 1958, Microfaciès de la région prérifaine et de la moyenne Moulouya (Maroc septentrional): Leiden, E. J. Brill, 41 pl., 97 pls., 1 table (International Sedimentary Petrographical Series 3).

Brief stratigraphic introduction and bibliography. Biotics, principally foraminifers, identified to genus and species.

SACAL, VINCENT, 1963, Microfaciès du Paléozoïque Saharien: Paris, Cie. Française des Pétroles, Notes et Mém., no. 6, 30 p., 100 figs., 4 charts.

Brief stratigraphic introduction and short bibliography. Full range of biotics beautifully illustrated; identifications mostly to major fossil groups.

SAMPÓ, M., 1969, Microfacies and microfossils of the Zagros area southwestern Iran: Leiden, E. J. Brill, 102 p., 105 pls., 6 figs. (International Sedimentary Petrographical Series 12).

Pre-Permian to Miocene microfacies in a 11,300 square mile area of Iran. Discussion of geology and stratigraphy followed by lists of microfacies and plates arranged in stratigraphic sequence. Author uses FOLK's classification. Many identifications to species. Fossils indexed.

References Cited

ADAMS, T. D., *et al.*, 1967, Stratigraphic significance of some oligosteginid assemblages from Lurestan Province, northwest Iran: Micropaleontology, v. 13, p. 55–67, 1 pl., 4 text-figs., 10 tables.

ALLOITEAU, JAMES, 1952, Madréporaires post-Paléozoïques, *in* PIVETEAU, JEAN, Traité de Paléontologie: Paris, Masson et Cie., v. 1, p. 539–684, 10 pls., 130 figs.

— 1957, Contribution à la systématique des Madréporaires fossiles: Centre Nat. Rech. Sci., v. 1, 462 p., 6 pls., 4 tables; v. 2, 20 pls., 286 figs.

ANDRES, DIETMAR, 1969, Ostracoden aus dem mittleren Kambrium von Öland: Lethaia, v. 2, p. 165–180, 12 figs.

ANDREWS, H. N., JR., 1961, Studies in Paleobotany: New York, John Wiley & Sons, Inc., 487 p.

ARKELL, W. J., *et al.*, 1957, Cephalopoda Ammonoidea, *in* MOORE, R. C., ed., Treatise on invertebrate paleontology Part L Mollusca 4: Geological Society of America and University of Kansas Press, 490 p., 558 figs.

ARMSTRONG, JOHN, 1969, The crossed-bladed fabrics of the shells of *Terrakea solida* (ETHERIDGE and DUN) and *Streptorhynchus pelicanensis* FLETCHER: Palaeontology, v. 12, p. 310–320, pls. 57–60, 1 text-fig.

BATHER, F. A., 1923, The shell of *Cornulites*: Geol. Mag., v. 60, p. 542–545, 1 fig.

BATHURST, R. G. C., 1964, The replacement of aragonite by calcite in the molluscan shell wall, *in* IMBRIE, JOHN, and NEWELL, NORMAN, eds., Approaches to paleoecology: New York, John Wiley & Sons, Inc., p. 357–376, 4 pls., 2 tables.

— 1966, Boring algae, micritic envelopes, and lithification of molluscan biosparites: Geol. Jour., v. 5, p. 15–32.

BEAVER, H. H., *et al.*, 1967, General characters Homalozoa-Crinozoa (except Crinoidea), *in* MOORE, R. C., ed., Treatise on invertebrate paleontology Part S Echinodermata 1: Geological Society of America and University of Kansas Press, 650 p., 400 figs.

BLOOM, WILLIAM, and FAWCETT, D. W., 1968, A textbook of histology (9th edition): Philadelphia, W. B. Saunders Company, 858 p.

BOARDMAN, R. S., 1960, Trepostomatous Bryozoa of the Hamilton Group of New York State: U. S. Geol. Survey Prof. Paper 340, 87 p., 22 pls., 27 figs.

BOARDMAMN, R. S., and CHEETHAM, A. H., 1969, Skeletal growth, intracolony variation, and evolution in Bryozoa: a review: Jour. Paleontology, v. 43, p. 205–233, pls. 27–30, 8 text-figs.

— and TOWE, K. M., 1966, Crystal growth and lamellar development in some Recent cyclostome Bryozoa: Geol. Soc. America Program 1966 Ann. Meeting, November 14–16, San Francisco, California, p. 20 (reprinted in Abstracts for 1966: Geol. Soc. America Spec. Paper 101, p. 20).

BØGGILD, O. B., 1930, The shell structure of the mollusks: Kgl. Danske Vidensk. Selsk. Skrifter, Naturvidensk. og Mathem. Afd. (Acad. Roy. Sci. Lettre Danemark, Mém. ser. 9, v. 2, p. 230–359, 15 pls., 10 figs.

BONET, FREDERICO, 1956, Zonificación microfaunística de las calizas Cretácicas del este de Mexico: Asoc. Mexicana Geólogos Petroleros Bol., v. 8, p. 389–488, 31 pls., 4 figs., 3 tables (reprinted XX Congreso Geol. Internatl., 102 p.).

BRYAN, W. H., and HILL, DOROTHY, 1940, Spherulitic crystallization as a mechanism of skeletal growth in the hexacorals: Royal Soc. Queensland Proc., v. 52, p. 78–91, 2 figs.

CAIN, J. D. B., 1968, Aspects of the depositional environment and palaeoecology of crinoidal limestone: Scottish Jour. Geology, v. 4, p. 191–208, 2 pls., 5 figs., 1 table.

CAMACHO, H. H., 1966, Invertebrados fósiles: Buenos Aires, Eudeba Editorial Universitaria de Buenos Aires, 707 p.

CAMPBELL, A. S., 1954a, Radiolaria, *in* MOORE, R. C., ed., Treatise on invertebrate paleontology Part D: Geological Society of America and University of Kansas Press, p. D11–D163, figs. 6–86.

— 1954b, Tintinnina, *in* MOORE, R. C., ed., Treatise on invertebrate paleontology Part D: Geological Society of America and University of Kansas Press, p. D166–D180, figs. 88–92.

CAREFOOT, T. H., 1965, Magnetite in the radula of the Polyplacophora: Proc. Malacol. Soc. London, v. 36, p. 203–212, pl. 10, 1 fig., 5 tables.

CARLSTRÖM, DIEGO, 1963, A crystallographic study of vertebrate otoliths: Biol. Bull., v. 125, p. 441–463, 5 figs., 1 table.

CAYEUX, LUCIEN, 1916, Introduction à l'étude pétrographique des roches sédimentaires: Mém. Ministère des Travaux Publics, Paris, Imprimerie Nationale, 2 v., 524 p., 56 pls., 80 figs.

— 1935, Les roches sédimentaires de France—roches carbonatées (calcaires et dolomites): Paris, Masson & Cie., 463 p., 26 pls., 9 figs.

CHAVE, K. E., 1954a, Aspects of the biogeochemistry of magnesium 1. Calcareous marine organisms: Jour. Geology, v. 62, p. 266–283, 16 figs., 3 tables.

CHAVE, K. E., 1954b, Aspects of the biogeochemistry of magnesium 2. Calcareous sediments and rocks: Jour. Geology, v. 62, p. 587–599, 4 figs., 9 tables.

— 1962, Factors influencing the mineralogy of carbonate sediments: Limnology and Oceanography, v. 7, p. 218–223, 2 figs., 7 tables.

CHEETAM, A. H., *et al.*, 1969, Wall structure and mineralogy of the cheilostome bryozoan *Metrarabdotos:* Jour. Paleontology, v. 43, p. 129–135, pl. 26, 1 text-fig., 2 tables.

COGAN, D. G., *et al.*, 1958, Crystalline calcium sulfate (gypsum) in scleral plaques of a human eye: Jour. Histochemistry and Cytochemistry, v. 6, p. 142–145, 4 figs.

COLOM, GUILLERMO, 1948, Fossil tintinnids: loricated Infusoria of the order Oligotricha: Jour. Paleontology, v. 22, p. 233–263, pls. 33–35, 14 figs., 1 table.

— 1955, Jurassic-Cretaceous pelagic sediments of the western Mediterranean zone and the Atlantic area: Micropaleontology, v. 1, p. 109–124, 5 pls., 4 text-figs.

CONKIN, J. E., 1961, Mississippian smaller Foraminifera of Kentucky, southern Indiana, northern Tennessee, and southcentral Ohio: Bull. Am. Paleontology, v. 43, p. 129–368, pls. 17–27, 43 figs., 23 charts, 1 map, 60 tables.

COOK, S. F., *et al.*, 1962, Histological studies on fossil bone: Jour. Paleontology, v. 36, p. 483–494, pls. 81–85.

COX, L. R., 1960, Gastropoda—General characteristics of Gastropoda, *in* MOORE, R. C., ed., Treatise on invertebrate paleontology Part I Mollusca 1: Geological Society of America and University of Kansas Press, p. I84–I169, figs. 51–88.

— *et al.*, 1969, Bivalvia, *in* MOORE, R. C., ed., Treatise on invertebrate paleontology Part N Mollusca 6: Geological Society of America and University of Kansas Press, 2 v., 952 p.

CUFFEY, R. J., 1967, Bryozoan Tabulipora carbonaria in Wreford Megacyclothem (Lower Permian) of Kansas: Univ. Kansas Paleont. Contrib. Bryozoa, art. 1, 96 p., 9 pls., 33 figs., 17 tables.

— 1970, Bryozoan-environment interrelationships—an overview of bryozoan paleoecology and ecology: Pennsylvania State Univ. Earth and Mineral Sci. Bull., v. 39, p. 41–45, 48, 4 figs.

CULLIS, C. G., 1904, The mineralogical changes observed in the cores of the Funafuti borings, *in* The atoll of Funafuti: London, Royal Soc. London, Rept. Coral Reef Comm., p. 392–420, Pl. F, figs. 24–69.

CUMMINGS, R. H., 1955, *Nodosinella* BRADY, 1876, and associated upper Paleozoic genera: Micropaleontology, v. 1, p. 221–238, 1 pl., 10 text-figs.

DECHASEAUX, COLETTE, 1952, Classe des lamellibranches, *in* PIVETEAU, JEAN, ed., Traité de Paléontologie: Paris, Masson et Cie., v. 2, p. 220–364, 215 figs.

DEFLANDRE, GEORGES, 1936, Tintinnoïdiens et Calpionelles. Comparaison entre les Tintinnoïdiens, Infusoires loriqués pélagiques des mers actuelles et les Calpionelles, microfossiles de l'époque secondaire: Bull. Soc. Français Microsc., v. 5, p. 112–122, 42 figs.

— 1952, Sous-embranchement des Actinopodes, *in* PIVETEAU, JEAN, ed., Traité de Paléontologie: Paris, Masson et Cie., v. 1, p. 303–313, figs. 1–53.

DEHORNE, YVONNE, 1920, Les stromatoporoïdes des terrains secondaires: Mém. pour servir à l'explication de la Carte géol. detaillée de la France, 170 p., 17 pls.

DENNELL, RALPH, 1960, Integument and exoskeleton, *in* WATERMAN, T. H., ed., The physiology of Crustacea: New York, Academic Press, v. 1, p. 449–472, 4 figs., 1 table.

DONNAY, GABRIELLE, and PAWSON, D. L., 1969, X-ray diffraction studies of echinoderm plates: Science, v. 166, p. 1147–1150, 2 figs., 1 table.

DUNCAN, HELEN, 1957, Bryozoans, *in* LADD, H. S., ed., Treatise on marine ecology and paleoecology: Geol. Society of America Mem. 67, v. 2. p. 783–799.

DURHAM, J. W., *et al.*, 1966, Asterozoa—Echinozoa, *in* MOORE, R. C., ed., Treatise on invertebrate paleontology Part U Echinodermata 3: Geological Society of America and University of Kansas Press, 695 p. 534 figs.

EASTON, W. H., 1960, Invertebrate paleontology: New York, Harper & Brothers, Publishers, 701 p.

ELLIOTT, G. F., 1958, Fossil microproblematica from the Middle East: Micropaleontology, v. 4, p. 419–428, 3 pls.

— 1960, Fossil calcareous algal floras of the Middle East with a note on a Cretaceous problematicum, *Hensonella cylindrica* gen. et sp. nov.: Quart. Jour. Geol. Soc. London, v. 115, p. 217–232, pl. 8.

— 1962, More microproblematica from the Middle East: Micropaleontology, v. 8, p. 29–44, 6 pls., 1 table.

ERBEN, H. K., *et al.*, 1969, Die frühontogenetische Entwicklung der Schalenstruktur ectocochleater Cephalopoden: Palaeontographica Abt. A, v. 132, p. 1–54, 15 pls., 12 figs., 3 tables.

FERAY, D. E., *et al.*, 1962, Biological, genetic, and utilitarian aspects of limestone classification, *in* HAM, W. E., ed., Classification of carbonate rocks: Am. Assoc. Petroleum Geologists Mem. 1, p. 20–32, 3 figs.

FISCHER, A. G., and TEICHERT, CURT, 1969, Cameral deposits in cephalopod shells: Univ. Kansas Paleont. Contr. Paper 37, 30 p., 4 pls., 8 figs.

— *et al.*, 1967, Electron micrographs of limestones and their nannofossils: Princeton, Princeton University Press, 141 p., 94 figs.

FISHER, D. W., 1962, Small conoidal shells of uncertain affinities, *in* MOORE, R. C., ed., Treatise on invertebrate paleontology Part W: Geological Society of America and University of Kansas Press, p. W98–W143, figs. 50–84.

FLEISCHER, MICHAEL, 1969, Occurrences of $CaCO_3 \cdot H_2O$ and its naming: Science, v. 166, p. 1309.

FLOWER, R. H., 1961, Part I Montoya and related colonial corals Part II Organisms attached to Montoya corals: New Mexico Bur. Mines & Min. Res. Mem. 7, 229 p., 52 pls., 10 text-figs., 1 table.

FLÜGEL, ERIK, and FLÜGEL-KAHLER, EVENTRAUD, 1968, Stromatoporoidea (Hydrozoa palaeozoica): Fossilium Catalogus I: Animalia, Pars 115–116, 681 p.

FOLK, R. L., 1965, Some aspects of recrystallization in ancient limestones, *in* PRAY, L. C., and MURRAY, R. C., eds., Dolomitization and limestone diagenesis: Soc. Econ. Paleontologists and Mineralogists Spec. Pub. 13, p. 14–48, 14 figs., 7 tables.

FRIEDMAN, G. M., 1964, Early diagenesis and lithification in carbonate sediments: Jour. Sed. Petrology, v. 34, p. 777–813, 53 figs.

FÜCHTBAUER, HANS, ed., 1969, Lithification of carbonate sediments, 1 and 2: Sedimentology (Special Issues), v. 12, p. 1–323.

GALLOWAY, J. J., 1957, Structure and classification of Stromatoporoidea: Bull. Am. Paleontology, v. 37, no. 164, p. 341–480, pls. 31–37, 1 table.

GARRETT, PETER, 1970, Phanerozoic stromatolites: noncompetitive ecologic restriction by grazing and burrowing animals: Science, v. 169, p. 171–173, 1 fig.

GAURI, K. L., and BOUCOT, A. J., 1968, Shell structure and classification of Pentameracea M'Coy, 1844: Palaeontographica, Abt. A, v. 131, p. 79–135, pls. 6–23, 31 text-figs., 2 tables.

— — 1970, *Cryptothyrella* (Brachiopoda) from the Brassfield Limestone (Lower Silurian) of Ohio and Kentucky: Jour. Paleontology, v. 44, p. 125–132, pls. 29–31, 2 text-figs.

GAVISH, ELIEZER, and FRIEDMAN, G. M., 1969, Progressive diagenesis in Quaternary to Late Tertiary carbonate sediments: sequence and time scale: Jour. Sed. Petrology, v. 39, p. 980–1006, 32 figs.

GLAESSNER, M. F., 1963, Major trends in the evolution of the Foraminifera, *in* KOENIGSWALD, G. H. R., *et al.*, eds., Evolutionary trends in Foraminifera: Amsterdam, Elsevier Publishing Company, p. 9–24, 1 table.

GOLL, R. M., 1968, Classification and phylogeny of Cenozoic Trissocyclidae (Radiolaria) in the Pacific and Caribbean Basins Part I: Jour. Paleontology, v. 42, p. 1409–1432, pls. 173–176, 9 text-figs.

— 1969, Classification and phylogeny of Cenozoic Trissocyclidae (Radiolaria) in the Pacific and Caribbean basins Part II: Jour. Paleontology, v. 43, p. 322–339, pls. 55–60, 2 text-figs.

GRAHAM, D. K., 1970, Scottish Carboniferous Lingulacea: Geol. Survey Great Britain Bull. 31, p. 139–184, pls. 14–20, 10 figs.

GREGUSS, PÁL, 1967, Fossil gymnosperm woods in Hungary from the Permian to the Pliocene: Budapest, Akadémiai Kiadó, 136 p., 86 pls., 14 maps.

GROSS, WALTER, 1967, Über Thelodontier-Schuppen: Palaeontographica, Abt. A, v. 127, p. 1–67, 7 pls., 15 figs.

HAM, A. W., 1969, Histology (6th edition): Philadelphia, J. B. Lippincott Company, 1037 p.

HÄNTZSCHEL, WALTER, *et al.*, 1968, Coprolites an annotated bibliography: Geol. Soc. America Mem. 108, 132 p., 11 pls., 6 figs., 3 tables.

HARRINGTON, H. J., 1959, General description of Trilobita, *in* MOORE, R. C., ed., Treatise on invertebrate paleontology Part O Arthropoda 1: Geological Society of America and University of Kansas Press, p. O38–O117, figs. 27–85.

HENNINGSMOEN, GUNNAR, 1965, On certain features of Palaeocope ostracodes: Geologiska Föreningens i Stockholm Förhandlingar, v. 86, p. 329–394, 16 figs.

HILL, DOROTHY, 1956, Rugosa, *in* MOORE, R. C., Treatise on invertebrate paleontology Part F Coelenterata: Geological Society of America and University of Kansas Press, p. F233–F324, figs. 165–219.

— 1965, Archaeocyatha from Antarctica and a review of the phylum: Trans-Antarctic Expedition 1955–1958, Sci. Repts. 10, Geology 3, 151 p., 12 pls., 25 figs.

HOFKER, JAN, 1962, Studien an planktonischen Foraminiferen: Neues Jahrb. Geol. Paläont. Abh., v. 114, p. 81–134, 85 figs.

— 1967, Hat die feinere Wandstruktur der Foraminiferen supragenerische Bedeutung?: Paläont. Zeitsch., v. 41, p. 194–198, pls. 19–21.

HOWELL, B. F., 1962, Worms, *in* MOORE, R. C., ed., Treatise on invertebrate paleontology Part W Miscellanea: Geological Society of America and University of Kansas Press, p. W144–W177, figs. 85–108.

HUDSON, R. G. S., 1956, Tethyan Jurassic hydroids of the family Milleporidiidae: Jour. Paleontology, v. 30, p. 714–730, pls. 75–77, 6 text-figs.

— 1958, *Actostroma* gen. nov., a Jurassic stromatoporoid from Maktesh Hathira, Israel: Palaeontology, v. 1, p. 87–98, pls. 15–17, 7 text-figs.

— 1959, A revision of the Jurassic stromatoporoids *Actinostromina*, *Astrostylopsis*, and *Trupetostromaria* Germovšek: Palaeontology, v. 2, p. 28–38, pls. 4–6.

— 1960, The Tethyan Jurassic stromatoporoids *Stromatoporina*, *Dehornella*, and *Astroporina*: Palaeontology, v. 2, p. 180–199, pls. 24–28, 6 text-figs.

JAANUSSON, VALDAR, 1966, Fossil brachiopods with probable aragonitic shells: Geol. Fören. Stockholm Förh., v. 88, p. 279–281, 1 fig.

JELETZKY, J. A., 1966, Comparative morphology, phylogeny, and classification of fossil coleoidea: Univ. Kansas Paleont. Contr. Mollusca, art. 7, 162 p., 25 pls., 15 figs.

JENSEN, J. A., 1966, Dinosaur eggs from the Upper Cretaceous North Horn Formation of central Utah: Brigham Young Univ. Geology Studies, v. 13, p. 55–67, 4 pls., 2 text-figs.

JOHNSON, J. H., 1943, Geologic importance of calcareous algae with annotated bibliography: Colorado School Mines Quart., v. 38, no. 1, 102 p., 23 figs., 2 tables.

— 1951, An introduction to the study of organic limestones: Colorado School Mines Quart., v. 46, no. 2, 185 p., 104 pls., 1 fig., 19 tables.

— 1957, Bibliography of fossil algae: 1942–1955: Colorado School Mines Quart., v. 52, no. 2, 92 p.

— 1961, Limestone-building algae and algal limestones: Golden, Colorado School of Mines, 297 p., 139 pls., 14 tables.

— 1967, Bibliography of fossil algae, algal limestones, and the geological work of algae, 1956–1965: Colorado School Mines Quart., v. 62, no. 4, 148 p.

KATO, MAKOTO, 1963, Fine skeletal structures in Rugosa: Jour. Fac. Sci. Hokkaido Univ., ser. 4., Geology and Mineralogy, v. 11, p. 571–630, 3 pls., 19 text-figs.

— 1968, Note on the fine skeletal structures in Scleractinia and in Tabulata: Jour. Fac. Sci. Hokkaido Univ., ser. 4, Geology and Mineralogy, v. 14, p. 51–56, 1 fig.

KENNEDY, W. J., and TAYLOR, J. D., 1968, Aragonite in rudists: Proc. Geol. Soc. London, no. 1645, p. 325–331, 3 figs.

— *et al.*, 1969, Environmental and biological controls on bivalve shell mineralogy: Biol. Rev., v. 44, p. 499–530, 4 pls., 13 figs.

KIELAN, ZOFIA, 1954, Les trilobites mésodévoniens des Montes de Sainte-Croix: Palaeontologia Polonica, no. 6, 50 p., 7 pls., 35 figs.

KNAPP, W. D., 1969, Declinida, a new order of Late Paleozoic inadunate crinoids: Jour. Paleontology, v. 43, p. 340–391, pls. 61, 62, 50 text-figs.

KNIGHT, J. B., *et al.*, 1960, Mollusca 1, *in* R. C. MOORE, ed., Treatise on invertebrate paleontology Part I: Geological Society of America and University of Kansas Press, 351 p., 216 figs.

KRANS, T. F., 1965, Études morphologiques de quelques Spirifères dévoniens de la Chaîne Cantabrique (Espagne): Leidse Geol. Medeleel., v. 33, p. 71–148, 16 pls., 71 figs.

LAND, L. S., 1967, Diagenesis of skeletal carbonate: Jour. Sed. Petrology, v. 37, p. 914–930, 15 figs., 5 tables.

LAPORTE, L. F., 1962, Paleoecology of the Cottonwood Limestone (Permian), northern Midcontinent: Geol. Society America Bull., v. 73, p. 521–544, 4 pls., 5 figs., 4 tables.

LAUBENFELS, M. W. DE, 1955, Porifera, *in* MOORE, R. C., ed., Treatise on invertebrate paleontology Part E: Geological Society of America and University of Kansas Press, E21–E122, figs. 14–89.

LECOMPTE, MARIUS, 1951, Les stromatoporoïdes du Dévonian moyen supérieur du bassin de Dinant: Mém. Inst. roy. Sci. nat. Belgique, v. 116, p. 1–215, pls. 1–35.

— 1952, Les stromatoporoïdes du Dévonian moyen et supérieur du bassin de Dinant: Mém. Inst. roy. Sci. nat. Belgique, v. 117, p. 216–359, pls. 36–70.

— 1956, Stromatoporoidea, *in* MOORE, R. C., ed., Treatise on invertebrate paleontology Part F Coelenterata: Geological Society of America and University of Kansas Press, p. F107–F144, figs. 86–114.

LEVINSON, S. A., 1961, Identification of fossil ostracodes in thin section, *in* MOORE, R. C., ed., Treatise on invertebrate paleontology Part Q Arthropoda 3: Geological Society of America and University of Kansas Press, p. Q70–Q73, fig. 31.

LINDSTRÖM, MAURITS, 1964, Conodonts: Amsterdam, Elsevier Publishing Company, 196 p., 64 figs., 5 tables.

LOEBLICH, A. R., JR., and TAPPAN, HELEN, 1964, Sarcodina chiefly "Thecamoidians" and Foraminiferida, *in* MOORE, R. C., ed., Treatise on invertebrate paleontology Part C Protista 2: Geological Society of America and University of Kansas Press, 2 v., 900 p., 653 figs.

— — 1966, Annotated index and bibliography of the calcareous nannoplankton: Phycologia, v. 5, p. 81–216.

— — 1968, Annotated index to the genera, subgenera and suprageneric taxa of the ciliate Order Tintinnida: Jour. Protozoology, v. 15, p. 185–192.

LOWENSTAM, H. A., 1962, Magnetite in denticle capping in Recent chitons (Polyplacophora): Geol. Soc. America Bull., v. 73, p. 435–438, 1 pl.

— 1963, Biologic problems relating to the composition and diagenesis of sediments, *in* DONELLY, T. W., ed., The earth sciences: Chicago, The University of Chicago Press, p. 137–195, 4 pls., 14 figs., 1 table.

— 1967, Lepidocrocite, an apatite mineral, and magnetite in teeth of chitons (Polyplacophora): Science, v. 156, p. 1373–1375, 3 figs.

— 1968, Weddellite in a marine gastropod and in Antarctic sediments: Science, v. 162, p. 1129–1130, 2 figs.

— and MCCONNELL, DUNCAN, 1968, Biologic precipitation of fluorite: Science, v. 162, p. 1496–1498, 1 fig.

LUCIA, F. J., 1962, Diagenesis of a crinoidal sediment: Jour. Sed. Petrology, v. 32, p. 848–865, 16 figs.

MACCLINTOCK, COPELAND, 1967, Shell structure of patelloid and bellerophontoid gastropods (Mollusca): Peabody Mus. Nat. Hist. Yale Univ., Bull. 22, 140 p., 32 pls., 128 text-figs., 10 tables.

MAJEWSKE, O. P., 1969, Recognition of invertebrate fossil fragments in rocks and thin sections: Leiden, E. J. Brill, 101 p., 106 pls., 8 tables (International Sedimentary Petrographical Series 13).

MASLOV, V. P., 1937, Atlas karbonatnykh porod chast' I Porodoobraziushchie organizmy (Atlas de roches Carbonatées Part I Organismes formant les roches): Moskva, Vsesouznyi Nauchno-Issledovatel'skiĭ Institut Mineral'nogo Syr'ya, 54 p., 55 pls.

— 1956, Iskopaemye isvestkovye vodorosli SSSR (Fossil calcareous algae of the USSR): Acad. Nauk S.S.S.R. Trudy Instituta Geol. Nauk, no. 160, 301 p., 86 pls., 136 figs., 9 tables, 4 charts.

— *et al.*, 1963, Vodorosli (Algae), *in* ORLOV, YU. A., Osnovy paleontologiĭ: Moskva Akad. Nauk SSSR, v. 14, p. 19–312, 22 pls.

MATTHEWS, R. K., 1966, Genesis of Recent lime mud in southern Honduras: Jour. Sed. Petrology, v. 36, p. 428–454, 8 figs., 5 tables.

MELIK, J. C., 1966, Hingement and contact margin structure of palaeocopid ostracodes from some Middle Devonian formations of Michigan, southwestern Ontario, and western New York: Univ. Michigan Contr. Mus. Paleontology, v. 20, p. 195–269, 24 pls., 4 figs.

MOORE, R. C., and JEFFORDS, R. M., 1968, Classification and nomenclature of fossil crinoids based on studies of dissociated parts of their columns: Univ. Kansas Paleont. Contr. Echinodermata, art. 9, 86 p., 28 pls., 6 figs., 4 tables.

— *et al.*, 1952, Invertebrate fossils: New York, McGraw-Hill Book Company, Inc., 766 p.

MÜLLER, A. H., 1957–1968, Lehrbuch der Paläozoologie: Jena, Gustav Fischer Verlag, 3 v.

MÜLLER, K. J., 1964, Ostracoda (Bradorina) mit phosphatischen Gehäusen aus dem Oberkambrium von Schweden: Neues Jahrb. Geol. Paläont. Abh., v. 121, p. 1–46, 5 pls., 2 figs., 3 tables.

— and MOSHER, L. C., 1969, Reports of post-Triassic conodonts: Geol. Soc. America Abstracts with Programs for 1969, part 6, p. 33.

MUTVEI, HARRY, 1967, On the microscopic shell structure in some Jurassic ammonoids: Neues Jahrb. Geol. Paläont. Abh., v. 129, p. 157–166, pl. 14, 4 text-figs.

NĚMEJC, FRANTIŠEK, 1959, Paleobotanika: Praha, Nakladalelství Československé Akademie Věd, v. 1, 402 p., 32 pls., 174 figs.

NISSEN, HANS-UDE, 1969, Crystal orientation and plate structure in echinoid skeletal units: Science, v. 166, p. 1150–1152, 4 figs.

NORLING, ERIC, 1968, On Liassic nodosariid Foraminifera and their wall structures: Sveriges Geologiska Undersökning, ser. C, no. 623, Årsbok 61, no. 8, 75 p., 9 pls., 12 text-figs., 5 tables.

NYE, O. B., 1969, Aspects of microstructure in post-Paleozoic *Cyclostomata:* Atti Soc. It. Sc. Nat. e Museo Civ. St. Nat. Milano: v. 108, p. 111–114.

OBERLING, J. J., 1964, Observations on some structural features of the pelecypod shell: Naturf. Gesell. Bern Mitt., n. ser., v. 20, p. 1–63, 5 pls., 3 figs., 1 table.

OKULITCH, V. J., 1955, Archaeocyatha, *in* MOORE, R. C., ed., Treatise on invertebrate paleontology Part E: Geological Society of America and University of Kansas Press, E1–E20, figs. 1–13.

ORLOV, YU. A., ed., 1958–1964, Osnovy paleontologii (Fundamentals of paleontology): Moscow, Akad. Nauk SSSR, 15 v.

ORME, G. R., and BROWN, W W. M., 1963, Diagenetic limestone fabrics in the Avonian limestones of Derbyshire and North Wales: Yorkshire Geol. Soc. Proc., v. 34, p. 51–66, pls. 7–13, 1 fig.

PALMER, A. R., 1964, An unusual Lower Cambrian trilobite fauna from Nevada: U. S. Geol. Survey Prof. Paper 483–F, p. F1–F13, 3 pls., 2 figs.

PECK, R. E., 1934, The North American trochiliscids, Paleozoic Charophyta: Jour. Paleontology, v. 8., p. 83–119, pls. 9–13, 2 figs.

— 1957, North American Mesozoic Charophyta: U.S. Geol. Survey Prof. Paper 294–A, 44 p., 8 pls., 7 figs.

PETERSON, N. M. A., 1966, Calcite: Rates of dissolution in a vertical profile in the central Pacific: Science, v. 154, p. 1542–1544, 2 figs.

PEYER, BERNHARD, 1968, Comparative odontology: Chicago, The University of Chicago Press, 347 p., 88 pls. and 8 color pls., 220 figs.

PIA, JULIUS, 1926, Pflanzen als Gesteinsbildner: Berlin, Gebrüder Borntraeger, 355 p., 166 text-figs.

PIETZNER, HORST, *et al.*, 1968, Zur chemischen Zusammensetzung und Mikromorphologie der Conodonten: Palaeontographica Abt. A, v. 128, p. 115–152, pls. 18–27, 10 figs., 8 tables.

PIVETEAU, JEAN, ed., 1952–1969, Traité de Paléontologie: Paris, Masson et Cie., 7 v.

PLAYFORD, P. E., and COCKBAIN, A. E., 1969, Algal stromatolites: deepwater forms in the Devonian of Western Australia: Science, v. 165, p. 1008–1010, 2 figs.

POKORNÝ, VLADIMÍR, 1958, Grundzüge der zoologischen Mikropaläontologie: Berlin, VEB Deutscher Verlag der Wissenschaften, v. 1, 582 p., 549 figs. (English translation by K. A. ALLEN and J. W. NEALE, 1963, Principles of zoological micropaleontology: Oxford, Pergamon Press, 652 p.).

PRAY, L. C., and MURRAY, R. C., eds., 1965, Dolomitization and limestone diagenesis a symposium: Soc. Econ. Paleontologists and Mineralogists Spec. Pub. 13, 180 p.

PURDY, E. G., 1968, Carbonate diagenesis: an environmental survey: Geologica Romana, v. 7, p. 183–228, 6 pls., 10 figs.

RAUP, D. M., 1961, The geometry of coiling in gastropods: Natl. Acad. Sci. Proc., v. 47, p. 602–609, 4 figs.

RAUP, D. M., 1966a, The endoskeleton, *in* BOOLOOTIAN, R. A., Physiology of Echinodermata: New York, John Wiley & Sons, Inc., p. 379–395, 6 figs., 1 table.

— 1966b, Geometric analysis of shell coiling: general problems: Jour. Paleontology, v. 40, p. 1178–1190, 10 text-figs.

— 1967, Geometric analysis of shell coiling: coiling in ammonoids: Jour. Paleontology, v. 41, p. 43–65, 19 text-figs.

REISS, ZEEV, 1963a, Reclassification of perforate Foraminifera: Israel Geol. Survey Bull. 35, 111 p., 8 pls.

— 1963b, Comments on wall structure of foraminifera: Micropaleontology, v. 9, p. 50–52.

REITLINGER, E. A., 1950, Foraminifery srednekammenougol'nikh otlozheniĭ tsentral'noi chasti Russkoi platformy (isklyuchaya semeistvo Fusulinidae) [Foraminifera of the middle Carboniferous deposits of the central part of the Russian platform (including the family Fusulinidae)]: Akad. Nauk SSSR, Trudy Inst. Geol. Nauk., no. 126 (Geol. Ser. no. 47), 126 p., 22 pls., 15 text-figs.

— 1958, K voprosu sistematiki i filogeniĭ nadsemeistva Endothyridea (On the question of the systematics and phylogeny of the superfamily Endothyridea): Voprosu Mikropaleontologiĭ Akad. Nauk SSSR, no. 2, p. 53–73, 4 figs.

REMANE, JÜRGEN, 1963, Les Calpionelles dans les couches de passage Jurassique-Crétacé de la fosse vocontienne: Univ. Grenoble lab. géologie fac. sci. Travaux, v. 39, p. 25–82, pls. 1–6, 18 figs.

— 1964, Untersuchungen zur Systematik und Stratigraphie der Calpionellen in den Jura-Kreide-Grenzschichten des vocontischen Troges: Palaeontographica, Bd. 123, Abt. A, p. 1–57, 6 pls., 18 figs.

REVELLE, ROGER, and FAIRBRIDGE, RHODES, 1957, Carbonates and carbon dioxide, *in* HEDGPETH, J. W., ed., Treatise on marine ecology and paleoecology: Geol. Soc. America Mem. 67, v. 1., p. 239–296, 8 figs., 8 tables.

RICHARDS, A. G., 1951, The integument of arthropods: Minneapolis, University of Minnesota Press, 411 p., 65 figs.

RIGBY, J. K., 1958, Two new upper Paleozoic hydrozoans: Jour. Paleontology, v. 32, p. 583–586, pl. 86, 3 text-figs.

ROSS, J. P., 1967, Fossil problematica from Upper Ordovician, Ohio: Jour. Paleontology, v. 41, p. 37–42, pls. 4–8, 2 text-figs.

RUCKER, J. B., 1967, Carbonate mineralogy of Recent cheilostome Bryozoa: Geol. Soc. America Program 1967 Ann. Meeting, November 20–22, New Orleans, Louisiana, p. 191–192 (reprinted in Abstracts for 1967: Geol. Soc. America Spec. Paper 115, p. 192).

— 1969, Skeletal mineralogy of cheilostome *Bryozoa:* Atti Soc. It. Sc. Nat. e Museo Civ. St. Nat. Milano, v. 108, p. 101–110, 2 figs., 1 table.

ST. JEAN, JOSEPH, JR., 1967, Maculate tissue in Stromatoporoidea: Micropaleontology, v. 13, p. 419–444, pls. 1–6, 1 text-fig.

SASS, D. B., 1967, Electron microscopy, punctae, and the brachiopod genus *Syringothyris* WINCHELL, 1863: Jour. Paleontology, v. 41, p. 1242–1246, pls. 167–169.

SCHENK, H. G., 1934, Literature on the shell structure of pelecypods: Mus. roy. Hist. nat. Belgique Bull., v. 10, no. 34, 20 p.

SCHMALZ, R. F., 1965, Brucite in carbonate secreted by the red alga Goniolithon sp.: Science, v. 149, p. 993–996, 2 figs., 1 table.

SCHMIDT, W. J., 1924, Die Bausteine des Tierkörpers in polarisiertem Lichte: Bonn, Verlag von Frederich Cohen, 528 p., 230 figs.

— 1951, Die Unterscheidung der Röhren von Scaphopoda, Vermetidae und Serpulidae mittels mikroskopischer Methoden: Mikroskopie, v. 6, p. 373–381, 14 figs.

— 1955, Die tertiären Würmer Österreichs: Österreichs Akad. Wissenschaften Mathematisch-naturwissenschaftliche Klasse Denkschriften, v. 109, no. 7, 121 p., 8 pls., 2 tables.

SCHOPF, T. J. M., 1969a, Paleoecology of ectoprocts (bryozoans): Jour. Paleontology, v. 43, p. 234–244, 5 text-figs., 1 table.

— 1969b, Generalizations regarding the phylum *Ectoprocta* in the deep-sea (200–6000 m): Atti Soc. It. Sc. Nat. e Museo Civ. St. Nat. Milano, v. 108, p. 152–154.

SCHOUPPÉ, ALEXANDER VON, and STACUL, PAUL, 1966, Morphogenese und Bau des Skelettes der Pterocorallia: Palaeontographica, Supplement-Band 11, 186 p., 6 pls., 132 figs., 8 tables.

SCHROEDER, J. H., *et al.*, 1969, Primary protodolomite in echinoid skeletons: Geol. Soc. America Bull., v. 80, p. 1613–1616, 3 figs.

SCHWAB, K. W., 1965, Microstructure of some Middle Ordovician conodonts: Jour. Paleontology, v. 39, p. 590–593, pls. 69, 70, 3 text-figs.

— 1966, Microstructure of some fossil and recent scolecodonts: Jour. Paleontology, v. 40, p. 416–423, pls. 53, 54, 3 text-figs., 1 table.

SHIMER, H. W., and SHROCK, R. R., 1944, Index fossils of North America: New York, John Wiley & Sons, Inc., 837 p., 303 pls., 5 text-figs.

SHROCK, R. R., and TWENHOFEL, W. H., 1953, Principles of invertebrate paleontology: New York, McGraw-Hill Book Company, Inc., 816 p.

SIEVER, RAYMOND, 1957, The silica budget in the sedimentary cycle: Amer. Mineralogist, v. 42, p. 821–841, 2 figs.

— 1962, Silica solubility, 0°–200° C., and the diagenesis of siliceous sediments: Jour. Geology, v. 70, p. 127–150, 6 figs., 1 table.

— and SCOTT, R. A., 1963, Organic geochemistry of silica, *in* BREGER, IRVING, ed., Organic geochemistry of silica: Oxford, Pergamon Press, p. 579–595, 6 figs.

SLITER, W. V., 1968, Shell-material variation in the agglutinated foraminifer *Trochammina pacifica* CUSHMAN: Tulane Studies Geology, v. 6, p. 80–84, 1 fig.

SMOUT, A. H., 1954, Lower Tertiary Foraminifera of the Qatar Peninsula: British Mus. (Nat. History), 96 p., 15 pls.

SOCHAVA, A. V., 1969, Yaitsa dinosavrov iz verchnego mel Gobi (Dinasaurian egg shells from the Upper Cretaceous deposits of the Gobi): Paleont. Zhur. 1969, no. 4, p. 76–88, pls. 11–12, 5 figs., 1 table.

SOKOLOV, B. S., *et al.*, 1962, Klass Anthozoa. Korallovye polipy, *in* ORLOV, YU. A., Osnovy paleontologii: Moskva, Akad. Nauk SSSR, v. 2, p. 192–430, 57 pls.

SORBY, H. C., 1879, Anniversary address of the president: Proc. Geol. Soc. London, v. 35, p. 56–95, 11 figs., 1 table.

SPANGENBERG, D. B., and BECK, C. W., 1968, Calcium sulfate dihydrate statoliths in *Aurelia:* Trans. Am. Microsc. Soc., v. 87, p. 329–335, 3 figs., 2 tables.

STANTON, R. J., JR., 1967, Radiosphaerid calcispheres in North America and remarks on calcisphere classification: Micropaleontology, v. 13, p. 465–472, 1 pl., 1 table.

STEARN, C. W., 1966, The microstructure of stromatoporoids: Palaeontology, v. 9, p. 74–124, pls. 14–19, 15 text-figs.

STØRMER, LEIF, 1930, Scandinavian Trinucleidae: Skrifter Utgitt av Det Norske Videnskaps-Akademi Oslo I. Mat.-Naturv. Klasse, no. 4, 111 p., 14 pls., 47 figs.

SUTER, D. J., and WOOLEY, S. E., 1968, Gallstone of unusual composition: calcite, aragonite, and vaterite: Science, v. 159, p. 1113–1114, 1 fig.

SWINCHATT, J. P., 1969, Algal boring: a possible depth indicator in carbonate rocks and sediments: Geol. Soc. America Bull., v. 80, p. 1391–1396, 1 pl., 2 figs.

TAPPAN, HELEN, and LOEBLICH, A. R., JR., 1968, Lorica composition of modern and fossil Tintinnida (Ciliate Protozoa), systematics, geologic distribution, and some new Tertiary taxa: Jour. Paleontology, v. 42, p. 1378–1394, pls. 165–171, 1 text-fig.

TASCH, PAUL, and SHAFFER, B. L., 1961, Study of scolecodonts by transmitted light: Micropaleontology, v. 7, p. 369–371, 1 pl.

TAVENER-SMITH, RONALD, 1969, Skeletal structure and growth in the Fenestellidae (Bryozoa): Palaeontology, v. 12, p. 281–309, pls. 52–56, 9 text-figs.

TAYLOR, J. D., *et al.*, 1969, The shell structure and mineralogy of the Bivalvia Introduction. Nuculacea—Trigonacea: British Mus. (Nat. History), Zoology, Supplement 3, 128 p., 29 pls., 77 text-figs., 16 tables.

TEICHERT, CURT, *et al.*, 1964, Mollusca 3, *in* MOORE, R. C., ed., Treatise on invertebrate paleontology Part K: Geological Society of America and University of Kansas Press, 519 p., 361 figs.

THOMPSON, D'A. W., 1942, On growth and form, 2nd ed.: Cambridge, Cambridge University Press, 1116 p., 554 figs.

THOMPSON, M. L., 1964, Fusulinacea, *in* MOORE, R. C., ed., Treatise on invertebrate paleontology Part C Protista 2: Geological Society of America and the University of Kansas Press, p. C358–C436, figs. 274–328A.

TOOMEY, D. F., 1967, Additional occurrences and extension of stratigraphic range of the problematical microorganism *Nuia:* Jour. Paleontology, v. 41, p. 1457–1460, pl. 185, 1 table.

— 1969, The biota of the Pennsylvanian (Virgilian) Leavenworth Limestone, Midcontinent region. Part 2: Distribution of algae: Jour. Paleontology, v. 43, p. 1313–1330, pls. 151–154, 1 text-fig., 2 tables.

TOWE, K. M., 1967a, Wall structure and cementation in *Haplophragmoides canariensis:* Contr. Cushman Found. Foram. Research, v. 18, p. 147–151, pls. 12, 13, 1 text-fig., 1 table.

— 1967b, Echinoderm calcite: single crystal or polycrystalline aggregate: Science, v. 157, p. 1048–1050, 8 figs.

— and CIFELLI, RICHARD, 1967, Wall ultrastructure in the calcareous foraminifera: crystallographic aspects and a model for calcification: Jour. Paleontology, v. 41, p. 742–762, pls. 87–99.

— and HARPER, C. W., JR., 1966, Pholidostrophiid brachiopods: origin of the nacreous luster: Science, v. 154, p. 153–155, 1 fig.

— and RÜTZLER, KLAUS, 1968, Lepidocrocite iron mineralization in keratose sponge granules: Science, v. 162, p. 268–269, 3 figs., 1 table.

TRAVIS, D. F., 1960, Matrix and mineral deposition in skeletal structures of the decapod Crustacea, *in* SOGNNAES, R. F., Calcification in Biological Systems: Amer. Assoc. Adv. Sci. Publ. 64, p. 57–116, 44 figs., 3 tables.

TREJO H., MARIO, 1960, La familia *Nannoconidae* y su alcance-estratigrafico en America (Protozoa, Incertae Saedis): Asoc. Mexicana Geólogos Petroleros Bol., v. 12, p. 259–314, 3 pls., 15 figs., 11 tables.

UTGAARD, JOHN, 1968a, A revision of North American genera of ceramoporoid bryozoans (Ectoprocta): Part I: Anolotichiidae: Jour. Paleontology, v. 42, p. 1033–1041, pls. 129–132.

— 1968b, A revision of North America genera of ceramoporoid bryozoans (Ectoprocta): Part II; *Crepipora, Ceramoporella, Acanthoceramoporella,* and *Ceramophylla:* Jour. Paleontology, v. 42, p. 1444–1455, pls. 181–184.

VOLOGDIN, A. G., 1962, Tip Archaeocyatha, *in* ORLOV, YU. A., ed., Osnovy paleontologii: Moskva, Akad. Nauk SSSR, v. 2, p. 89–144, 9 pls., 128 figs.

WALL, DAVID, and DALE, BARRIE, 1968, Quaternary calcareous dinoflagellates (Calciodinellidae) and their natural affinities: Jour. Paleontology, v. 42, p. 1395–1408, pl. 172, 3 text-figs.

WEBER, J. N., and KAUFMAN, J. W., 1965, Brucite in the calcareous alga Goniolithon: Science, v. 149, p. 996–997, 1 fig.

WELLS, J. W., 1956, Scleractinia, *in* MOORE, R. C., Treatise on invertebrate paleontology Part F Coelenterata: Geological Society of America and the University of Kansas Press, p. F328–F444, figs. 222–339.

WESTBROEK, PETER, 1967, Morphological observations with systematic implications on some Paleozoic Rhynchonellida from Europe, with special emphasis on the Uncinulidae: Leidse Geol. Mededeel., v. 41, p. 1–82, 14 pls., 81 figs.

WEYL, P. K., 1967, The solution of carbonate materials in sea water: Exploration Production Research Division, Shell Oil Company, Pub. 428, p. 178–228.

WILLGALLIS, ALEXANDER, 1969, Untersuchung des chemischen Aufbaus von mittelkambrischen Ostracodenschalen: Lethaia, v. 2, p. 181–183, 1 fig.

WILLIAMS, ALWYN, 1968, Evolution of the shell structure of articulate brachiopods: Palaeont. Assoc. Spec. Papers in Palaeontology No. 2, 55 p., 24 pls., 27 text-figs., 1 table.

— and ROWELL, A. J. 1965, Morphology, *in* MOORE, R. C., ed., Treatise on invertebrate paleontology Part H Brachiopoda: Geological Society of America and University of Kansas Press, p. H57–H138, figs. 59–138.

WILLIAMS, ALWYN, and WRIGHT, A. D., 1970, Shell structure of the Craniacea and other calcareous inarticulate brachiopods: Palaeont. Assoc. Spec. Papers in Palaeontology No. 7, 51 p., 15 pls., 17 text-figs.

WILSON, E. C., 1969, No new *Ungdarella* (Rhodophycophyta) in New Mexico: Jour. Paleontology, v. 43, p. 1245–1247, pl. 146, 1 text-fig.

WINLAND, H. D., 1968, The role of high Mg calcite in the preservation of micritic envelopes and textural features of aragonite sediments: Jour. Sed. Petrology, v. 38, p. 1320–1325, 7 figs.

WISE, S. W., JR., 1970, Scleractinian coral exoskeletons: surface microarchitecture and attachment scar patterns: Science, v. 169, p. 978–980, 2 figs.

WOOD, ALAN, 1949, The structure of the wall of the test in the Foraminifera; its value in classification: Quart. Jour. Geol. Society London, v. 104, p. 229–255, pls. 13–15.

YAVORSKY, V. I., 1955, Stromatoporoidea Sovetskogo Souza. Chast 1: Vses. Nauchno-Issled. Geol. Inst., Trudy, n. ser., no. 8, p. 1–173, 89 pls.

— 1957, Stromatoporoidea Sovetskogo Souza. Chast 2: Vses. Nauchno-Issled. Geol. Inst., Trudy, n. ser., no. 18, p. 1–168, 43 pls.

— 1961, Stromatoporoidea Sovetskogo Souza. Chast 3: Vses. Nauchno-Issled. Geol. Inst., Trudy, n. ser., no. 44, p. 1–144, 38 pls.

— 1963, Stromatoporoidea Sovetskogo Souza. Chast 4: Vses. Nauchno-Issled. Geol. Inst. Trudy, n. ser., no. 87, p. 1–160, 31 pls.

— 1967, Stromatoporoidea Sovetskogo Souza. Chast 5: Vses. Nauchno-Issled. Geol. Inst., Trudy, n. ser., no. 148, 56 p., 29 pls., 1 fig.

ZHURAVLEVA, I. T., 1960, Arkheotsiaty sibirskoi platformy (Archeocyathids of the Siberian Platform): Moskva, Akademiĭ Nauk SSSR, Institut Geologiĭ i Geofiziki Sibirskogo Otdeleniya Paleontologicheskiĭ Institut, 344 p., 33 pls., 147 figs., 26 tables.

4. Plates

We attempted to select for photography samples that included both world-wide geographic and stratigraphic coverage as well as those illustrating the fossil groups under consideration. Our efforts to obtain world-wide coverage are only modestly successful as 40 percent of our samples are from North America, and South America is not represented. About 1000 thin sections were prepared from more than 1000 available samples, mostly supplied by those geologists cited in the Preface. Nearly 1000 photomicrographs were obtained from about 400 thin sections. From these photomicrographs, 280 pictures from 200 thin sections were selected for the plates reproduced herein.

Plates 1 to 63 correspond with the discussion of biotic groups in the text. Most of the remaining plates represent full-plate examples of rocks from Recent to Cambrian age. These were added to give the reader a greater opportunity to see fossil constituents in limestones and to illustrate some typical biotic associations. Many more plates would be necessary to present a full sampling of all the microfacies of carbonate and related rocks of each geologic period.

We recognize that many geologists first will examine the plates in order to compare their unknowns directly with our photographs. Then they may want to read the plate captions and, as a last resort, refer to the tables on the geometry of biotic constituents or even read the text. Those wishing to examine the illustrations for a specific group can refer to the index. Most of the plates also are cited in the text on individual biotic groups.

We have attempted to select the best possible pictures. However, the reader should not be misled into believing either that all limestones (or all sedimentary rocks) are fossiliferous or that all debris is so well preserved. Many limestones are monotonous lime muds, and many fossiliferous limestones have been altered or are embedded in lime muds so that excellent photomicrographs are impossible to obtain. The most striking pictures are frequently from well-washed biocalcarenites. However, we have prepared a few plates showing poorly preserved debris. The effective interpetation of such limestones depends in large measure on the more readily obtained knowledge and understanding gained from better preserved rocks. In this respect, the papers by CAROZZI and his students (TEXTORIS and CAROZZI, 1964; 1966; and references cited therein) on various dolomitized rocks in the Silurian reefs of Indiana, Illinois and Wisconsin are instructive.

It is a tribute to man's ability to recognize patterns in the natural world that as few as 280 pictures will permit identification to major biotic group of a large proportion of fossil debris. This is strikingly illustrated by considering the size of this sample relative to the number of fossils or fossil fragments in carbonate rocks. The plates contain perhaps 10^4 identifiable fossil fragments. The number of pieces of biotic debris in sedimentary rocks depends on the volume of carbonate rocks, the average grain size of biotic debris, and the amount of fossiliferous limestone. The amount of carbonate rock is calculated from data presented by POLDERVAART (1955). Assuming the average grain volumes cited in Table 18, the maximum number of grains varies from about 10^{16} to 10^{20}. Not all of these grains would be biotic debris but this is an upper limit based on identifiable fossil fragments. Assuming that only one percent of the volume of carbonate rocks is fossil debris will reduce the limits by two orders of magnitude (10^{14} to

Table 18. *Calculation of Number of Fossil Grains in Carbonate Rocks*

A. Volume of Carbonate Sediment

Lithosphere	Total volume $\times 10^6$ km^3	Carbonate (%)	Carbonate volume $\times 10^6$ km^3
Deep oceanic	80.4	71.9	58.8
Continental shields	52.5	16	8.4
Young folded belts	126	22	27.7
Suboceanic			
Shelf	120	22	26.4
Hemipelagic	252	22.5	55.9
		Total	177.2

B. Particle Size and Abundance

Particle Volume in mm	Maximum no. of particles $\times 10^{14}$
1	177.2
0.5	1,417.6
0.25	11,340.8
0.125	90,726.4
0.0625	725,811.2

10^{18}). The average volume of shell secreted by fossils is not known but 1 mm^3 is perhaps a conservative figure in view of the large number of microfossils. Consequently, the sample in this atlas represent of the order of one part in ten billion (10^{10}). RONOV and YAROSHEVSKY (1969) calculated slightly different volumes of carbonate sediments, but the number of fossil grains based on this newer data differs by less than 20 percent from the values obtained from POLDERVAART's data.

Our thin sections were prepared by standard techniques to standard thickness, about 0.03 mm. Many thin sections were fine ground with 1000 powder before and after mounting. Special studies require thinner sections for which special techniques are needed, but these techniques were not utilized in this study. Well-prepared thin sections are initially the most useful petrographic preparations in the study of biotic constituents as they permit examination with polarized light, which is useful in differentiating some fossil groups. However, for individual studies well-prepared peels or polished surfaces can be a practical and effective substitute. Careful preparation can produce peels yielding very fine resolution of textures. However, each rock sample presents a somewhat different problem. Peels, polished surfaces, and thin sections are also effective on bore hole cuttings, particularly those mounted in a plastic medium. KUMMEL and RAUP (1965) have reviewed various thin section and peel techniques utilized in the study of fossils.

References Cited

(Including citations on plate descriptions)

JOHNSON, J. H., 1964, The Jurassic algae: Colorado School Mines Quart., v. 59, no. 2, 129 p., 45 pls., 1 table.

KUMMEL, BERNHARD, and RAUP, D. M., eds., 1965 Handbook of paleontological techniques: San Francisco, W. H. Freeman and Company, 852 p.

MACKAY, I. H., 1952, The shell structure of the modern mollusks: Colorado School Mines Quart., v. 47, no. 2, p. 1–27, 6 pls., 1 fig.

POLDERVAART, ARIE, 1955, Chemistry of the earth's crust, *in* POLDERVAART, ARIE, Crust of the earth (a symposium): Geol. Soc. America Spec. Paper 62, p. 119–144, 1 fig., 22 tables.

RADIOČIĆ, RAJKA, 1966, Microfaciès du Jurassique des Dinarides externes de la Yougoslavie: Geologija Razprave in Poročila, v. 9, p. 5–377, 165 pls., 11 tables.

RIGBY, J. K., 1958, Two new upper Paleozoic hydrozoans: Jour. Paleontology, v. 32, p. 583–586, pl. 86, 3 text-figs.

RONOV, A. B., and YAROSHEVSKY, A. A., 1969, Chemical composition of the earth's crust, *in* HART, P. J., ed., The earth's crust and upper mantle: Am. Geophys. Union Geophys. Mon. 13, p. 37–57, 3 figs., 11 tables.

TEXTORIS, D. A., and CAROZZI, A. V., 1964, Petrography and evolution of Niagaran (Silurian) reefs, Indiana: Am. Assoc. Petroleum Geologists, v. 48, p. 397–426, 2 pls., 24 figs.

— — 1966, Petrography of a Cayugan (Silurian) stromatolite mound and associated facies, Ohio: Am. Assoc. Petroleum Geologists Bull., v. 50, p. 1375–1388, 1 pl., 4 figs., 2 tables.

Plate Titles

Plates and Plate Descriptions

PLATE 1. TINTINNINES

1/10 mm
x 20
x 40
x 60
x 80
x100

Calpionella alpina LORENZ, in a dark, clotted, finely recrystallized matrix. Note horseshoe-shaped longitudinal cross sections and circular and elliptical transverse sections. Tintinnine-bearing limestones are best known in the Upper Jurassic and Lower Cretaceous of the Mediterranean region. These limestones usually are interpreted as deepwater deposits. Calpionellid zone B, Upper Tithonian, Upper Jurassic, east of Col de Cabre, Drôme, France. Remane Slide Collection, CCa6, ×100.

PLATE 2. TINTINNINES

1/10 mm

x 20

x 40

x 60

x 80

x100

Longitudinal (U-shaped) and transverse (circles and ellipses) sections of tintinnines. Some longitudinal sections show characteristic flanges at mouth of shell which gives cross sections a horseshoe shape. Some disarticulated ostracode valves. Other unidentifiable recrystallized fossil debris is present. Dark finely recrystallized matrix. Calpionellid zone D, uppermost Berriasian, Upper Jurassic, Clue de Taulanne, northwest of Castellane, Basses-Alpes, France. Remane Slide Collection, CTa3, ×100.

PLATE 3. RADIOLARIANS AND "LOMBARDIA"

1/10 mm
x 20
x 40
x 60
x 80
x100

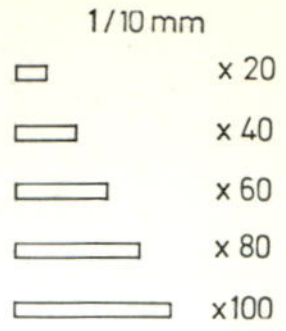

Fig. 1. Globular radiolarians and much smaller cross sections of the tintinnine, *Calpionella alpina* LORENZ. Radiolarian skeletons partially replaced by calcite. Mud matrix. Calpionellid zone B, Upper Tithonian, Upper Jurassic, Charens, Drôme, France. Remane Slide Collection CHR–4, ×40.

Fig. 2. Longitudinal (upper left) and transverse (center) sections of plates showing spines. Skeletal framework extincts at one position under cross nicols. Notwithstanding the extinction data, we originally classified the transverse section as a radiolarian. European microfacies workers would refer these plates to "Lombardia", a designation which includes isolated plates of the planktonic crinoid *Saccoma*. Scaglia Formation, Cretaceous, near Cagli, 50 kilometers north-northeast of Perugia, Urbino, Marche, Italy. IU 8099–154, ×80.

Fig. 3. Closely packed globular and conical (left center) radiolarian skeletons. Note meshwork (a) exhibited by many smaller cross sections. A few skeletons (b) are partially replaced by calcite. Dark matrix probably stained black by organic compounds. Lower Jurassic, 4 kilometers north-west of Valdorbia, Perugia, Umbria, Italy. IU 8099–358, ×80.

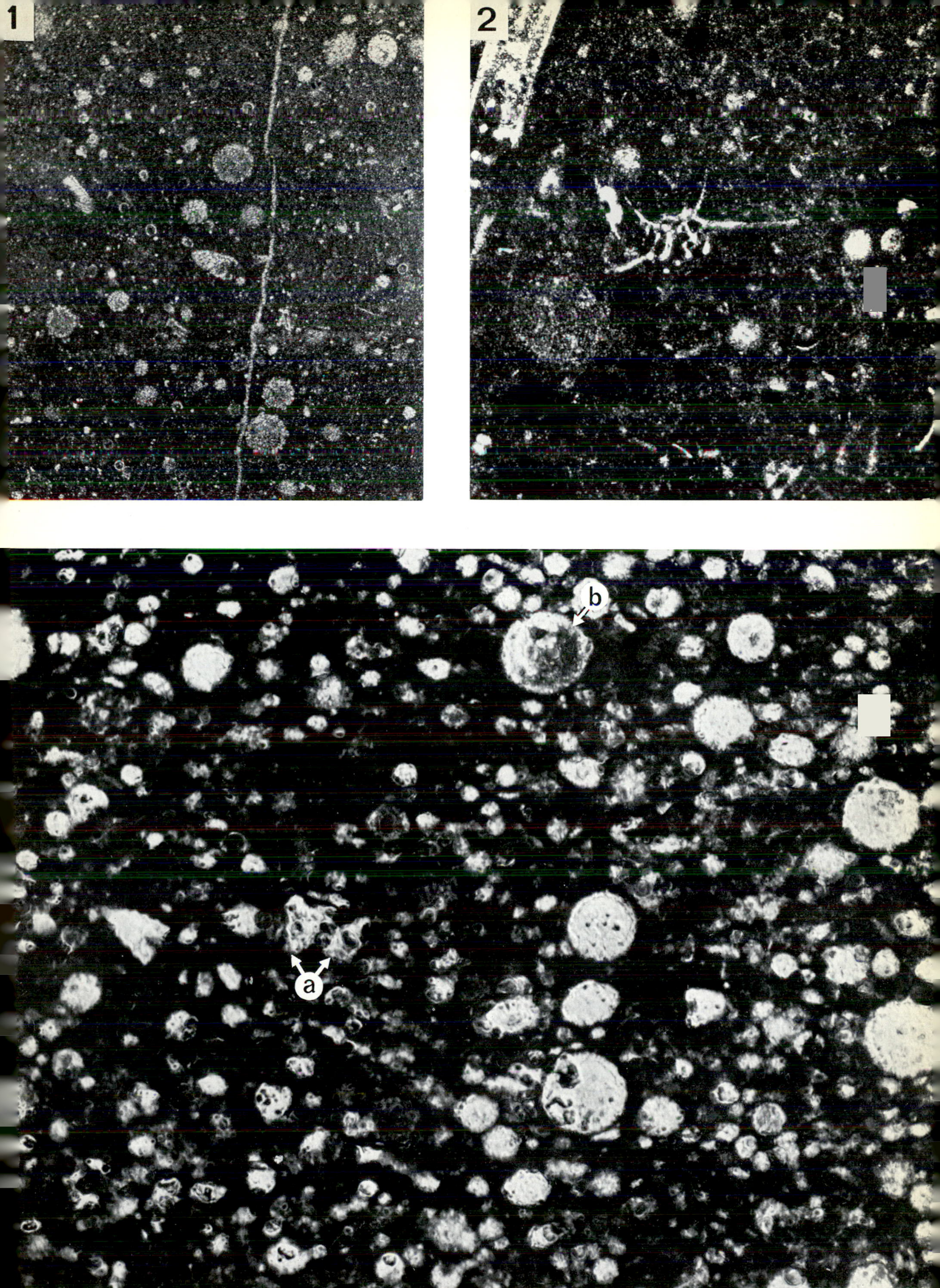
1
2
b
a

PLATE 4. FORAMINIFERS

1/10 mm
x 20
x 40
x 60
x 80
x100

Fig. 1. Cross sections of large miliolid foraminifers (*Marginopora* sp.) having numerous chambers displaying complex patterns and a gastropod fragment (top). Recent, Chireerete Island, Bikini Atoll, Marshall Islands, Pacific Ocean. IU 8099–476, ×40.

Fig. 2. Foraminifer having a partially agglutinated wall. Matrix is pelletal mud, calcite cement, and finely broken fossil debris. Note five-sided echinoderm-columnal having central perforation (lumen). Deese Formation, Middle Pennsylvanian, Dry Branch, Buckhorn Creek, sec. 26, T. 1 S., R. 3 E., Murray County, Oklahoma, U.S.A. IU 8099–510, ×40.

Fig. 3. Foraminifer at left has porous shell structure but one at right has layered lamellar walls. Probable altered molluscan fragment at upper center and barely visible fragment of *Halimeda*, a codiacean alga, at lower right. As fig. 1.

Fig. 4. Foraminifer sectioned parallel to the plane of coiling shows porous shell wall and numerous chambers. Recent, between Rongelap and Bokujanto Islands, Rongelap Atoll, Marshall Islands, Pacific Ocean. IU 8099–489, ×20.

Fig. 5. Biserial foraminifer. Fine-grained microstructure of wall indistinguishable from matrix, which is lime mud and finely comminuted fossil debris. Middle Jurassic, Pollino Mountain, 8 kilometers north of Castrovillari, Cosenza, Italy. IU 8099–357, ×40.

Fig. 6. Complex chambered pattern in foraminifer, which lies next to a recrystallized coral. Fine-grained microstructure of wall indistinguishable from matrix of lime mud and finely broken fossil debris. As fig. 5, ×20.

Fig. 7. Multichambered foraminifer has a fine-grained wall which is difficult to distinguish from mud matrix of finely comminuted fossil debris. As fig. 5.

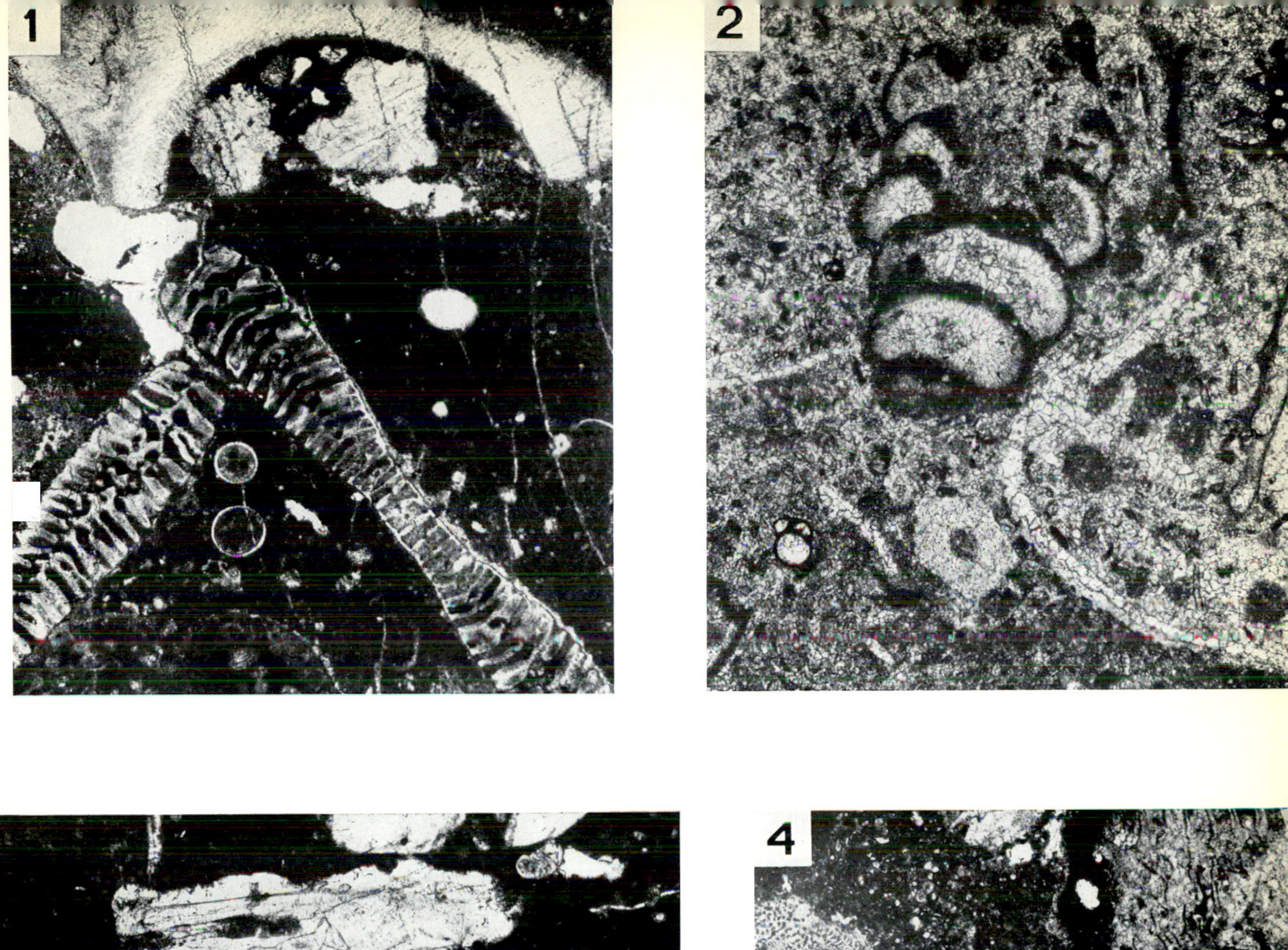
1
2

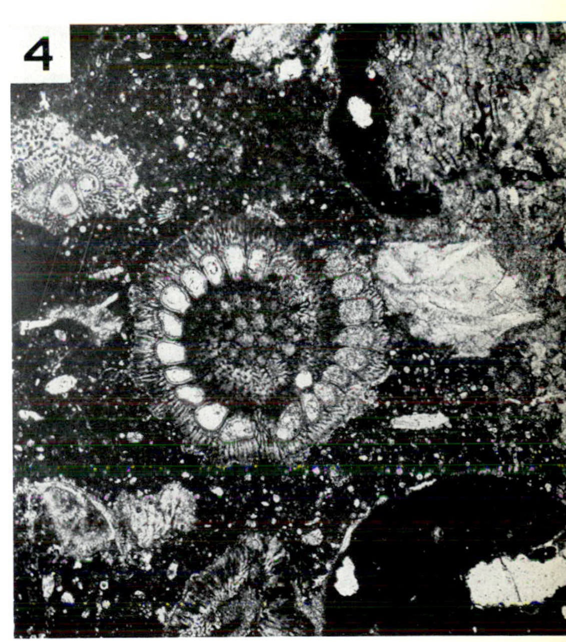
4

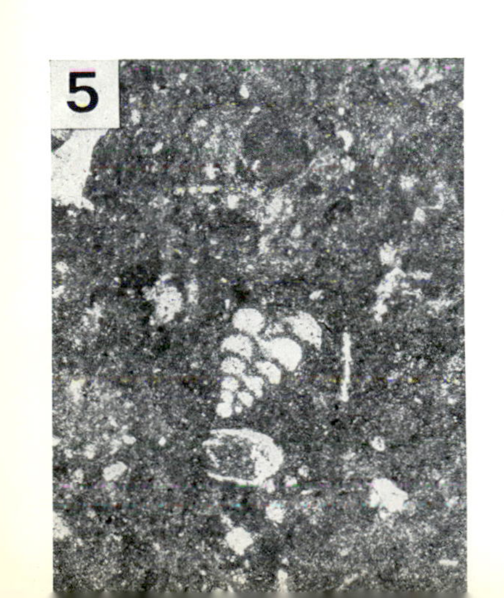
5

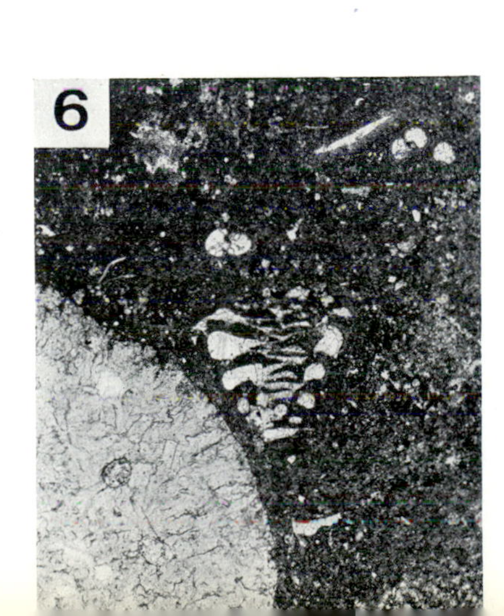
6

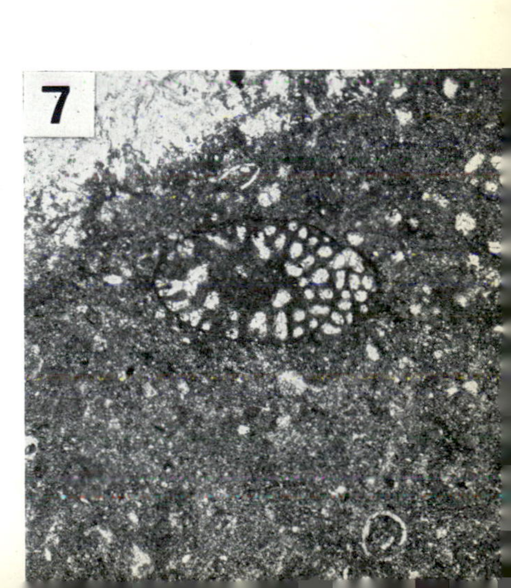
7

1/10 mm
x 20
x 40
x 60
x 80
x100

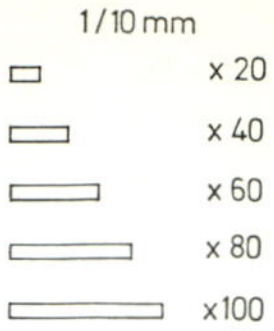

Fig. 1. Framework of large fusulinids all cut in approximately transverse section. Interstitial material consists of smaller foraminifers, echinodermal plates, and sparry calcite cement. Ibukiyama Limestone, Middle Permian, Shiga Province, Japan. IU 8099–955, ×20.

Fig. 2. Large nummulitid foraminifer in upper center. Two additional specimens at left center. Coralline algae in calcareous mud, but magnification is too small to show cellular internal structure of algal fragments. Ramose bryozoan at right-center margin. Upper Eocene, near highest bed on Lou Cachaou, Biarritz, Basses Pyrénées, France. IU 8099–2, ×20.

Fig. 3. Nearly transverse section of fusulinid in partially recrystallized calcareous mud. Deese Formation, Lower Pennsylvanian, Dry Branch, Buckhorn Creek, sec. 26, T. 1 S., R. 3 E., Murray County, Oklahoma, U.S.A. IU 8099–510, ×40.

Fig. 4. *Bradyina*, a coiled endothyrid foraminifer (lower left) exhibiting a vesicular or alveolar wall. Fusulinid at right. As fig. 3.

Fig. 5. Well-sorted quinqueloculine (foraminiferal) limestone. Note transverse (a) and longitudinal (b) sections. Dark fine-grained appearance is typical of porcellaneous shell microstructure. Sparry calcite cement. Unidentified sparry calcite grain at upper center. Asmari Formation, Miocene, Tang-e Do Roodeh, northeast flank of Kuh-e Gurpi, Khuzestan, Iran. IU 8099–293, ×20.

Fig. 6. Limestone as in figure 5, but oblique section of large miliolid at top center. As fig. 5.

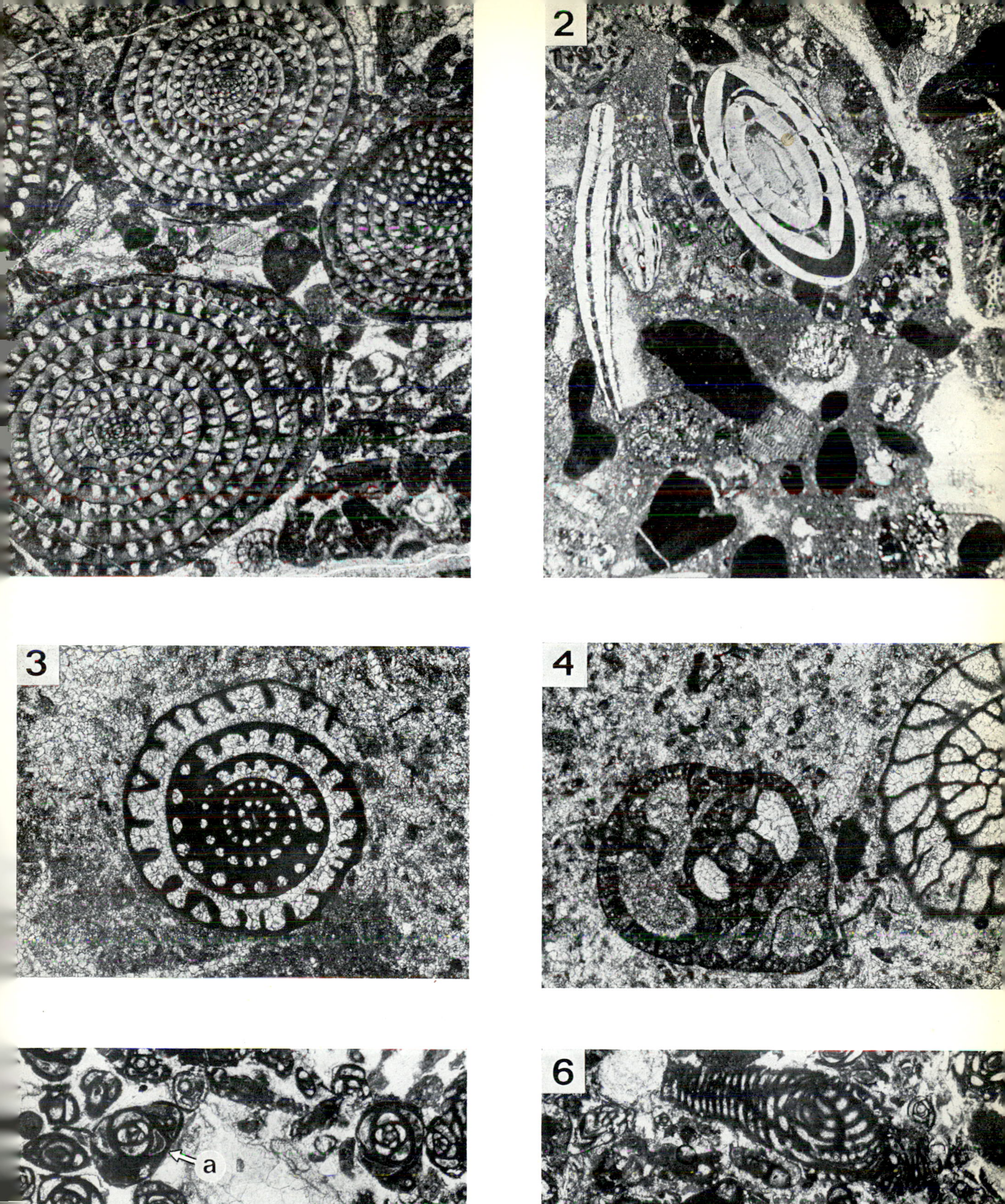
2
3
4

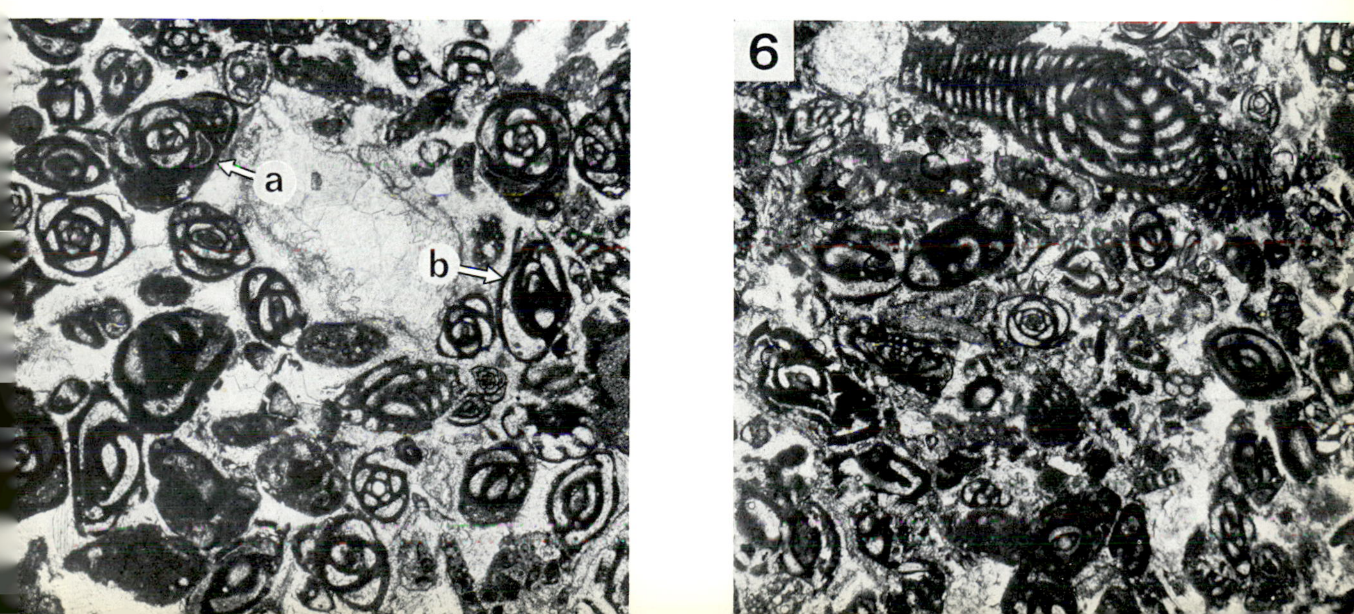
a
b
6

PLATE 6. FORAMINIFERS

1/10 mm
x 20
x 40
x 60
x 80
x100

Fig. 1. Schwagerinid type of fusulinid wall microstructure. Fine pores perpendicular to wall surfaces produce a columnar pattern. Septa between chambers folded or plicated so that they appear vesicular in cross sections. Chambers filled with sparry calcite and pelletal mud. Gerster Formation, Permian, Montello Canyon, Leach Mountains, T. 39 N., R. 68 E., Elko County, Nevada, U.S.A. IU 8099–562B, ×40.

Fig. 2. Transverse section of fusulinid showing septa and initial chamber of shell (proloculus). Skeletal microstructure poorly exhibited and partially replaced by silica (clear white areas in walls). Pennsylvanian, Shazaid Heraz Valley, Iran. IU 8099–309, ×40.

Fig. 3. Fusulinid with altered skeletal wall. Large calcite-filled fracture at right cuts fusulinid. Akasaka Limestone, Socio Stage, Middle Permian, Kinshozan, Akasaka-Cho, Fuha-gun, Gifu Prefecture, Honshu Island, Japan. IU 8099–1017, ×20.

Fig. 4. Fusulinid limestone containing smaller foraminifers and echinodermal plates. Akiyoshi Limestone, Lower Permian, Yamaguchi Prefecture, Honshu Island, Japan. IU 8099–1001, ×20.

Fig. 5. Transverse (right) and oblique sections of fusulinid in mud and finely comminuted fossil debris. Fusulinid on right sectioned through large initial chamber (proloculus). As fig. 1, ×20.

Fig. 6. Fusulinid skeletal limestone. Some echinodermal plates. As fig. 4.

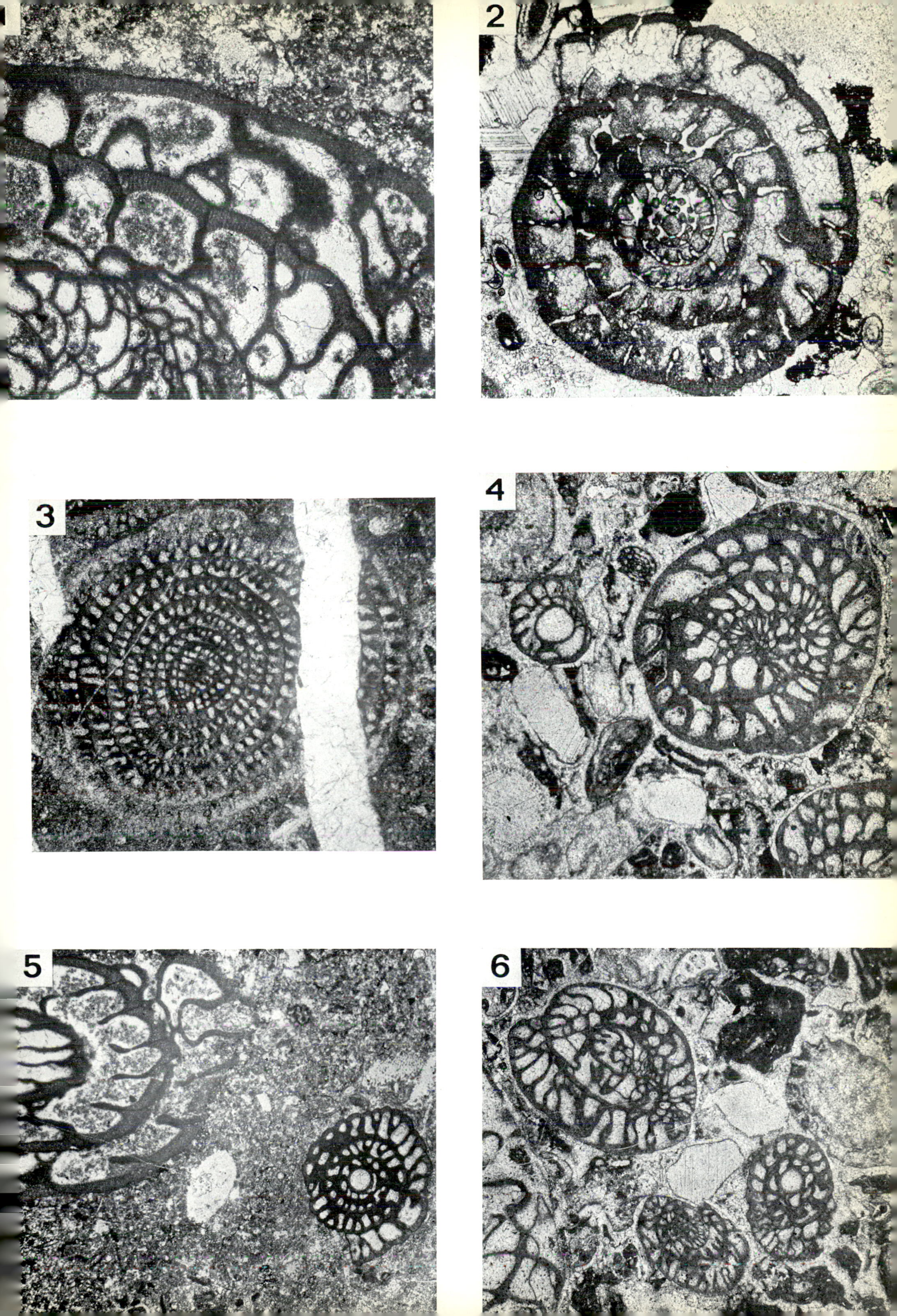

2
3
4
5
6

1/10 mm
x 20
x 40
x 60
x 80
x100

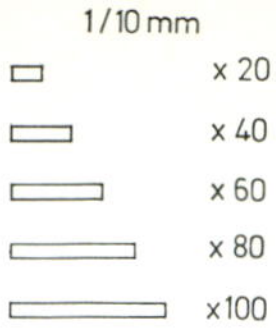

Fig. 1. Foraminiferal calcarenite. Cemented only at points of grain contact. Compare open packing with Pls. 83 and 96. Foraminiferal walls show dark mud-filled perforations (a), some resulting from activities of borers, and radial calcite fibers perpendicular to shell wall (b). Recent, Rongelap Atoll, Tufa Island, Marshall Islands, Pacific Ocean. IU 8099–473, ×40.

Fig. 2. Globerigerinid limestone. Mud and calcite matrix. Globigerinids filled with clear calcite cement. Many chambers filled by single calcite crystal. Note the different kinds of cross sections of this planispirally coiled foraminifer. Greenhorn Limestone, Lower Turonian, Upper Cretaceous, Boyes Hill, 4 miles west of Delphos, SW$\frac{1}{4}$ sec. 14, T. 9 S., R. 5 W., Ottawa County, Kansas, U.S.A. IU 8099–104, ×40.

a
b
a
b
a

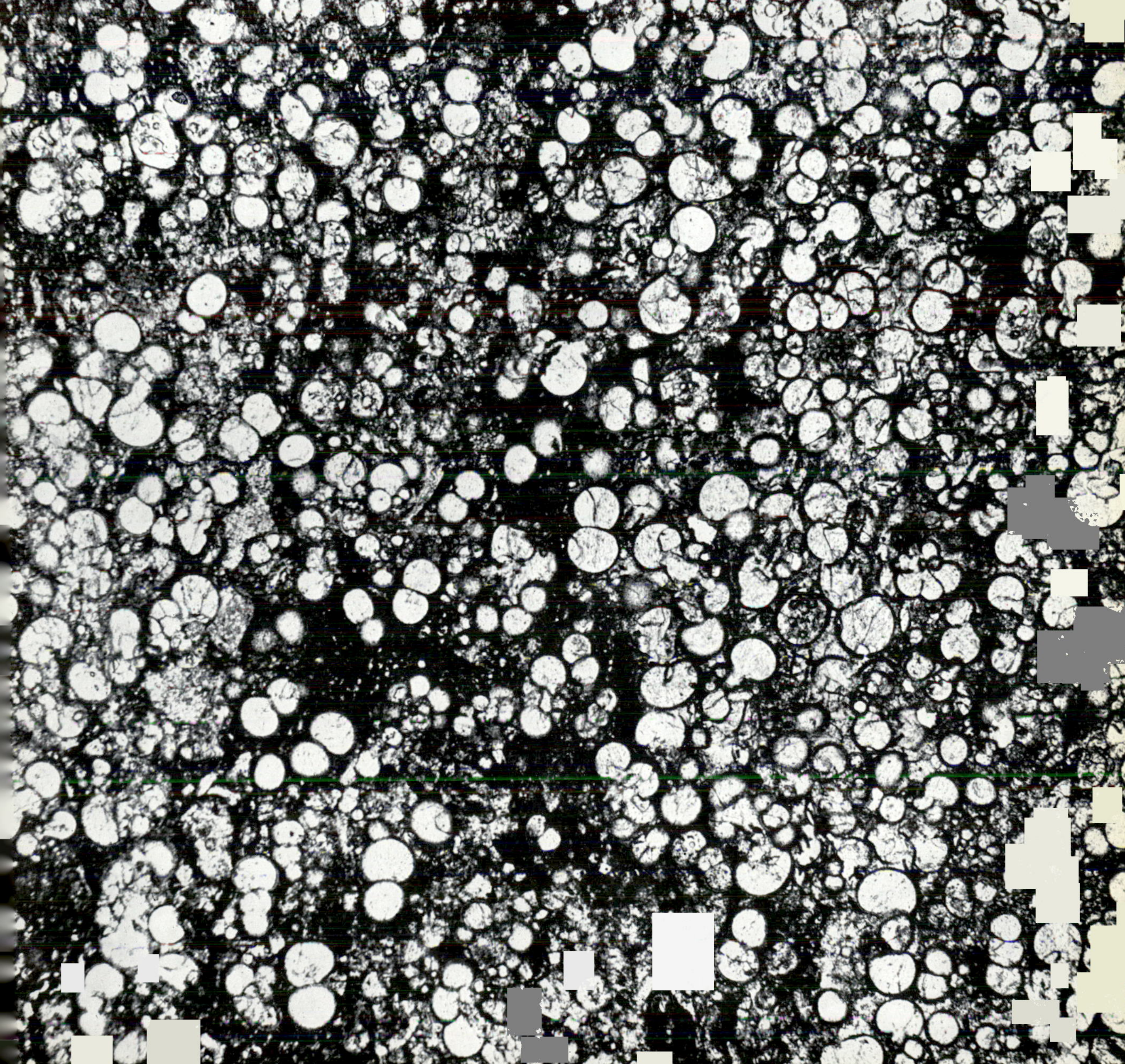

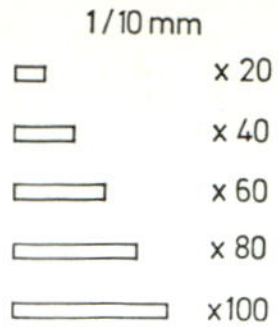

Fig. 1. Cross sections of large benthonic foraminifers: discocyclinid (a), perforate coiled rotalid foraminifer (b). Silty micrite matrix, quartz grain (c), echinoderm fragment (d), and coated grain (e). Upper Eocene, near highest bed on Lou Cachaou, Biarritz, Basses Pyrénées, France. IU 8099–2, ×40.

Fig. 2. Globigerinid limestone. Coiled shells are cut in all planes and not only exhibit coiling but also one to multichambered linear arrays. Note perforate shell structure in transverse and tangential sections of larger chambers (a). Dark mud matrix. Greenhorn Limestone, Lower Turonian, Upper Cretaceous, Boyes Hill, 4 miles west of Delphos, SW$\frac{1}{4}$ sec. 14, T. 9 S., R. 4 W., Ottawa County, Kansas, U.S.A. IU 8099–104, ×80.

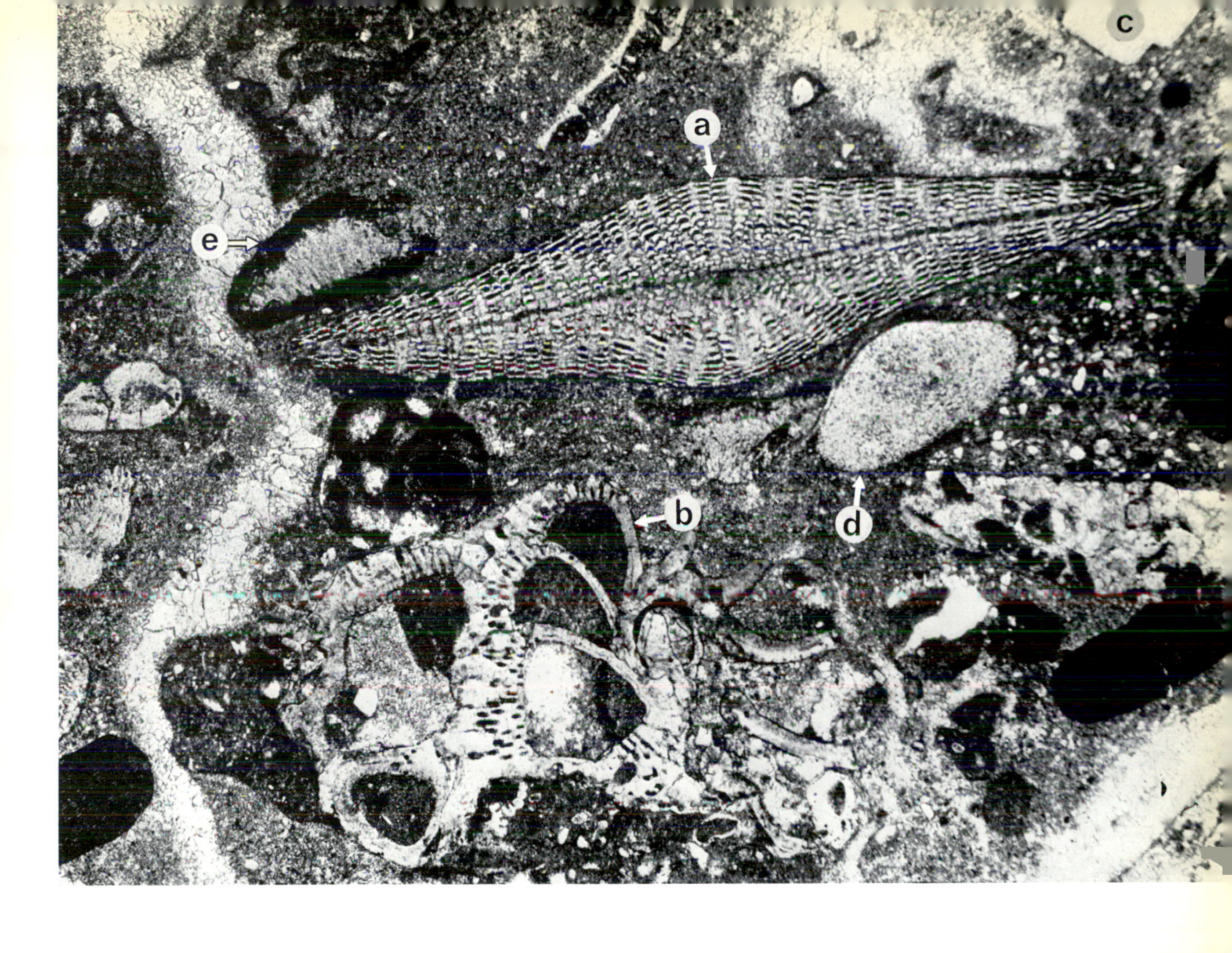

a
b
c
d
e

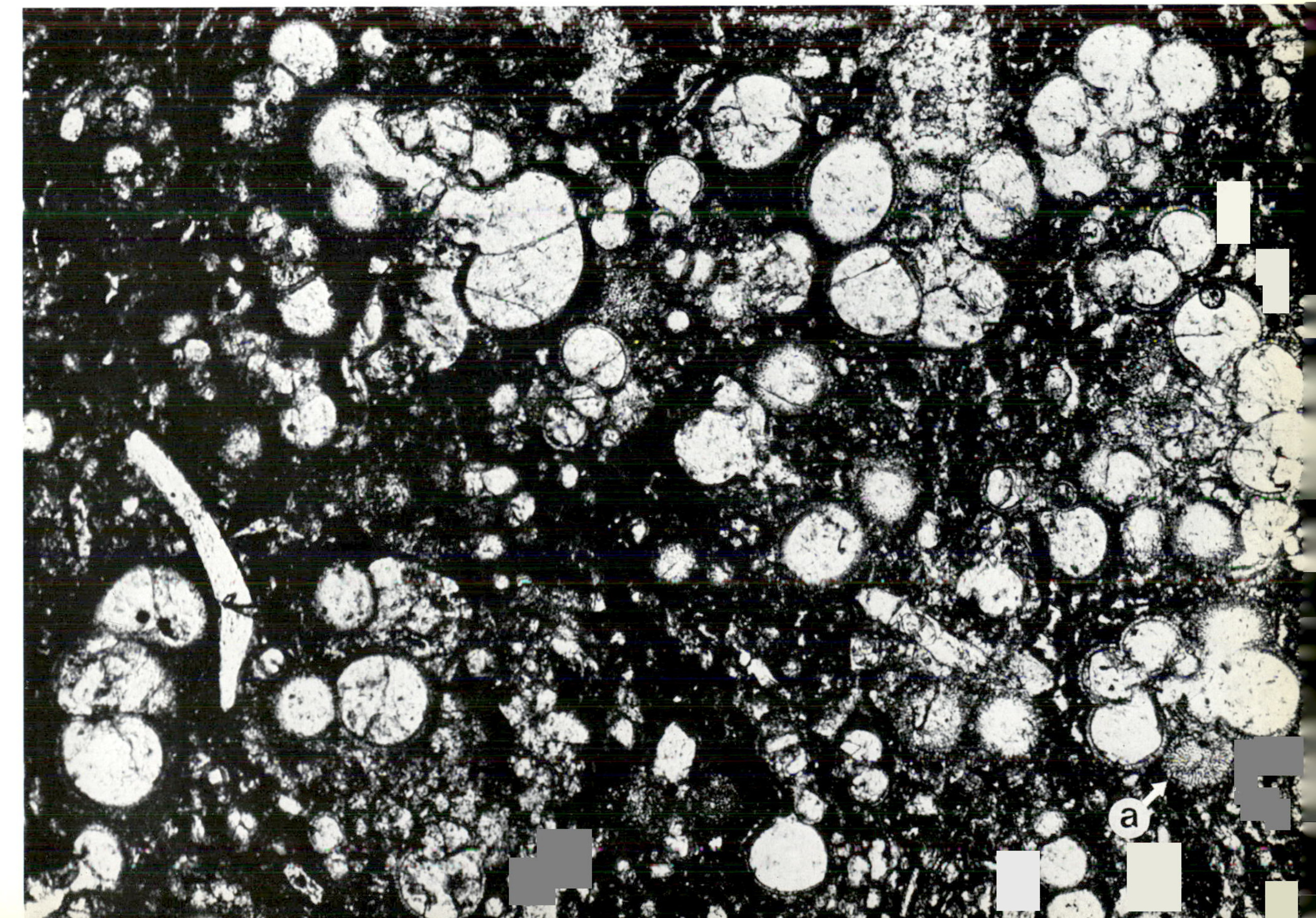

a

PLATE 9. FORAMINIFERS

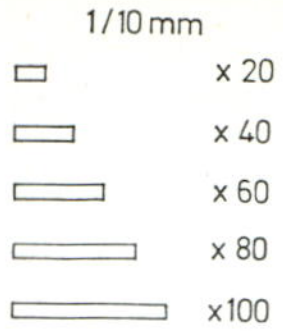

1/10 mm
x 20
x 40
x 60
x 80
x100

Fig. 1. Algal-foraminiferal limestone. Conical multichambered foraminifer in center of figure is dictyoconid. Coiled rotalids are also present. Fragments showing arched rows of cells are articulating coralline algae. Some pre-existing rock fragments and altered coral are present. Matrix is calcareous mud. Prominent angular quartz grain at lower right. Upper Eocene, near highest bed on Lou Cachaou, Biarritz, Basses Pyrénées, France. IU 8099–2B, ×40.

Fig. 2. Foraminiferal limestone. Note dark mud infilling suggesting redeposition from original environment. Shell wall exhibits radial fibrous pattern in large rotalid foraminifer at right. A few small molluscan fragments (a) and quartz grains (b) are present. Some very dark fragments are coralline algae in which the structure is not apparent at this magnification. Silty dark matrix. As fig. 1.

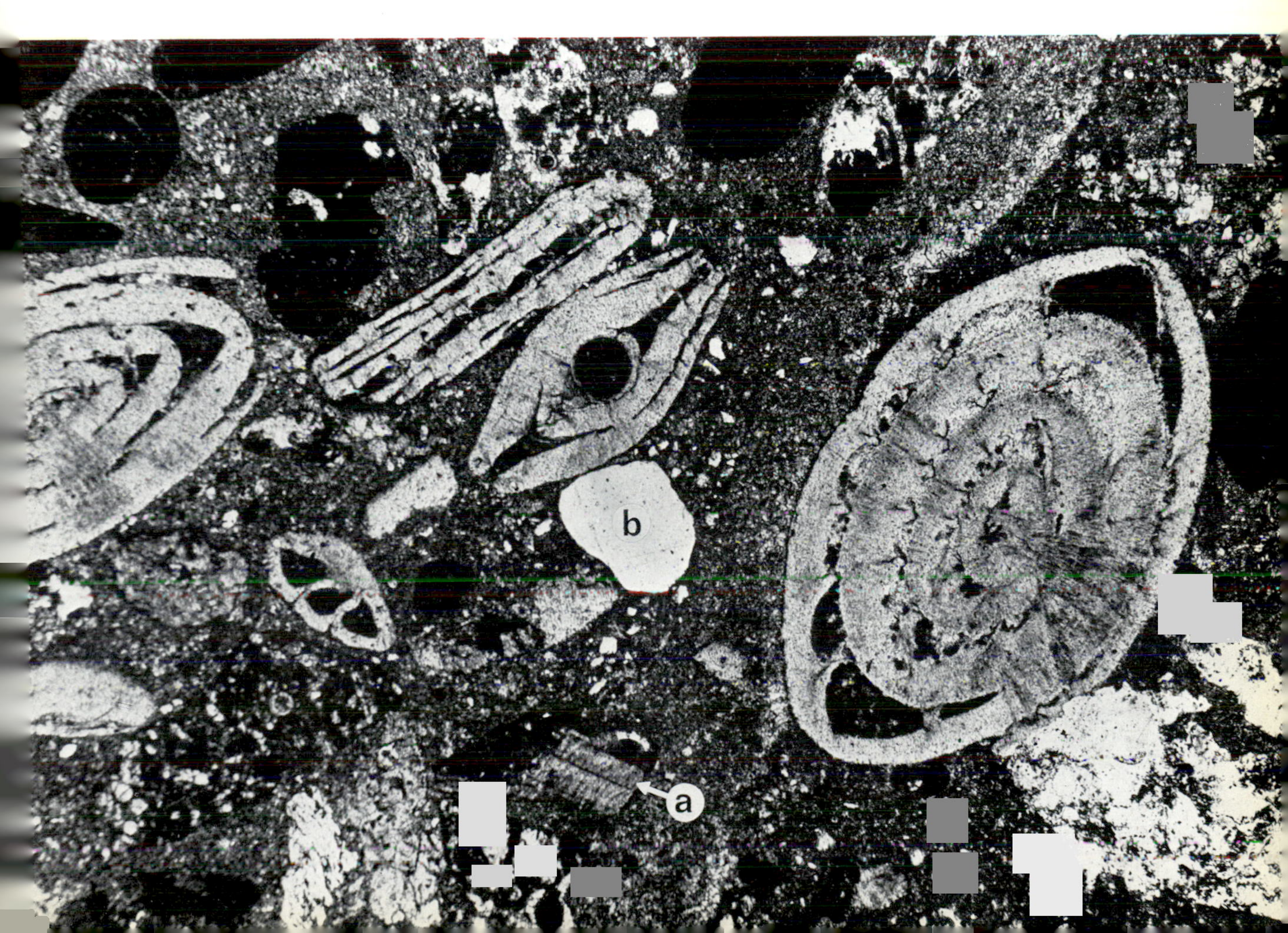
b
a

PLATE 10. FORAMINIFERS

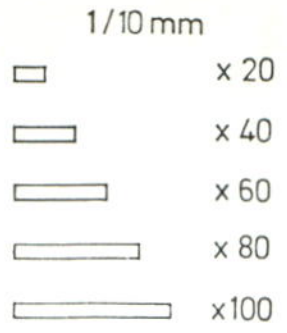

1/10 mm

x 20

x 40

x 60

x 80

x 100

Fig. 1. Closely packed discocyclinids in oblique and transverse sections. Jahrum Formation, Eocene, Kuh-e, Gurpi, Tang-e Do Roodeh, Kuzestan, Iran. IU 8099–294, ×40.

Fig. 2. Foraminiferal limestone. Large broken discocyclinid (a) cut in transverse section perpendicular to axis of coiling. Smaller foraminifers of various types and an echinodermal plate showing syntaxial rim cement (b) are also present. Possible faecal material at upper left shows elongate, partially beaded, variably constricted rods and strings (? *Aggregatella* sp.). As fig. 1.

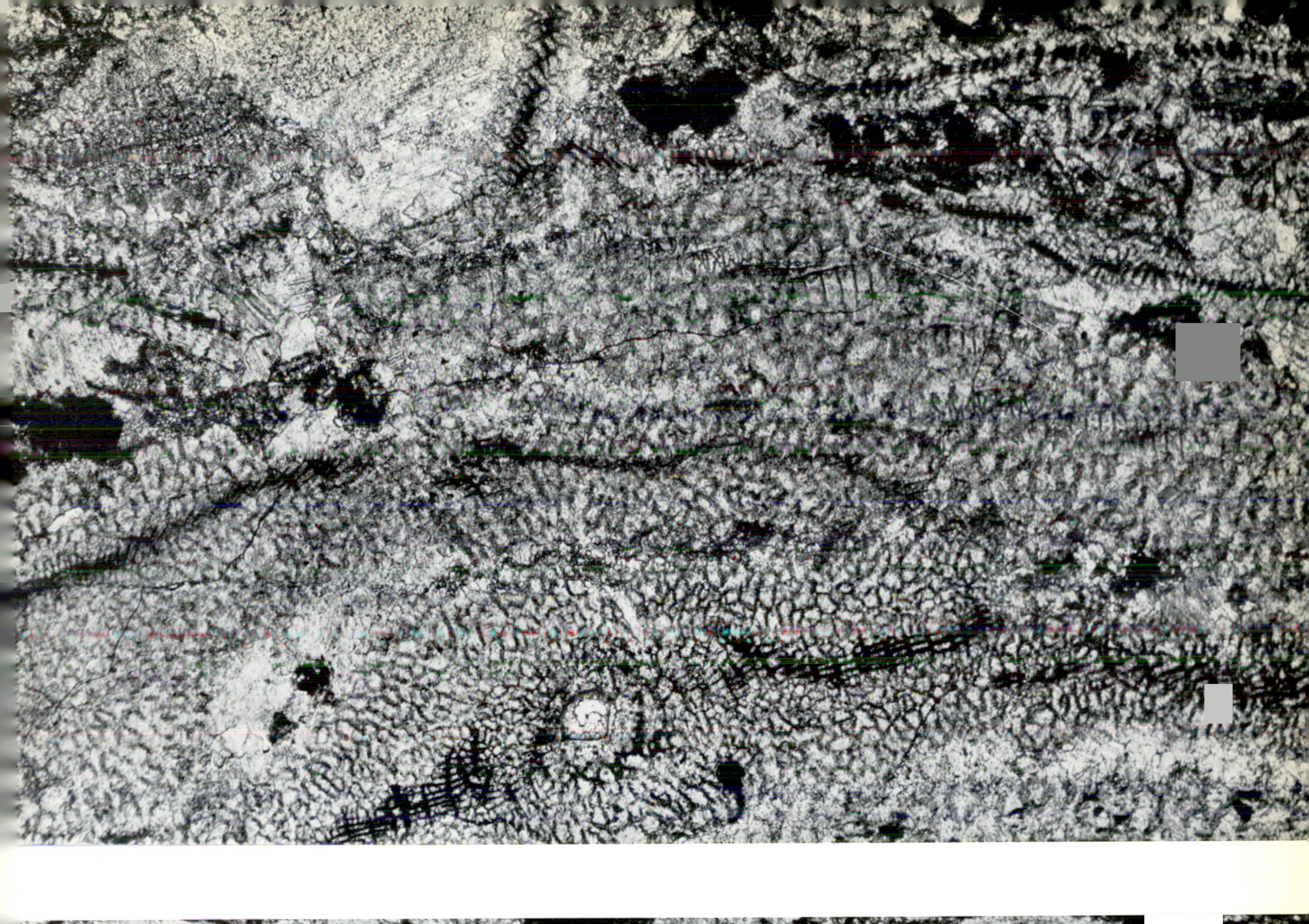

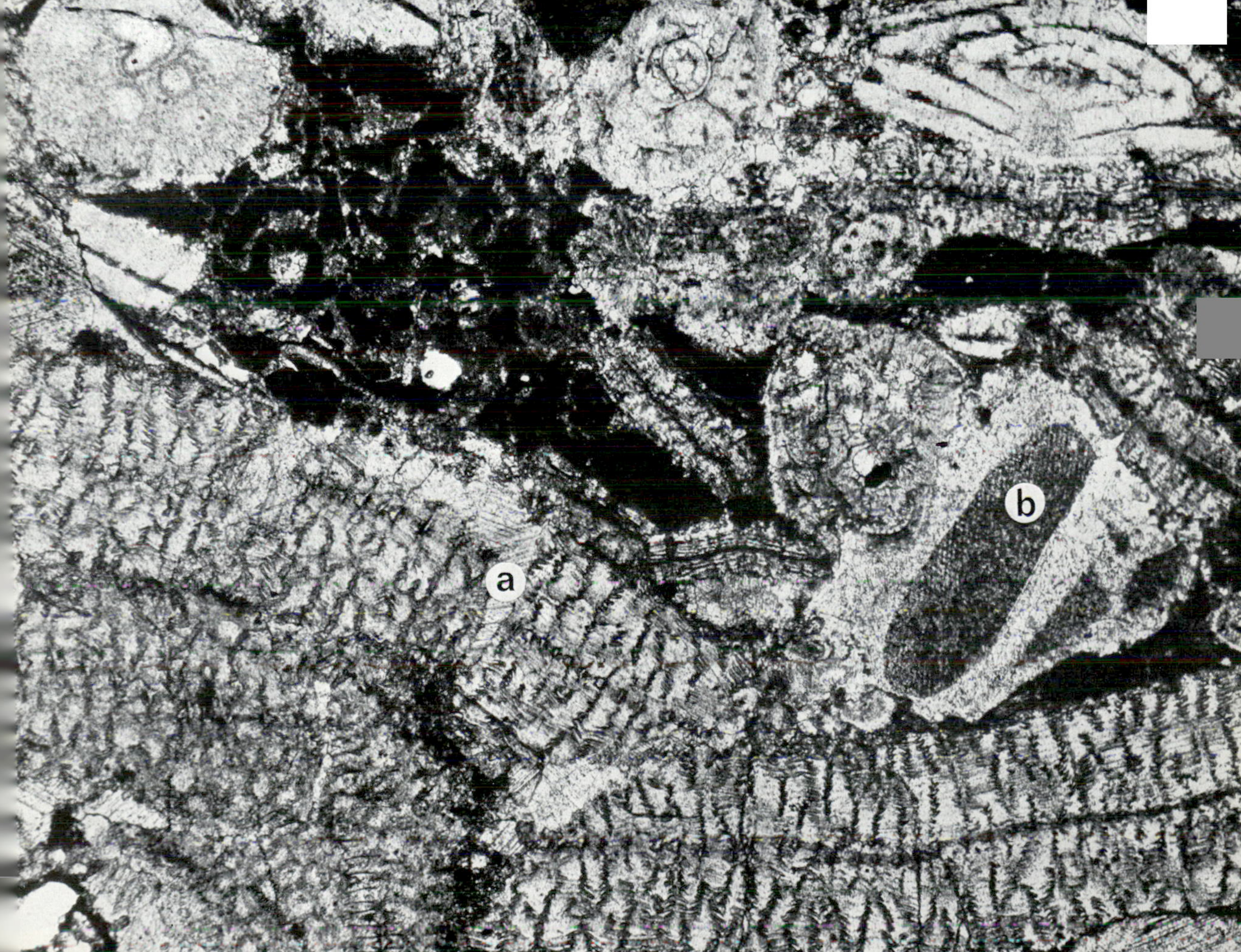
a
b

1/10 mm

x 20

x 40

x 60

x 80

x100

Fig. 1. Large orbitolinids cut in various directions. Silty light-gray muddy matrix. Cretaceous, San Mango Piemonte, Salerno, Italy. IU 8099–370, ×40.

Fig. 2. Foraminiferal limestone having sparry calcite matrix. Foraminifers are principally coiled quinqueloculines (a) or biserial forms (b) and are cut in various planes. Large partially calcite-filled voids probably are dissolved fragments of molluscan shells. Pisté Member, Chichén Itzá Formation, Middle Eocene, on Federal Route 180, 5.2 kilometers northwest of Pisté, Yucatán, México. IU 8099–126B, ×40.

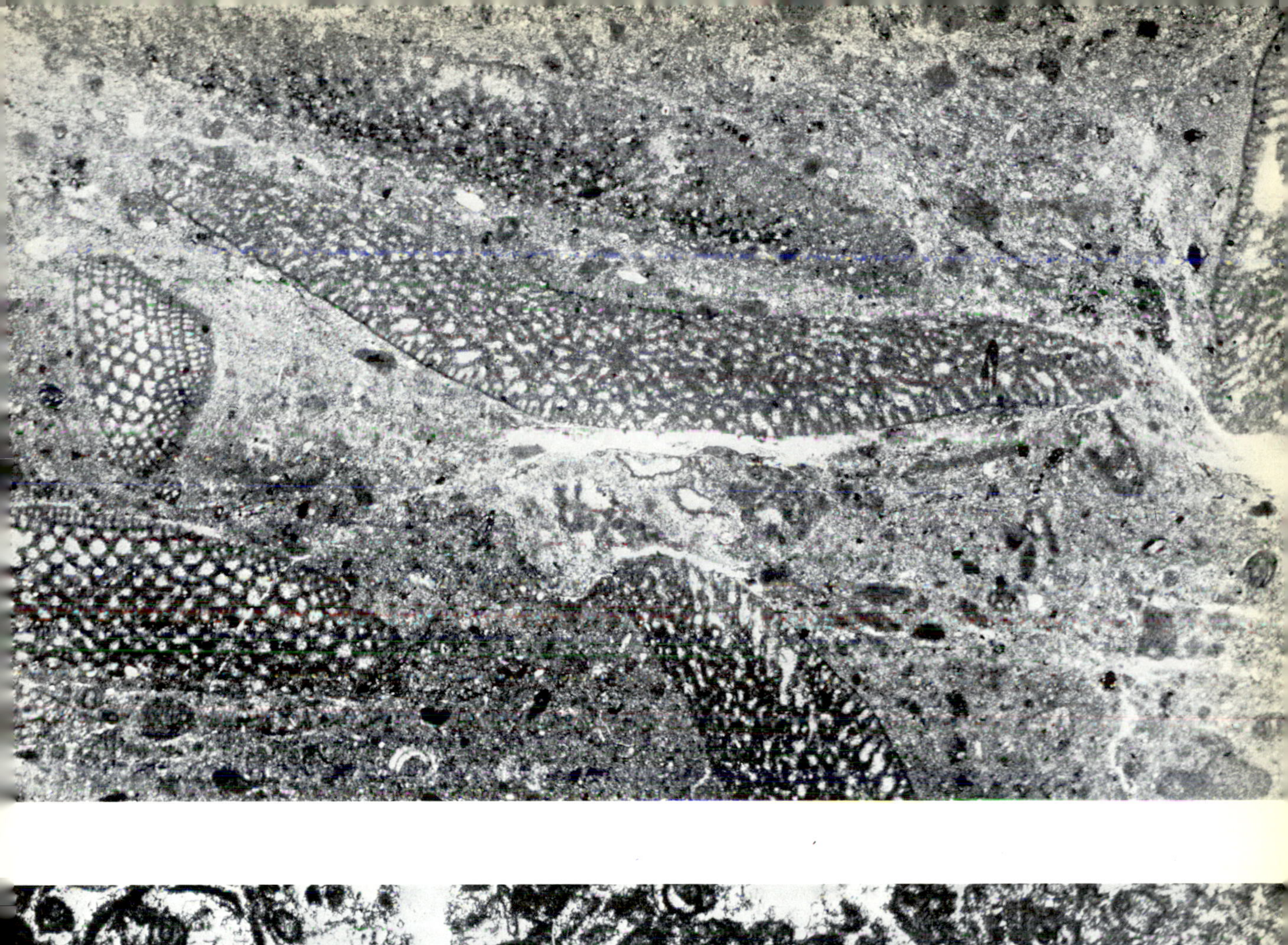

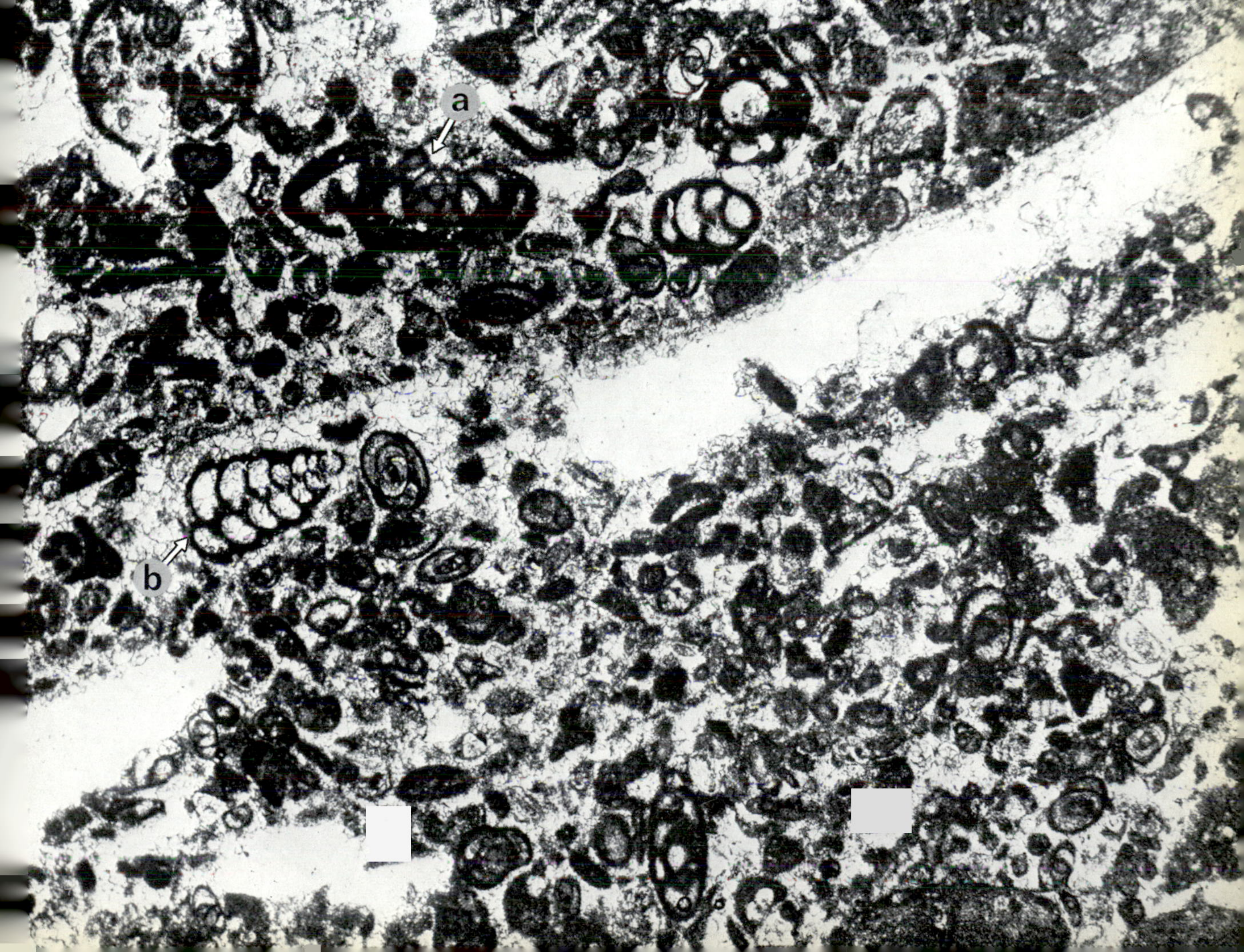
a
b

1/10 mm

x 20

x 40

x 60

x 80

x100

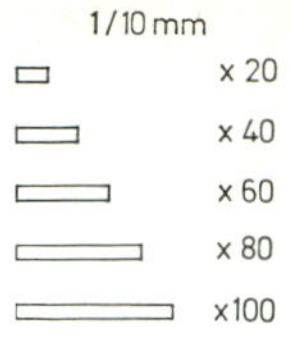

Fig. 1. Large nummulitid foraminifers. One is cut through the initial chamber (proloculus). Note the thick radial fibrous walls. Lower foraminifer shows some abrasion of external surface. Matred Formation, Eocene, Negev, Israel. IU 8099–645, ×20.

Fig. 2. Keeled globorotalid infilled with calcite and mud. Dense pelagic mud contains numerous very small foraminiferal shells. Lower Cretaceous, Santa Maria a Vico, Caserta, Italy. IU 8099–372, ×80.

Fig. 3. Longitudinal sections of flat coiled nummulitids (*Heterostegina* sp.) in dense mud. Smaller globigerinids in matrix. Oligocene, region of the city of Simferopol, Crimea, U.S.S.R. IU 8099–685, ×40.

2

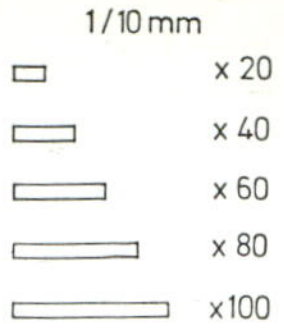

1/10 mm

x 20

x 40

x 60

x 80

x100

Fig. 1. Peel of large nummulitid shows a transverse section. Compare detail in this peel with that shown in thin section, Pl. 12–1. Wadi Tamet Formation (subsurface), Middle Eocene, Eastern Sirte Basin, Cyrenaica, Libya. IU 8099–698AP, ×15

Fig. 2. Peel of fusulinid limestone with a silty micritic matrix. Fractures and incipient stylolites shown by fine linear bubble trains and dark lines. Compare detail in this peel with that shown in thin sections, Pl. 6. Pequop Formation, Permian, north of Moorman Ranch, South Butte Mountains, sec. 9, T. 17 N., R. 59 E., White Pine County, Nevada, U.S.A. IU 8099–564AP, ×15.

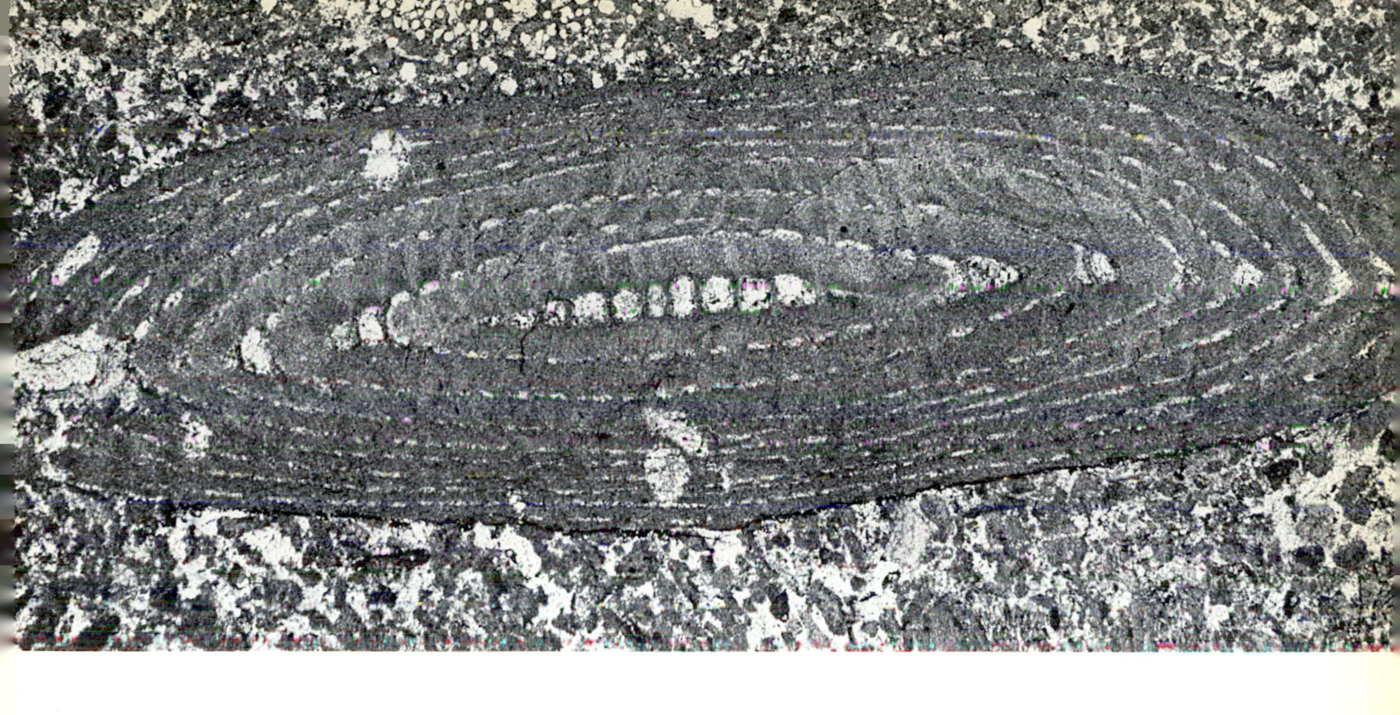

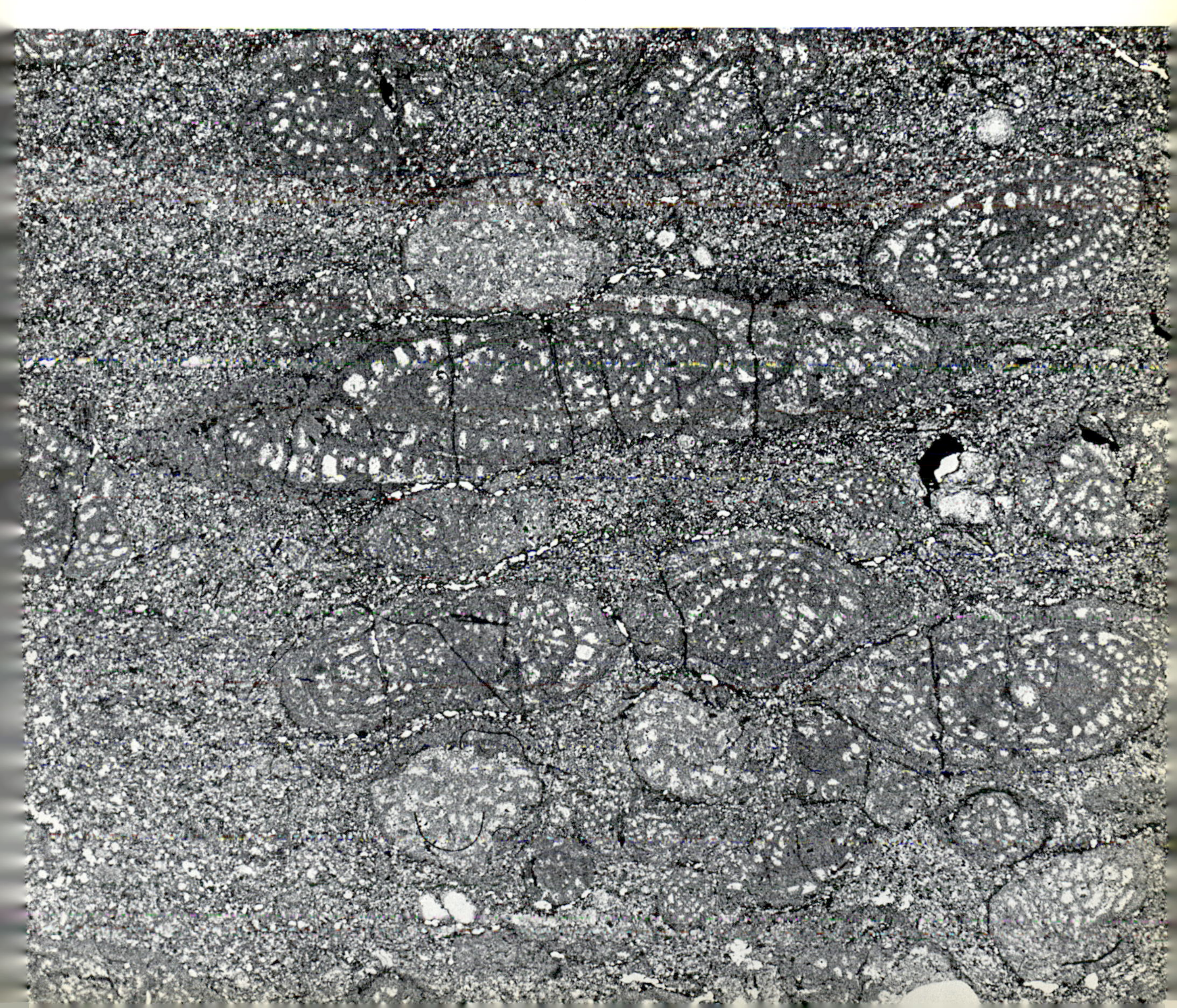

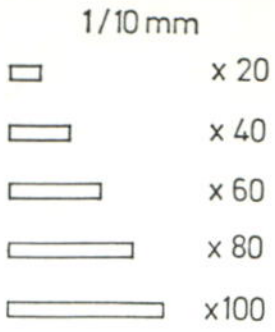

1/10 mm
x 20
x 40
x 60
x 80
x100

Fig. 1, 2. Micrite containing sponge spicules and a few small coiled foraminifers. Miocene, Pietraroia Benevento, Italy. IU 8099–375, ×20.

Fig. 3. Multirayed spicules in micrite matrix. Du Noir Formation, Middle Cedaria Zone, Upper Cambrian, Sheep Mountain, Wind River Range, T. 38 N., R. 108 W., Sublette County, Wyoming, U.S.A. IU 8099–626, ×20.

Fig. 4. Three-dimensional framework of sponge spicules. Note three kinds of cross sections of spicules: 1) transverse sections of rays of spicules, 2) sections displaying multirayed character of spicules, and 3) longitudinal sections along interlocking spicule rays that form canals, linear frameworks, or ladders. Micrite matrix. *Nevadocoelia wistae* Bassler, Middle Ordovician, Toquima Range, Nye County, Nevada, U.S.A. Rigby Collection, USNM 2(53), ×20.

Fig. 5. Three-dimensional framework of interlocking multirayed sponge spicules (dark pattern) in chert nodule. Framework partially obliterated by white chert. Note ladderlike structure of fused spicular elements. *Astylospongia praemorsa* (Goldfuss), Brownsport Formation, Middle Silurian, Perry County, Tennessee, U.S.A. Rigby Collection, CMNH 10972 (1), ×20.

Fig. 6. Three-dimensional framework of interlocking multirayed sponge spicules in chert nodule. Fused framework inhibited disarticulation of skeleton. Chert is white. Note colloform structure at lower right. *Aulocopium cylindraceum* Roemer, Brownsport Formation, Middle Silurian, Perry County, Tennessee, U.S.A. Rigby Collection, CMNH 10948, ×40.

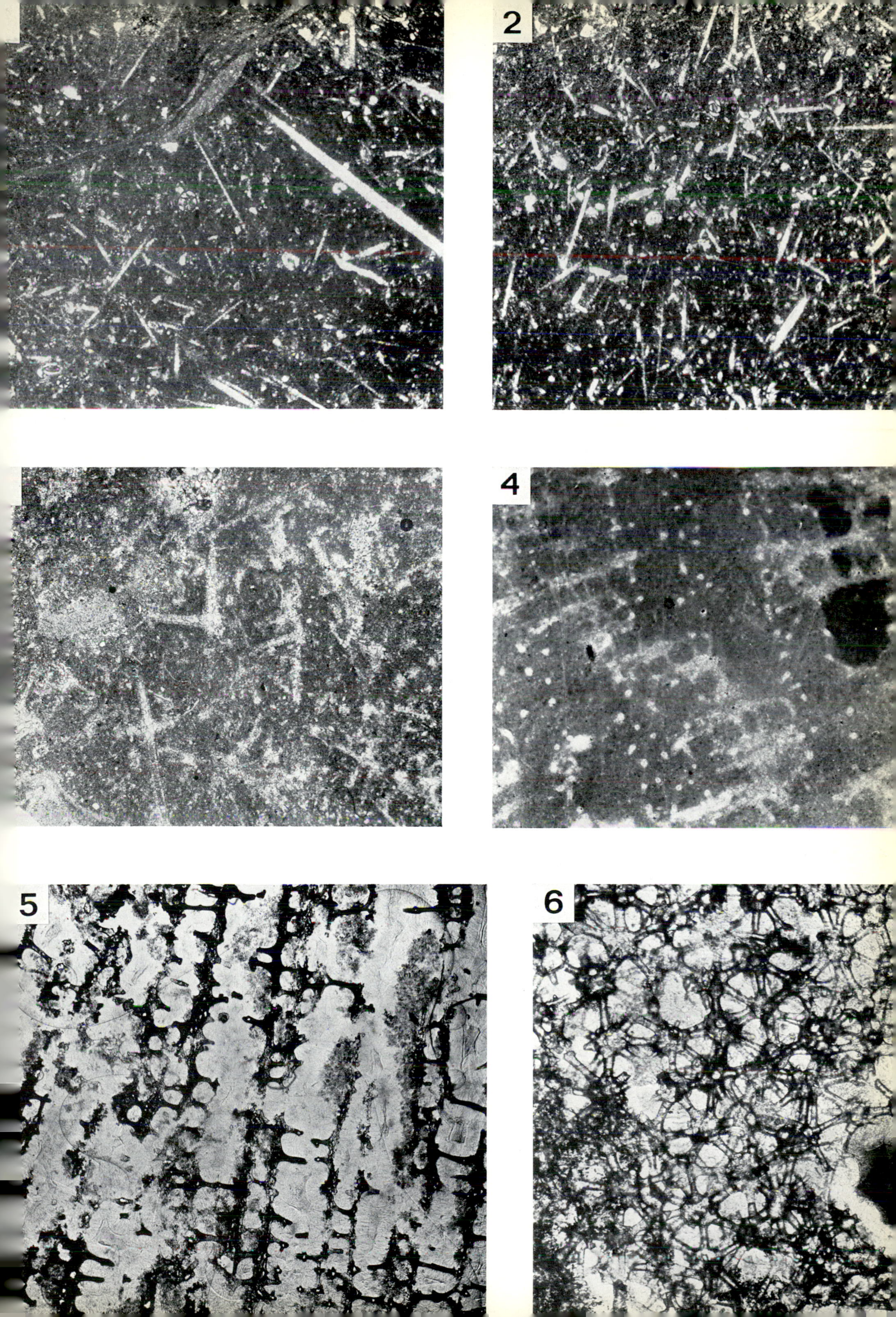
2
4
5
6

1/10 mm
x 20
x 40
x 60
x 80
x 100

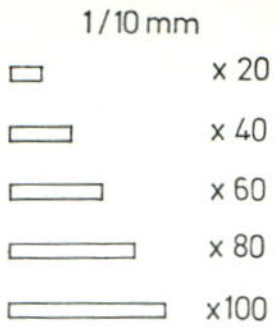

Fig. 1. Transverse and longitudinal sections of calcareous sponge spicules. Some cross sections exhibit internal canal. Beersheva Formation, Upper Jurassic, Makhtesh Ramon, Israel. IU 8099–640B, ×40.

Fig. 2. Calcareous sponge spicules in transverse and longitudinal section. Note several multirayed spicules having expanded ends. Toarcian, Lower Jurassic, Coupe de Bandol, Var, Provence, France. IU 8099–796, ×20.

Fig. 3. Muddy spiculite. Some siliceous spicules exhibit mud-filled interiors in transverse and longitudinal section. Area in upper left with few spicules may be a burrow filling as it shows crude subconcentric layering (spreite). Pennsylvanian, NW$\frac{1}{4}$ SW$\frac{1}{4}$ sec. 27, T. 2 S., R. 5 E., Johnston County, Oklahoma, U.S.A. IU 8099–450, ×40.

2

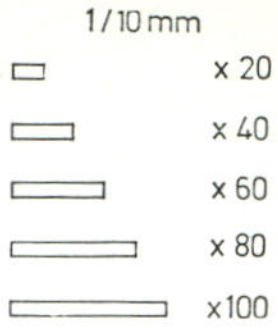

1/10 mm
x 20
x 40
x 60
x 80
x 100

Fig. 1. Transverse section of archaeocyathid in recrystallized matrix. Fine-grained wall is not perforate, and septa (parieties) are poorly developed in favor of vesicular internal wall. Mural Limestone, Lower Cambrian, Cariboo Mountains, Latitude 53° 54′ North, Longitude 120° 57′ West, British Columbia, Canada. IU 8099–224B, ×20.

Fig. 2. Longitudinal section of archaeocyathid in recrystallized matrix. Note dark very fine-grained wall. Cambrian, Kemerovsk District, Chrebet Salair, U.S.S.R. IU 8099–674, ×20.

Fig. 3. Closely packed prisms from pelecypod shells. Prisms exhibit unit extinction under cross polarizers. Compare with pls. 34 and 37–2. Packing and character of prisms simulates some spiculites (Pl. 15–1), but these calcareous prisms lack central canals of siliceous spicules. Limestone, Gemuk Group, Permian, $1^1/_2$ miles northwest of Goodnews Lake, Alaska, U.S.A. IU 8099–949, ×40.

Fig. 4. Transverse section of archaeocyathid in fine-grained matrix. Note dark very fine-grained wall, perforations in outer wall and well-developed septa (parieties) extending to vesiculose inner wall. As fig. 1, IU 8099–224, ×20.

Fig. 5. Fossil contains 1) a calcite-filled internal network, 2) a dense fine-grained nondistinctive wall microstructure, and 3) an asymmetrical arched structure suggesting attachment. Initially considered a sponge by us (note simulated spicular pattern of calcite infilling at right), but probably an oblique section of a large dicyclinid foraminifer (*Orbitopsella* sp.). Compare with Radoičić (1966, Pl. 122, fig. 1). Jurassic, Marsica Mountains, Abruzzo, Italy. IU 8099–165, ×40.

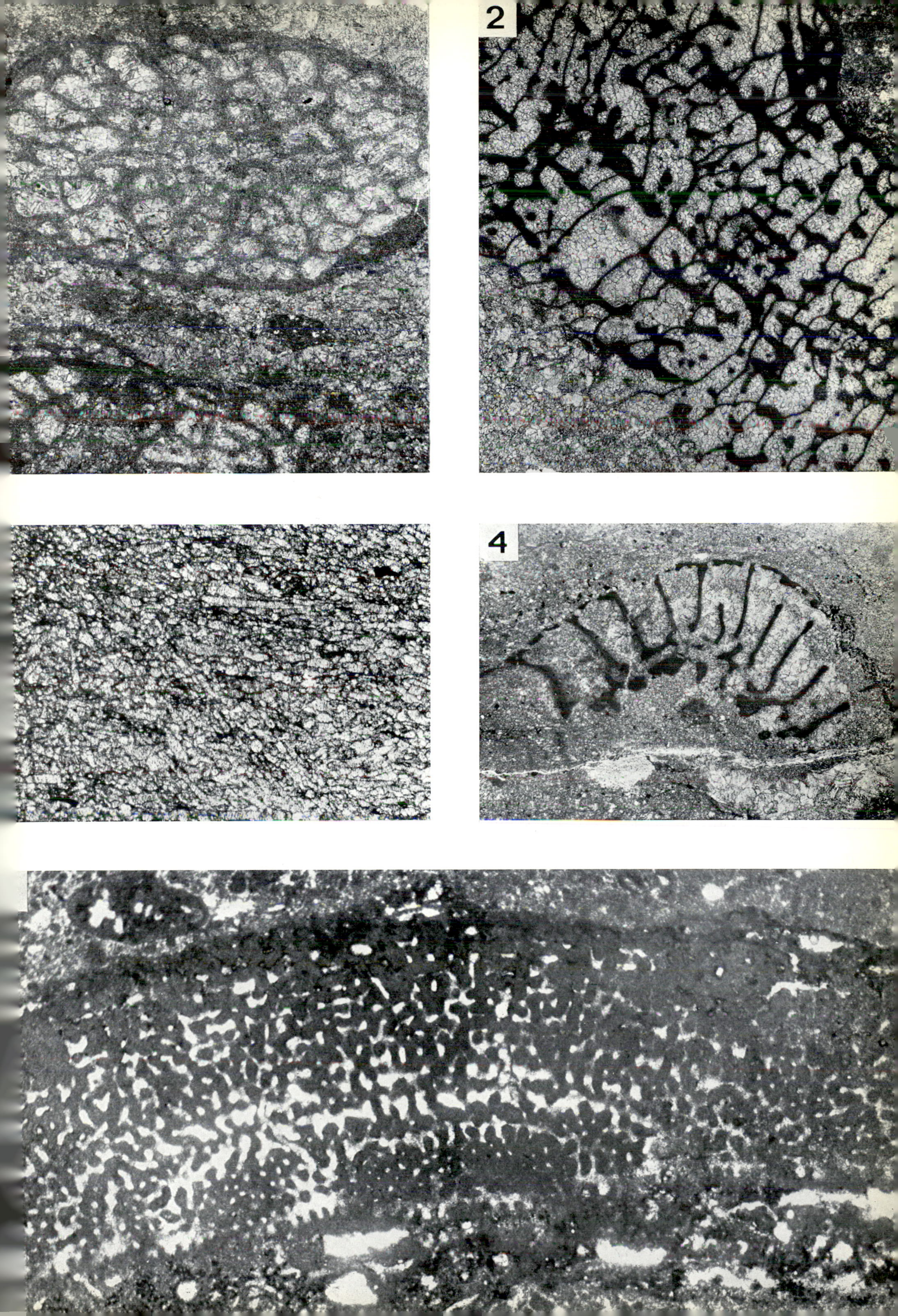

2
4

1/10 mm
x 20
x 40
x 60
x 80
x100

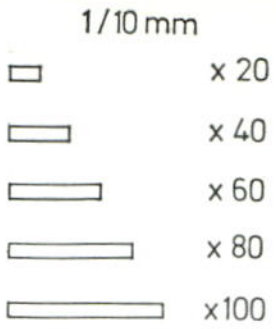

Fig. 1. Coral infilled with mud, microstructure altered. Thin rim cement between grains. Foraminifer at lower right. Mariana Limestone, Pleistocene, 1.9 kilometers northwest of Sakiburg Junction on north-central Guam Plateau, Guam, Mariana Islands, Pacific Ocean. IU 8099–480, ×20.

Fig. 2. Transverse section of a portion of the wall of a cup coral exhibiting fine microstructure and septa projecting inward from outer wall. Pelletal matrix between septa. Blue Fiord Formation, Lower Cretaceous, Bathurst Island, District of Franklin, Northwest Territories, Canada. IU 8099–135B, ×40.

Fig. 3. Oblique cross section of coral showing broad thin tabulae in center and short septa along outer wall. Internal structures partially destroyed by silicification along outer wall. Matrix of fine pelletal limestone. Lower Carboniferous, near Moorcock, about 7 kilometers northwest of Hawes, West Riding, Yorkshire, England. IU 8099–63, ×20.

Fig. 4. Transverse section of cup coral showing well-developed septa. Original shell microstructure destroyed, probably by recrystallization of original aragonitic shell. Dark very fine mud, matrix. Upper Cretaceous, Faxe Limestone quarry, Sjaelland, Denmark. IU 8099–445, ×20.

Fig. 5. Transverse section of cup coral encrusted at left by bryozoan. Original shell microstructure destroyed. Compare with fig. 6. Urgonian-Barremian, Lower Cretaceous, Coupe de St. Chamas, Bouches du Rhône, France. IU 8099–777, ×20.

Fig. 6. Transverse section of cup coral showing septa. Original shell microstructure destroyed probably by leaching and subsequent infilling of carbonate cements (calcite grain size increases from walls inward). Small dark-walled quinqueloculine foraminifers (upper right and lower left) and bryozoan cross sections (upper right and lower left) are present. As fig. 5.

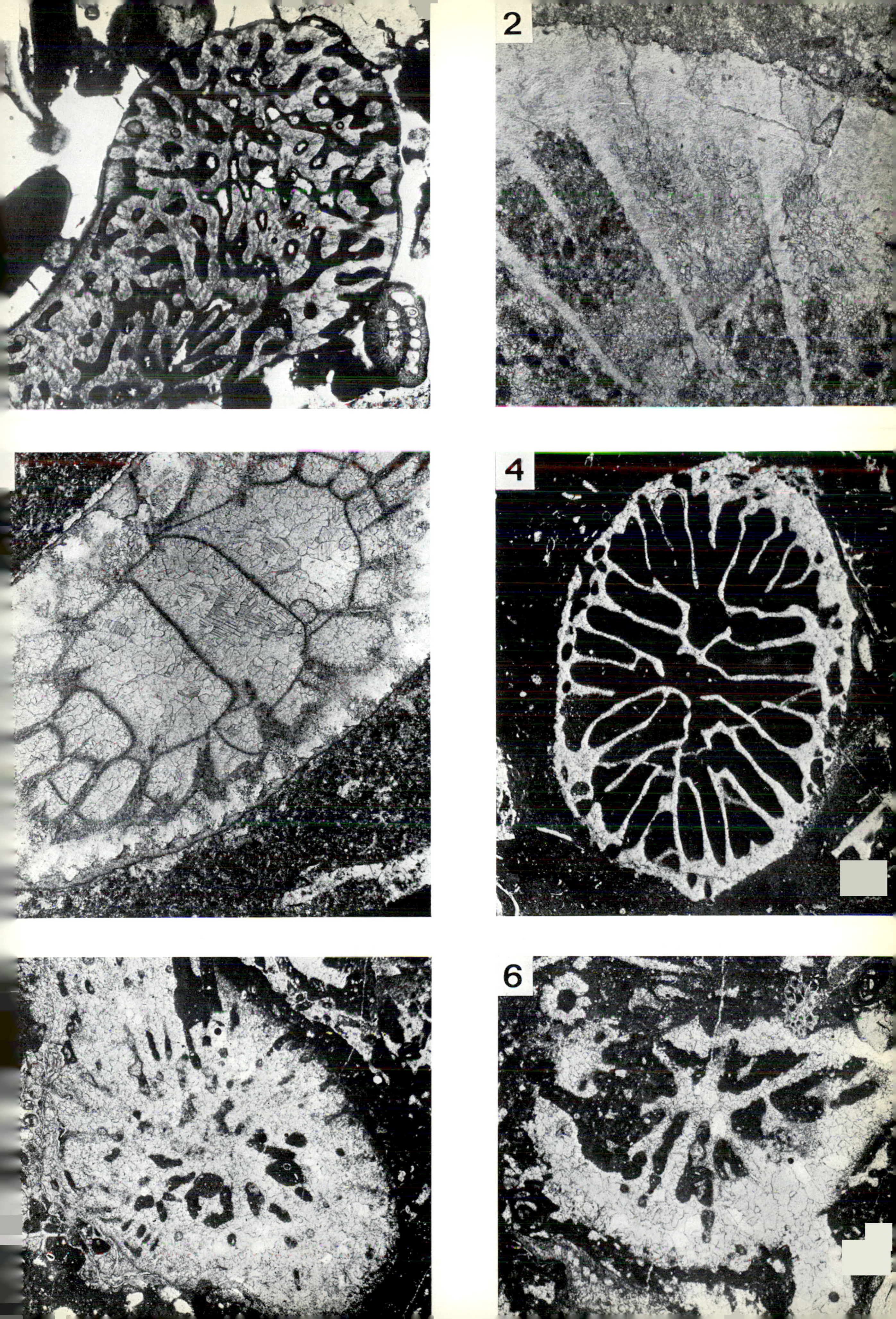
2
4
6

1/10 mm
x 20
x 40
x 60
x 80
x100

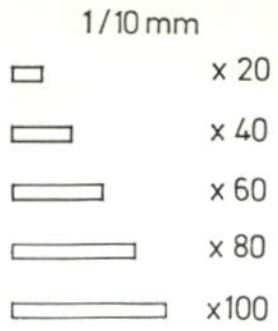

Fig. 1. Transverse section of cup coral exhibiting porous radial septa. Wall structure altered to calcite spar probably from original aragonite. Lower Cretaceous, Surduc Valley, Cariului Mountains, Padurea, Romania. IU 8099–409, ×20.

Fig. 2. Transverse section of cup corals showing solid radial septa and recrystallized wall structure. Pleistocene, near Villa Cisneros, Spanish Sahara. IU 8099–495, ×20.

Fig. 3. Longitudinal section of cup coral. Note porous septa. As fig. 1.

Fig. 4. Longitudinal section of cup coral. Note porous septa. As fig. 1.

Fig. 5. Transverse section showing internal dissepiments and thick fibrous outer wall. Buchan Caves Limestone, Middle Devonian, Heaths Quarry south of Buchan, Victoria, Australia. IU 8099–469, ×20.

Fig. 6. Coral with radiating septa. Pleistocene, Eniwetok Atoll, Marshall Islands, Pacific Ocean. IU 8099–474, ×20.

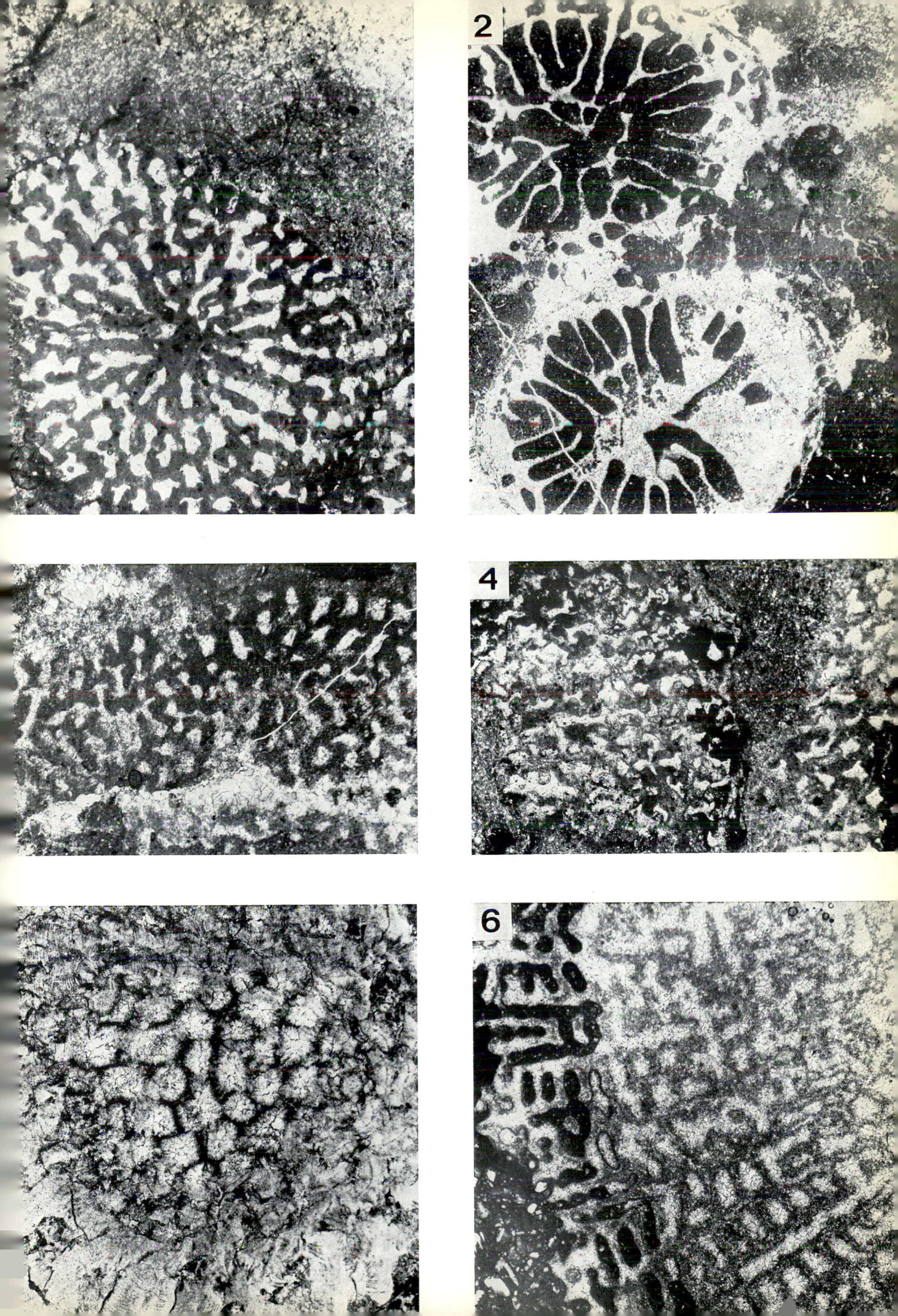
2
4
6

1/10 mm
x 20
x 40
x 60
x 80
x100

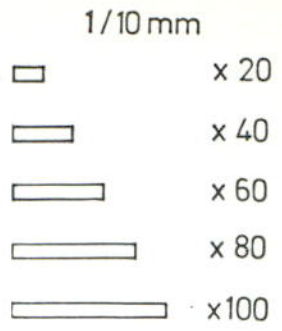

Fig. 1. Cross section of septa (a) and tabulae (b) of coral. Shell microstructure altered and shell partially silicified. Note pelletal matrix containing fine echinodermal debris and foraminifer. Lower Carboniferous, near Moorcock, about 7 kilometers northwest of Hawes, West Riding, Yorkshire, England. IU 8099–63, ×40.

Fig. 2. Colonial coral (*Syringopora*) exhibiting loosely packed corallites (a) and connecting tubes (b). Shell walls largely replaced by silica. Matrix is pelletal and calcite cement. Tournaisian, Lower Carboniferous, Usva River, Kizelovsk Region, Ural Mountains, U.S.S.R. IU 8099–679, ×20.

Fig. 3. Cross section of small cup coral. Note radial septa. Matrix consists of poorly defined debris, principally molluscan. Mastrojanni Formation, Middle Miocene, near Aschi, Abruzzo, Italy. IU 8099–175, ×40.

Fig. 4. Cross section of small cup coral (a), *Heterophyllia,* in brachiopod-echinodermal-pelletal limestone. Lower Carboniferous, Lathkill Dale, near Monyash, Derbyshire, England. IU 8099–39, ×20.

Fig. 5. Portion of a coral exhibiting septa, tabulae (a) and dissepiments (curved plates). Sparry calcite filling interior of coral skeleton. Pelletal mud at top. As fig. 4, 8099–39B, ×20.

Fig. 6. Septa (a) and tabulae (b) of portion of a coral. Note thick septal walls, partially replaced by silica, exhibiting tenuous fibrous microstructure perpendicular to the septal surfaces. Elongate microstructural units perpendicular to septal walls display unit extinction under crossed polarizers. The same microstructure appears on the upper surface of a few tabulae. As fig. 1, 8099–63B, ×20.

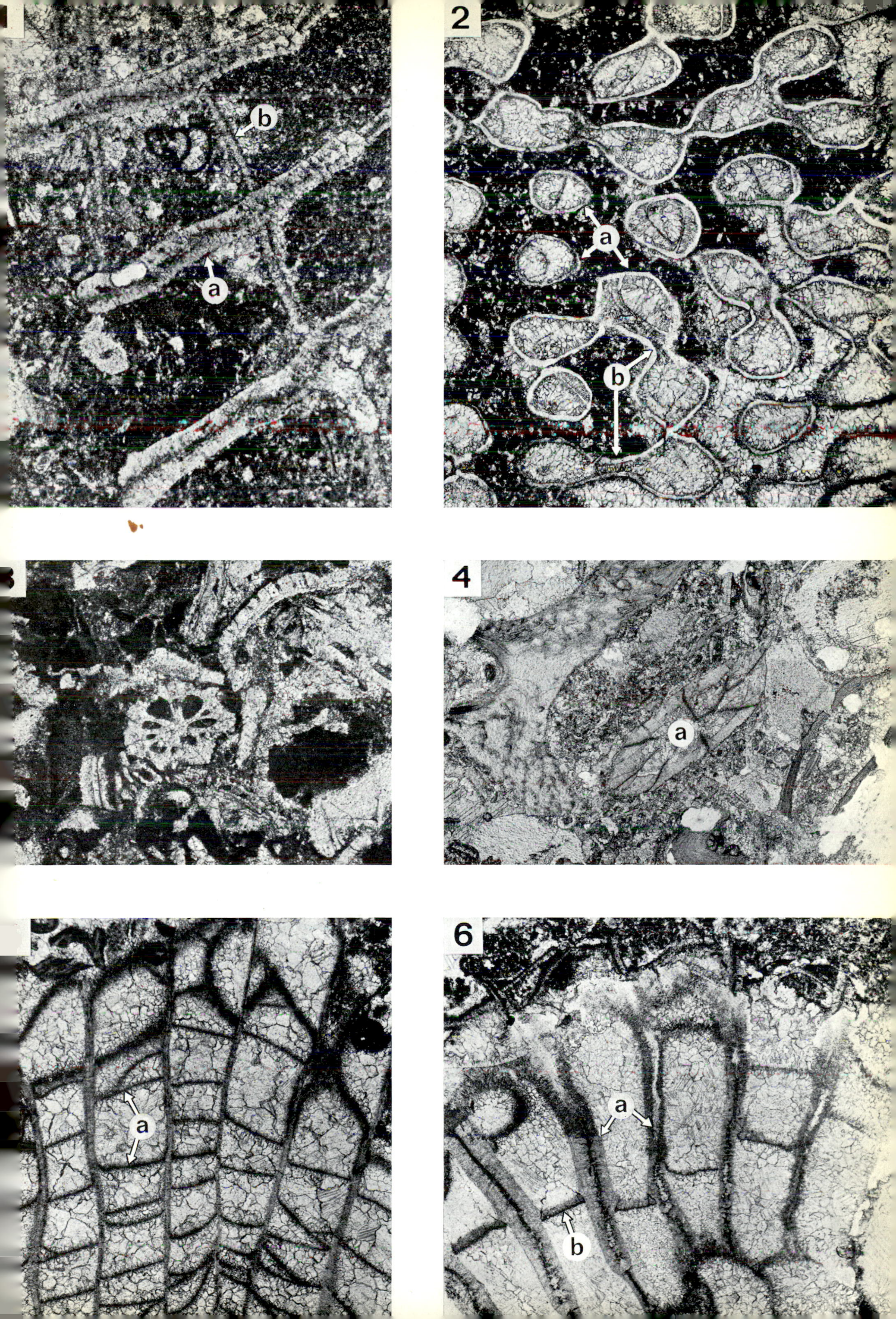

b
a
2
a
b
4
a
a
6
a
b

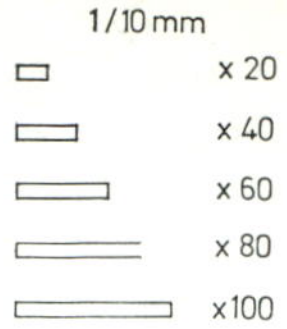

PLATE 20. CORALS AND ALGAE (?)

1/10 mm
x 20
x 40
x 60
x 80
x100

Fig. 1. Coral in dark mud matrix. Structural portions of skeleton are difficult to distinguish from clear calcite infilling. Shell microstructure probably altered. Middle Jurassic, Pollino Mountain, 8 kilometers north of Castrovillari, Cosenza, Italy. IU 8099–357, ×20.

Fig. 2. Colonial coral having poorly preserved microstructure in the walls. Individual corallites have dark incrustations around walls and were infilled with clear calcite spar. Buchan Caves Limestone, Middle Devonian, Heaths Quarry south of Buchan, Victoria, Australia. IU 8099–469, ×20.

Fig. 3. Questionable algae (a) exhibiting thick dense fine-grained walls and generally thin elongate voids (b), ovate or round in transverse section. Forms such as this have been referred to the hydrozoans (coelenterate relatives of the corals). See Pls. 63–3 and 73–2. Pelletal matrix containing calcite spar. Middle Jurassic, Montagna, Spaccata Dam, south of Alfedena, L'Aquila, Italy. IU 8099–373, ×40.

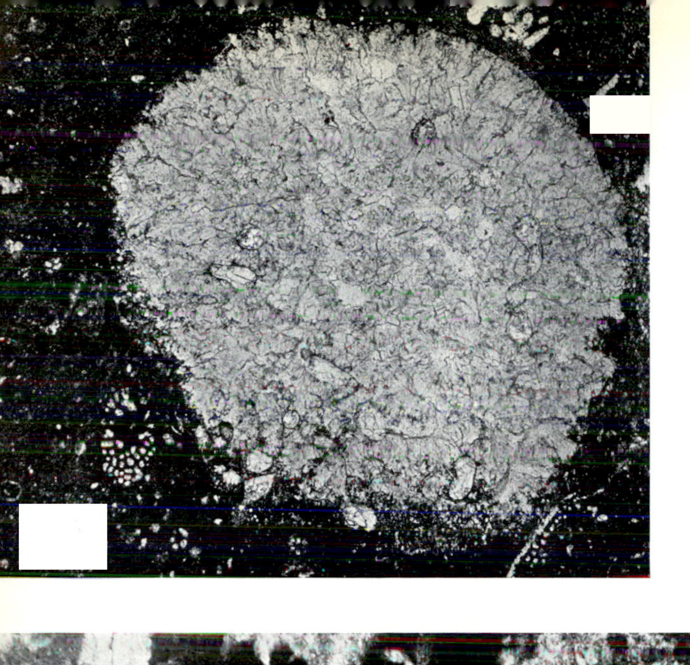

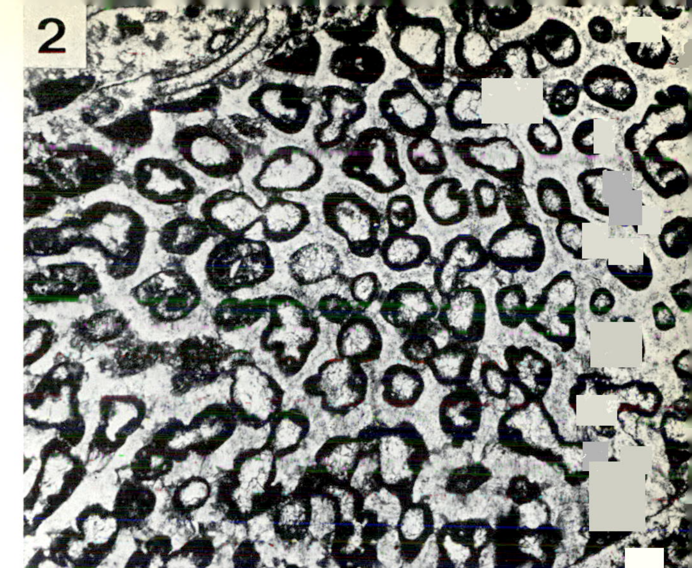
2

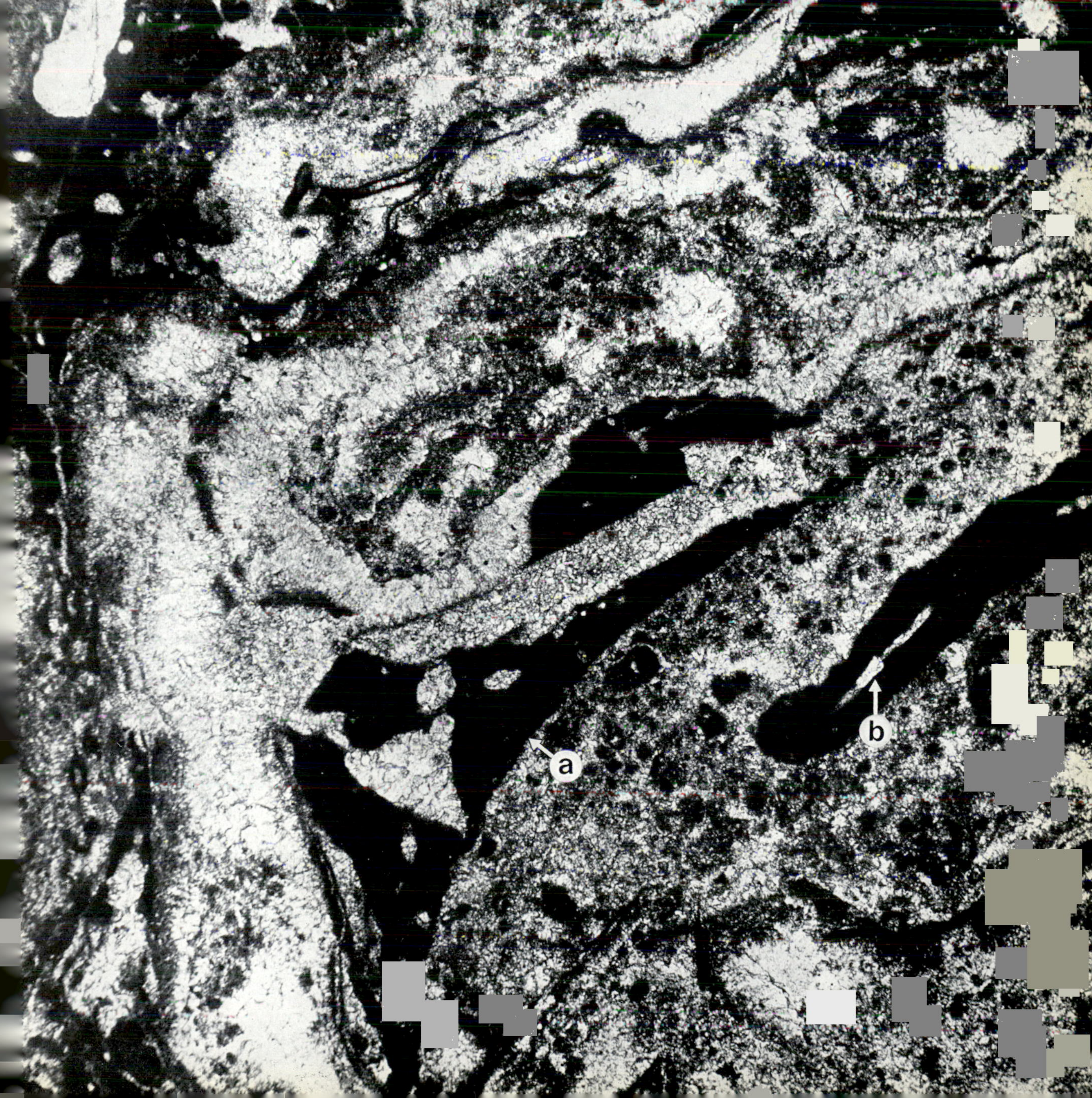
a
b

1/10 mm
x 20
x 40
x 60
x 80
x100

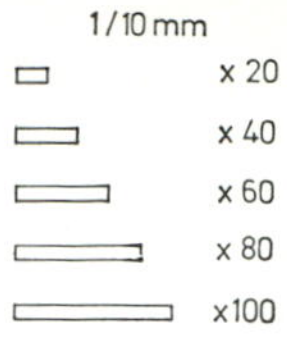

Fig. 1. *Anostylostroma columnare* (PARKS) in tangential section exhibiting circular structure of up-turned laminae in column and individual pillars (a) that appear as dark dots. Jeffersonville Limestone, Middle Devonian, Charlestown, Clark County, Indiana, U.S.A. IU S1–22, 23 (tangential), ×20.

Fig. 2. *Anostylostroma columnare* (PARKS) in longitudinal section. Laminae and pillars are clearly defined and form a generally rectangular grid. As fig. 1, IU S1–22, 23 (longitudinal), ×20.

Fig. 3. *Gerronostroma excellens* GALLOWAY and ST. JEAN in tangential section. Dark pillars contrast sharply with clear sparry infilling. Jeffersonville Limestone, Middle Devonian, Falls of the Ohio, Jeffersonville, Clark County, Indiana, U.S.A. IU 6289 (S1–31, 32, tangential), ×20.

Fig. 4. *Stromatoporella solitaria* (NICHOLSON), a longitudinal section that shows upturned laminae in a column. Pillars are dark and fine grained. Logansport Limestone, Middle Devonian, Pipe Creek Falls, 10 miles southeast of Logansport, Cass County, Indiana, U.S.A. IU S1–26, 27 (longitudinal), ×20.

Fig. 5. *Gerronostroma excellens* GALLOWAY and ST. JEAN, longitudinal section illustrating laminae and pillars less clearly defined than above. Partly recrystallized. As fig. 3, IU S1–31, 32 (longitudinal), ×20.

Fig. 6. *Stromatoporella solitaria* NICHOLSON in oblique section exhibiting pillars and laminae. As fig. 4. IU S1–26, 27 (oblique section), ×20.

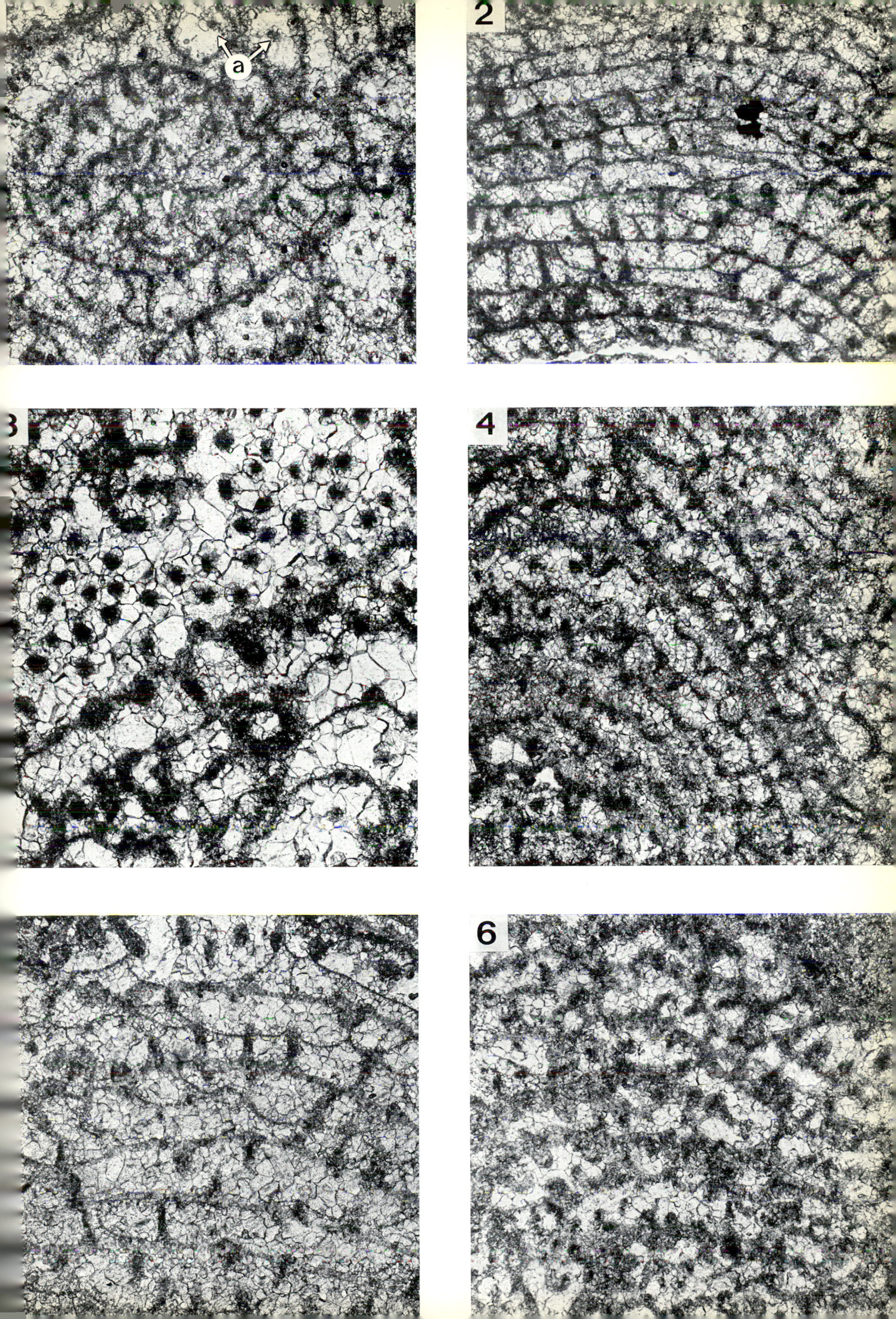
a
2
3
4
6

1/10 mm
x 20
x 40
x 60
x 80
x100

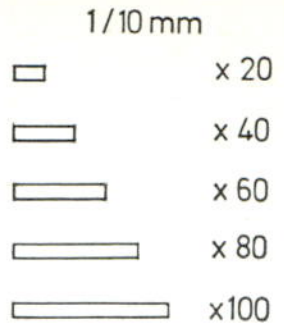

Fig. 1. *Aulacera plummeri* GALLOWAY and ST. JEAN, stromatoporoid showing in longitudinal section vesicular or cytose plates. Muddy wall structure has a dense dark median line. Galleries filled by clear calcite cement. Liberty Formation, Upper Ordovician, Wilson Creek, 2 miles southwest of Deatsville, Nelson County, Kentucky, U.S.A. IU S1–7, 8, 9 (longitudinal), ×20.

Fig. 2. *Trupetostroma iowense* PARKS, stromatoporoid exhibiting in tangential section the circular pattern of upturned laminae and associated pillars. Cedar Valley Limestone, Upper Devonian, Floyd County, Iowa, U.S.A. IU 6290 (S1–33, 34, tangential), ×20.

Fig. 3. *Gerronostroma excellens* GALLOWAY and ST. JEAN, stromatoporoid displaying in longitudinal section laminae, pillars and spotted (maculate) tissue (a). Galleries are infilled by clear calcite cement. Jeffersonville Limestone, Falls of the Ohio, Jeffersonville, Clark County, Indiana, U.S.A. IU 6289 (S1–31, 32, longitudinal), ×40.

Fig. 4. *Labechia huronensis* (BILLINGS), stromatoporoid showing vesicular tissue and pillars. Whitewater Formation, Upper Ordovician, Muscatatuck State Farm, Butlerville, Jennings County, Indiana, U.S.A. IU S1–11, 12 (longitudinal), ×20.

Fig. 5. *Millepora alcicornis* Linnaeus, a modern hydrozoan (a relative of the corals) showing in tangential section two sizes of skeletal tubes, dark skeletal tissue, and small "septa" around apertures of large tubes ("corallites"). Recent, Florida, U.S.A. IU 6293 (S1–40), ×20.

Fig. 6. *Stromatoporella solitaria* NICHOLSON, a stromatoporoid having generally linear arrangement of pillars. Logansport Limestone, Middle Devonian, Pipe Creek Falls, 10 miles southeast of Logansport, Cass County, Indiana, U.S.A. IU S1–26, 27 (tangential), ×40.

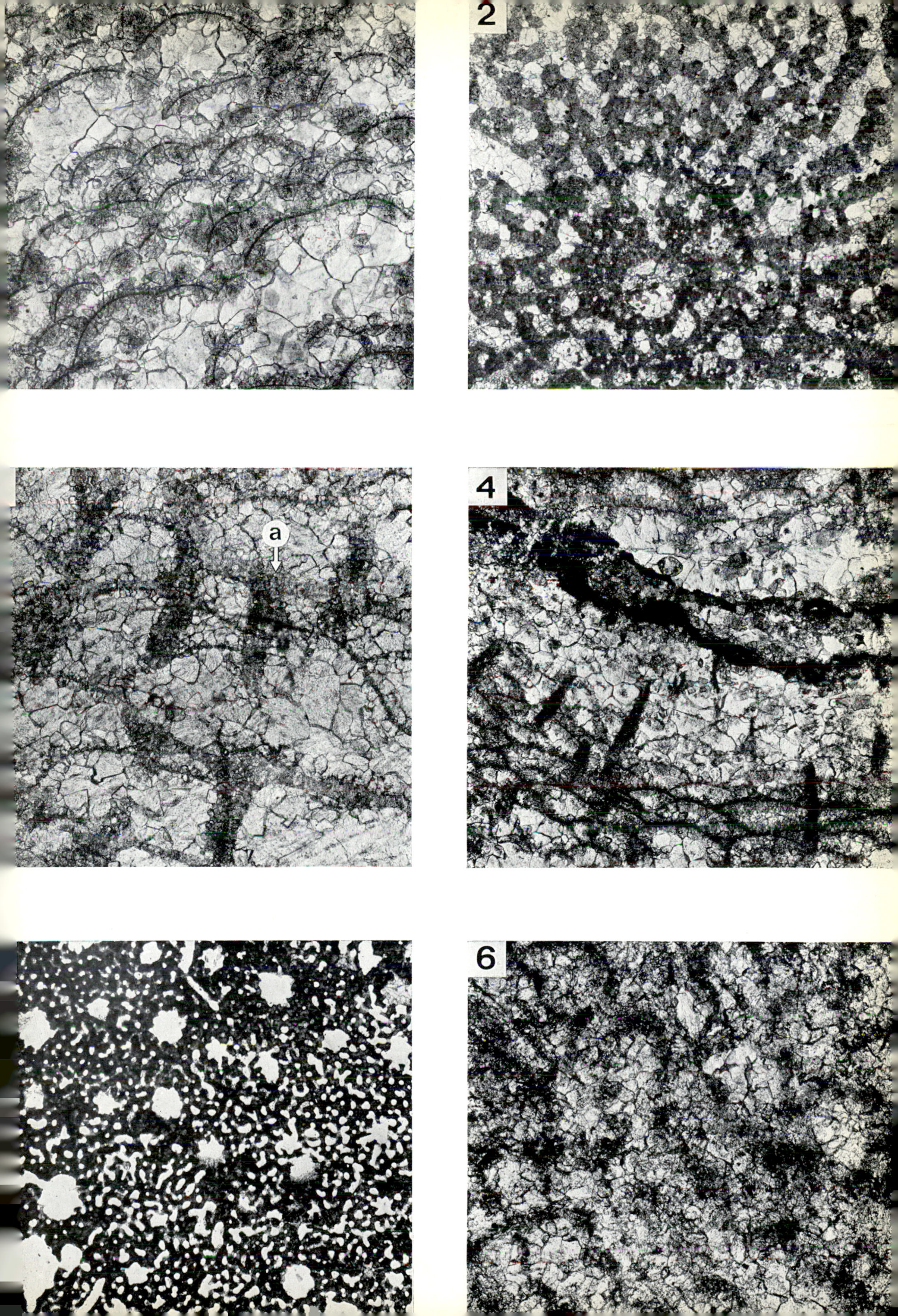

2
a
4
6

1/10 mm

x 20

x 40

x 60

x 80

x 100

Fig. 1. Laminated wall structure of Paleozoic trepostomatous bryozoan appears fibrous in longitudinal section. Zooecia contain a few transverse partitions (diaphragms), and smaller tubes (mesopores) have many partitions. Discontinuity in structure (a) represents plane of rejuvenation. Zooecia filled with fine spar. Echinoderm plates at left show syntaxial rim cement. Brassfield Limestone, Lower Silurian, Center N$\frac{1}{2}$ SW$\frac{1}{4}$ sec. 6, T. 5 N., R. 11 E., Jefferson County, Indiana, U.S.A. IU 8099–290, ×40.

Fig. 2. Mud- and spar-filled, thin-walled bryozoan. Zooecia (a) lack transverse partitions (diaphragms), which are numerous in the smaller tubes (b, mesopores). Upper Caradocian, Middle Ordovician, Rausjaer, Asker, Akerhus, Oslo region, Norway. IU 8099–812, ×20.

Fig. 3. Tangential section of bryozoan in mud matrix. Note thick walls between zooecia. Laminated walls contain numerous very small micracanthopores. Upper Pequop Formation, Permian, North Cherry Creek Mountains, sec. 12, T. 29 N., R. 63 E., Elko County, Nevada, U.S.A. IU 8099–570, ×20.

Fig. 4. Tangential and longitudinal sections of partially mud-filled bryozoans exhibiting recrystallized wall structure. Lower Carboniferous, between Kali and Arhlad, Gourara region, Sahara Desert, Algeria. IU 8099–667, ×20.

Fig. 5. Tangential section displaying poorly developed laminated wall microstructure. Campanian-Maestrictian, Upper Cretaceous, Dau-Port-Marant, Charente Maritime, France. IU 8099–33, ×40.

Fig. 6. Longitudinal section of hollow ramose form. Laminated wall microstructure is poorly shown. Note the difference in wall thickness between the inner (immature, endozone) and outer (mature, exozone) regions of the colony. Fine pelletal matrix contains a few small foraminifers. As fig. 5, ×20.

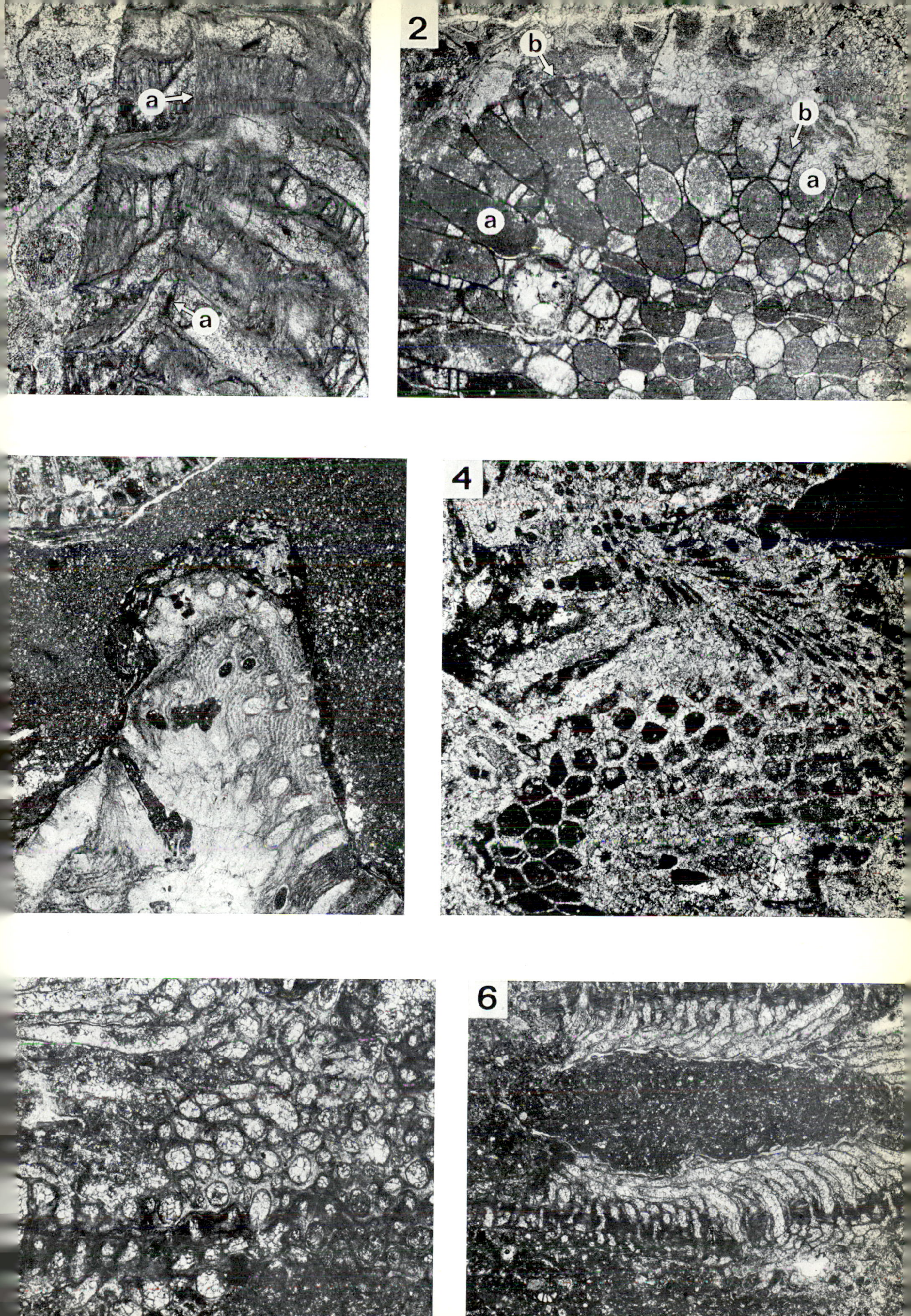
a
a
2
b
b
a
a
4
6

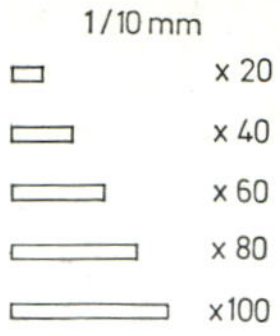

1/10 mm

x 20

x 40

x 60

x 80

x100

Fig. 1. Cross section of fenestrate bryozoan frond. Note clear granular layer lining zooecia and thick dark laminated outer shell wall. Brachiopods, punctate (a) and impunctate (b), gastropods (c), echinoderm debris (d), and fistuliporoid bryozoan (e) also present. Sparry calcite cement. Note that some pelletal lime mud is present and that the gastropod is filled with quartz-bearing dark mud. This fragment was reworked from another environment before final deposition and lithification. Bethel Formation, Chester Series, Upper Mississippian, Hardin County, Kentucky, U.S.A. IU 8099–P211, ×40.

Fig. 2. *Fistulipora*, an encruster exhibiting small cystose or vesicular tissue between tubes (zooecia). Bryozoologists (ectoproctologists) will recognize from the orientation of the vesicles (convex downward) that the colony is overturned. Vesicles and zooecia are filled principally by sparry calcite cement. Matrix is quartz-bearing pelletal mud. Geopetal surface at lower center indicates that direction of "up" is to the northeast in the photograph. As fig. 1, IU 8099–P212, ×40.

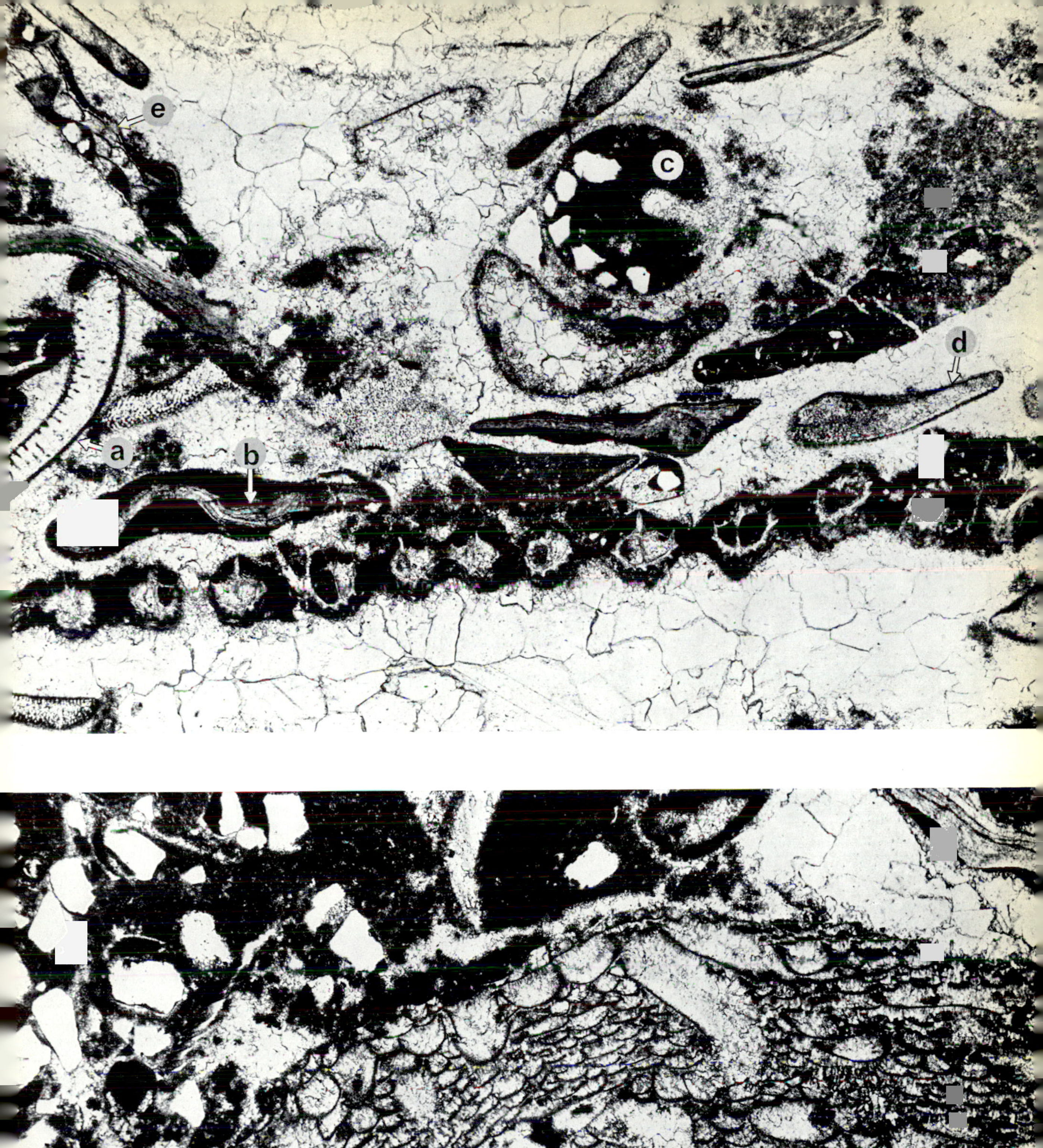
e
c
d
a
b
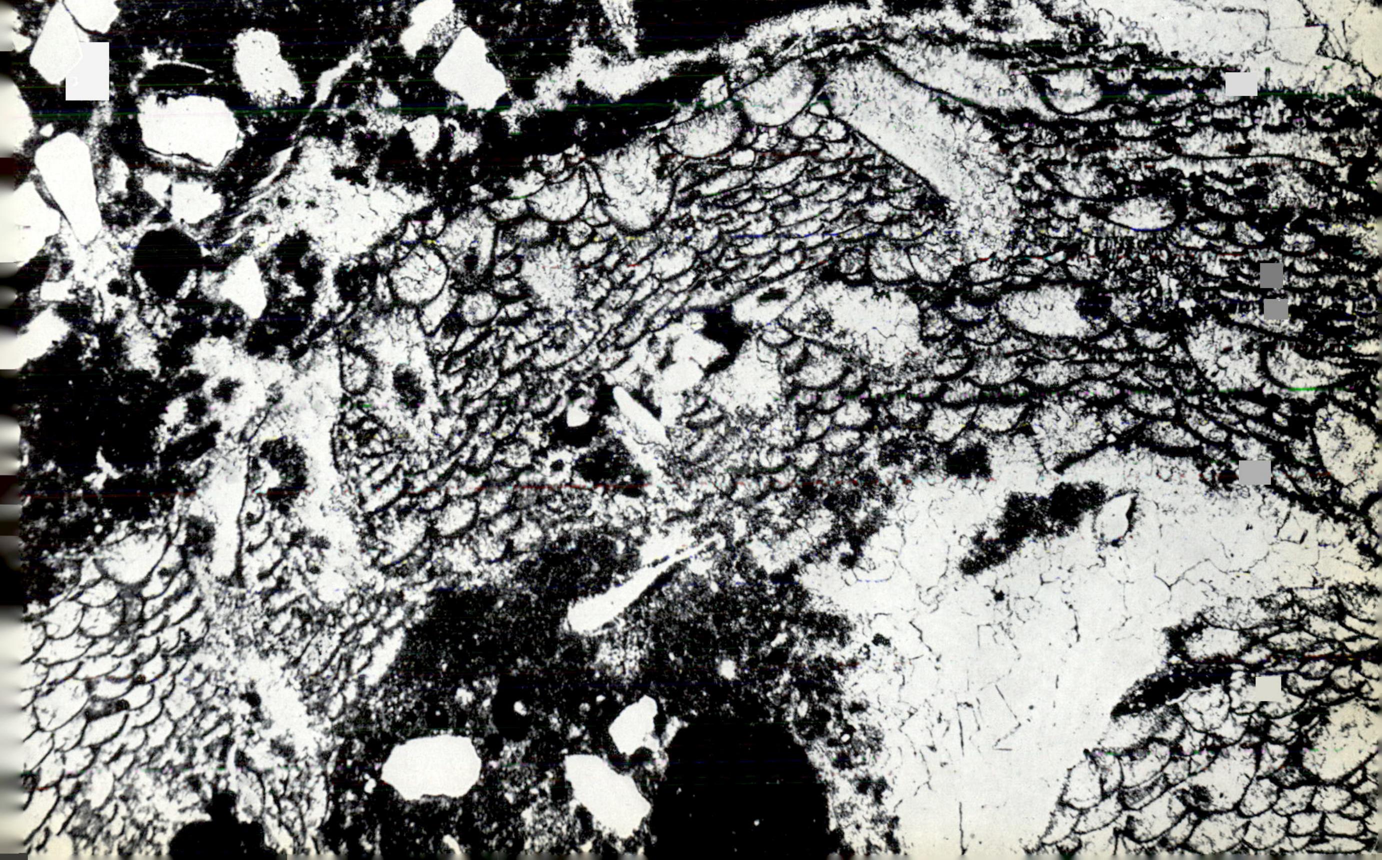

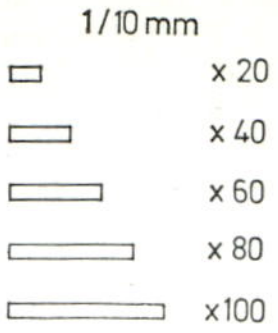

1/10 mm
x 20
x 40
x 60
x 80
x100

Fig. 1. Ramose specimens in transverse (a) and longitudinal (b) sections. Rock has unusually high porosity. Upper Danian, Upper Cretaceous, Faxe Quarry, Stevns Klint, Sjaelland, Denmark. IU 8099–764, ×40.

Fig. 2. Bifoliate form. Wall structure recrystallized and zooecia partially infilled with micrite. Matrix is dark micrite and fragmental molluscan sand grains. Geopetal fabric indicates "up" is to northwest in photograph. Carillo Puerto Formation, Pliocene, on Federal Highway 180, 3.9 kilometers southeast of Kantunil, Yucatán, México. IU 8099–125B, ×20.

Fig. 3. Bryozoan encrusting brachiopod. Zooecial tubes show diaphragms (flat plates) and cystiphragms (curved plates). Note fibrous shell structure of brachiopods. Smaller fragments of brachiopods and echinoderms in mud matrix at upper right. Recrystallized micrite at lower left. Upper Ordovician, southeastern Indiana, U.S.A. IU 8099–100B, ×40.

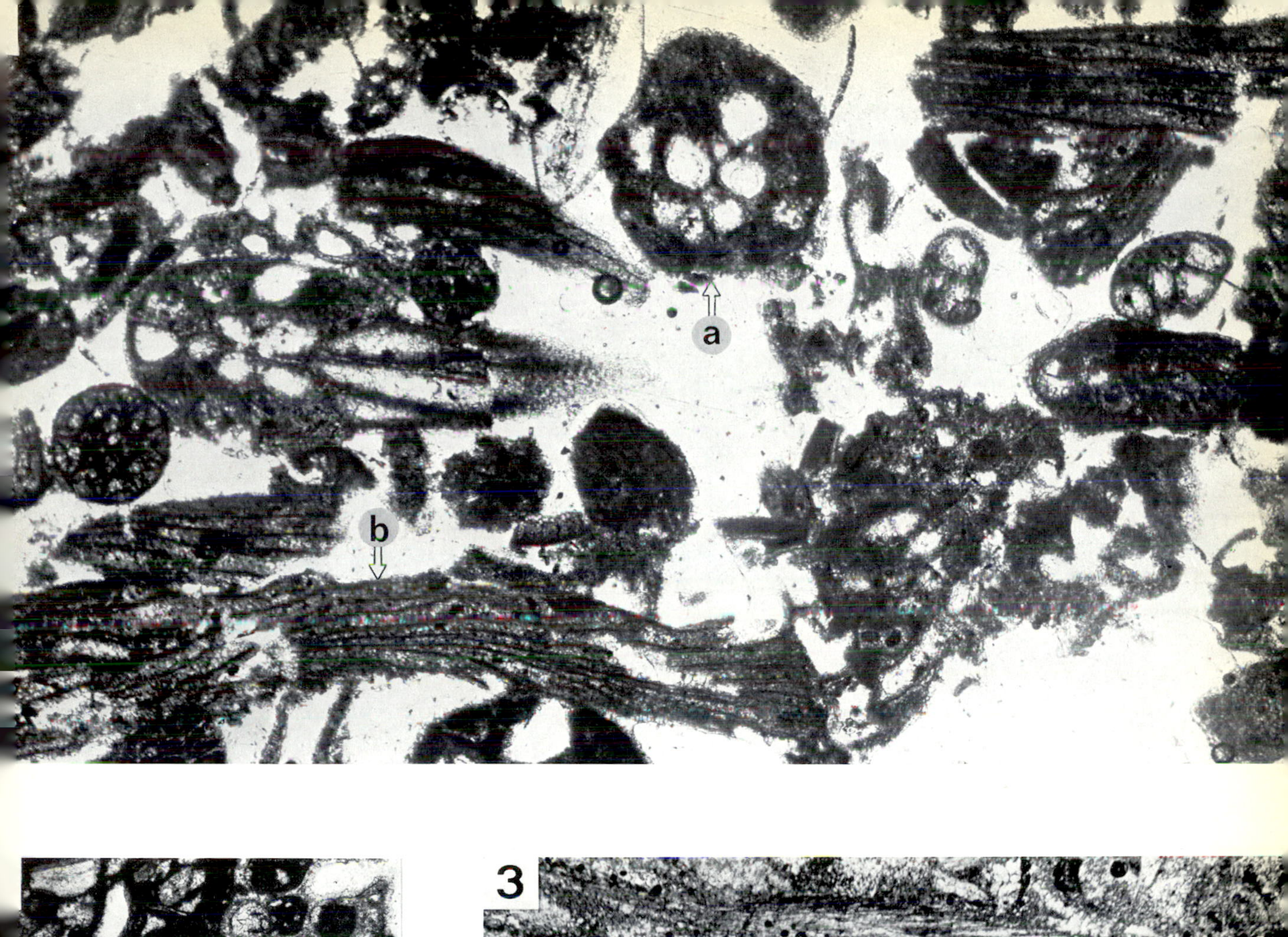
a
b

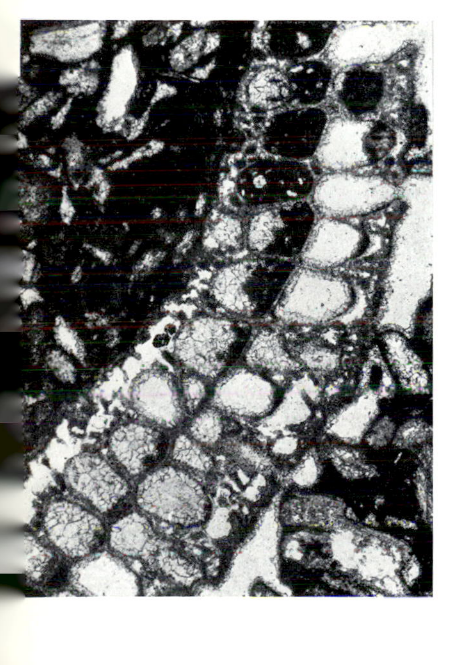

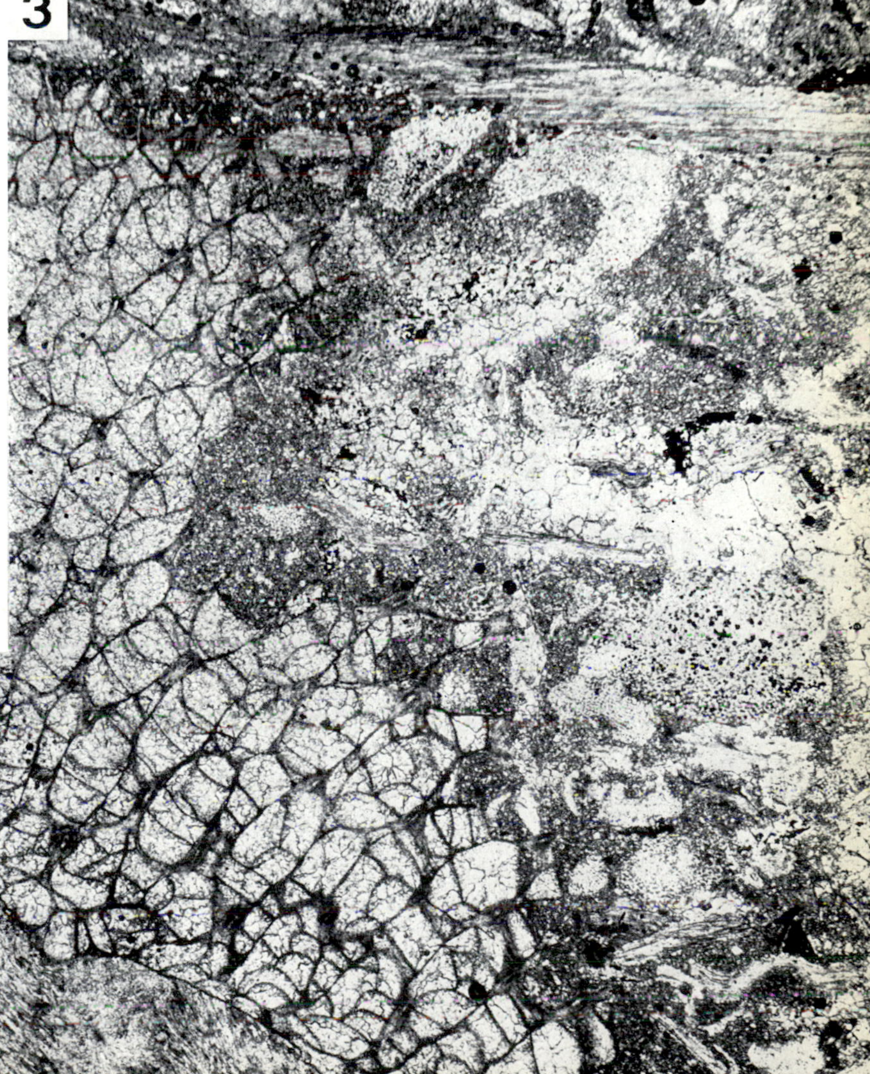
3

PLATE 26. BRYOZOANS

1/10 mm

x 20

x 40

x 60

x 80

x100

Fig. 1. Muddy, poorly sorted skeletal limestone, showing conspicuous bryozoans in transverse, tangential and longitudinal sections. Biotic debris contains an echinoid spine (a) and molluscan debris (b). Dark mud matrix. Fracture filling crosses bottom of figure. Miocene, Cusano Mutri, Benevento, Italy. IU 8099–366, ×40.

Fig. 2. Prominent longitudinal sections of bryozoans at right in a micritic foraminiferal-bryozoan limestone. Fracture filling at lower left contains large calcite crystals that exhibit twinning. As fig. 1.

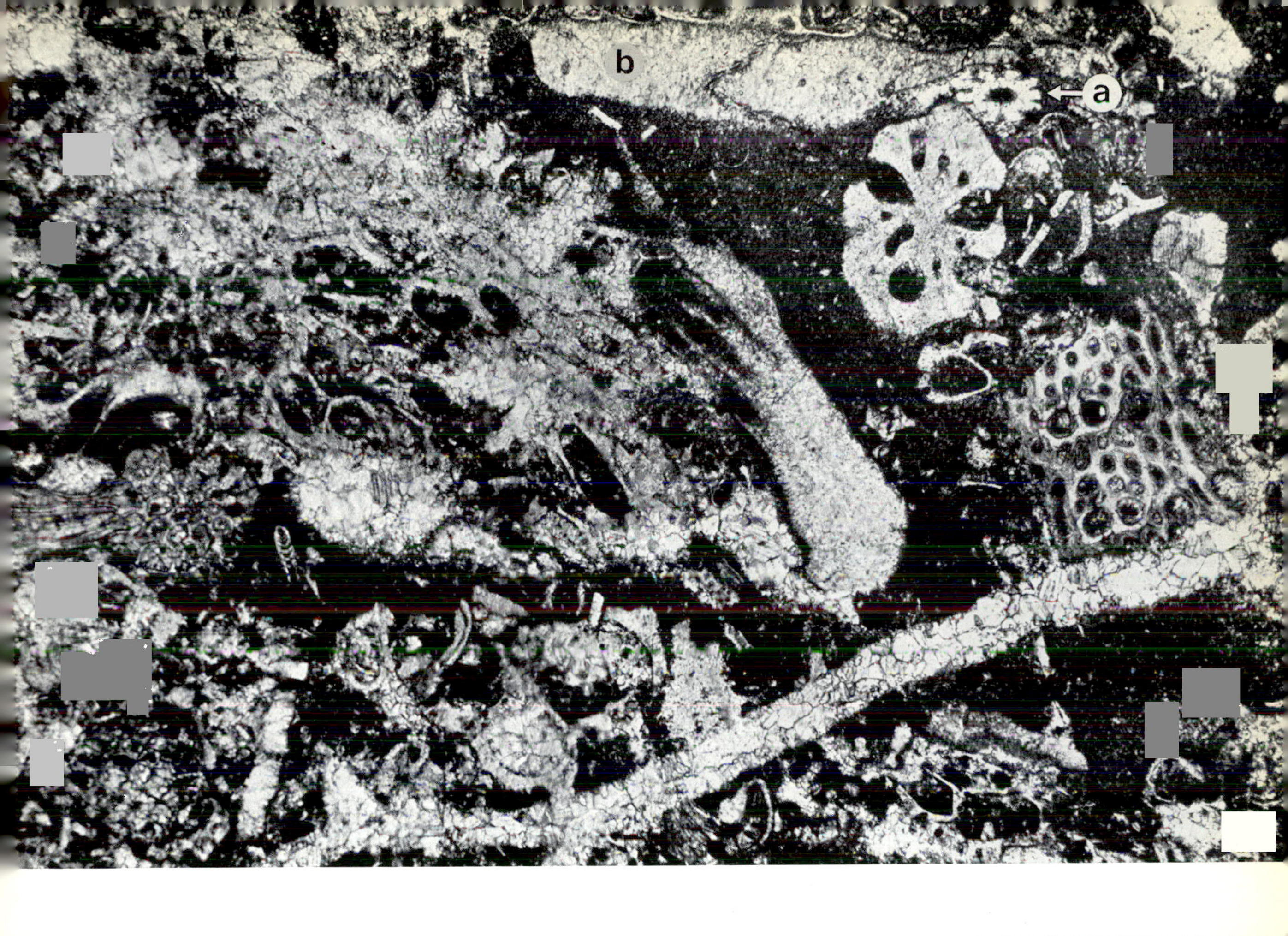
b
a

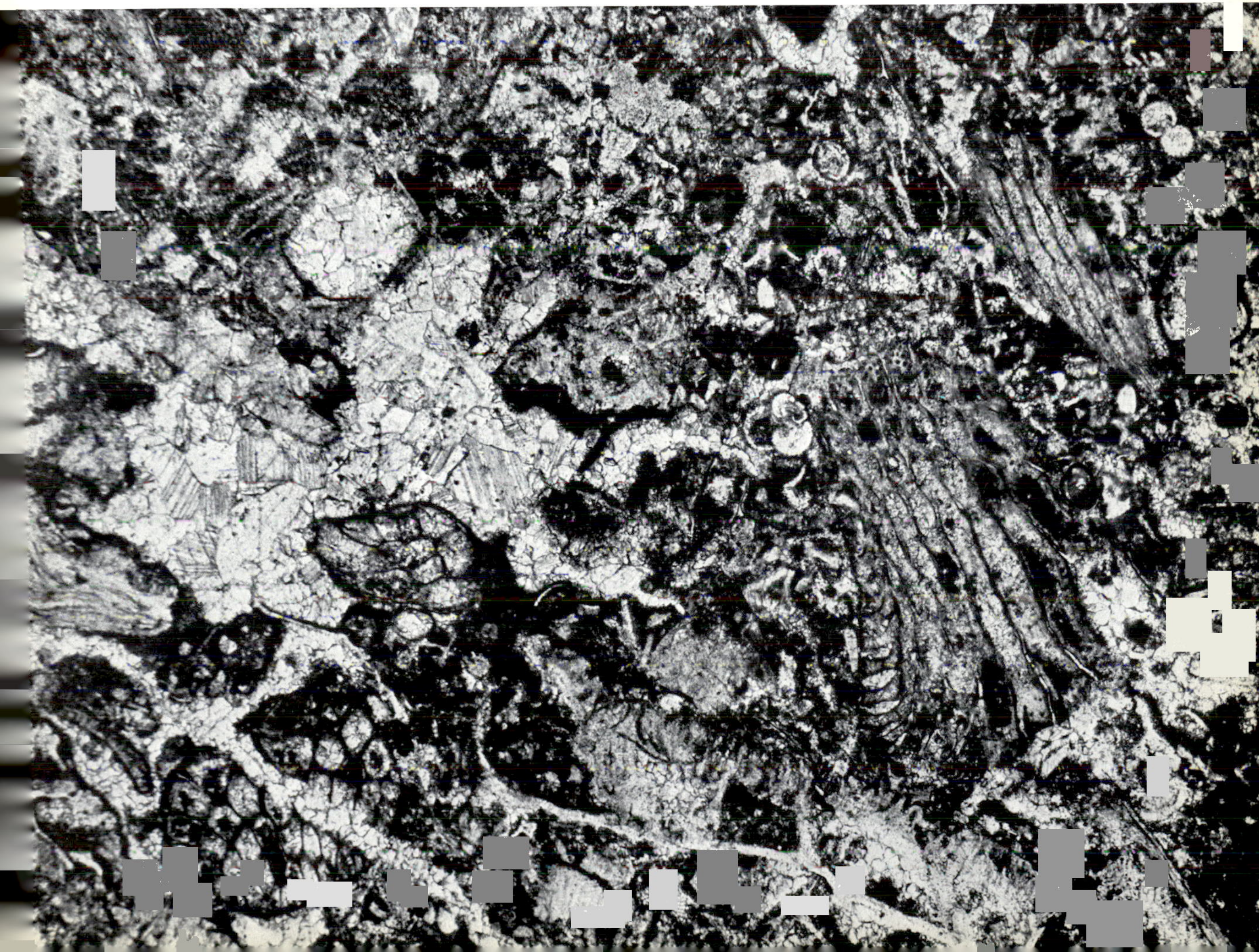

1/10 mm
x 20
x 40
x 60
x 80
x100

Fig. 1. Colony under cross nicols showing the laminated walls of the outer region (exozone) of the zoarium and the characteristic cone-in-cone structure of the laminae. Berriedale Limestone, Lower Permian, Mount Nausau, near Hobart, Tasmania, Australia. IU 8099–893, ×40.

Fig. 2. Ramose specimen as observed in tangential (upper right) and longitudinal (lower left) sections. Note numerous small acanthopores (dark dots) surrounding each zooecial opening in tangential section. Matrix consists of pelletal mud and calcite spar. Echinodermal plates, foraminifers, brachiopods, and fenestrate bryozoan debris are present. Lower Carboniferous, Lathkill Dale, near Monyash, Derbyshire, England. IU 8099–39B, ×40.

1/10 mm

x 20

x 40

x 60

x 80

x100

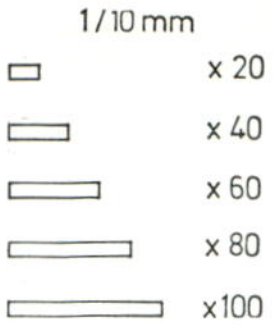

Fig. 1. Cross section of bifoliate form. Dark wall has been altered probably from original laminated structure. Matrix is fine biotic sand of echinodermal debris in calcite. Sparry calcite cement. Echinodermal plates commonly have syntaxial rim cement. Jeffersonville Limestone, Middle Devonian, active quarry, sec. 34, T. 7 N., R. 8 E., North Vernon, Jennings County, Indiana, U.S.A. IU 8099-80, ×40.

Fig. 2. Bryozoans largely infilled with calcite spar but having pelletal muds adhering to the colony surfaces as well as extending into areas of the matrix. Bryozoan wall thick in outer zoarium (exozone) and thin in inner zone (endozone). Dark line in wall marks boundary between adjacent zooecia in colony. Laminated walls exhibit faint cone-in-cone texture at base of photograph. Zooecia contain numerous transverse partitions (diaphragms). Zone 9a, Upper Wenlockian, Middle Silurian, road section between Vik and Sundvollen, Ringerike area, Buskerud, northwest of Oslo, Norway. IU 8099-808, ×40.

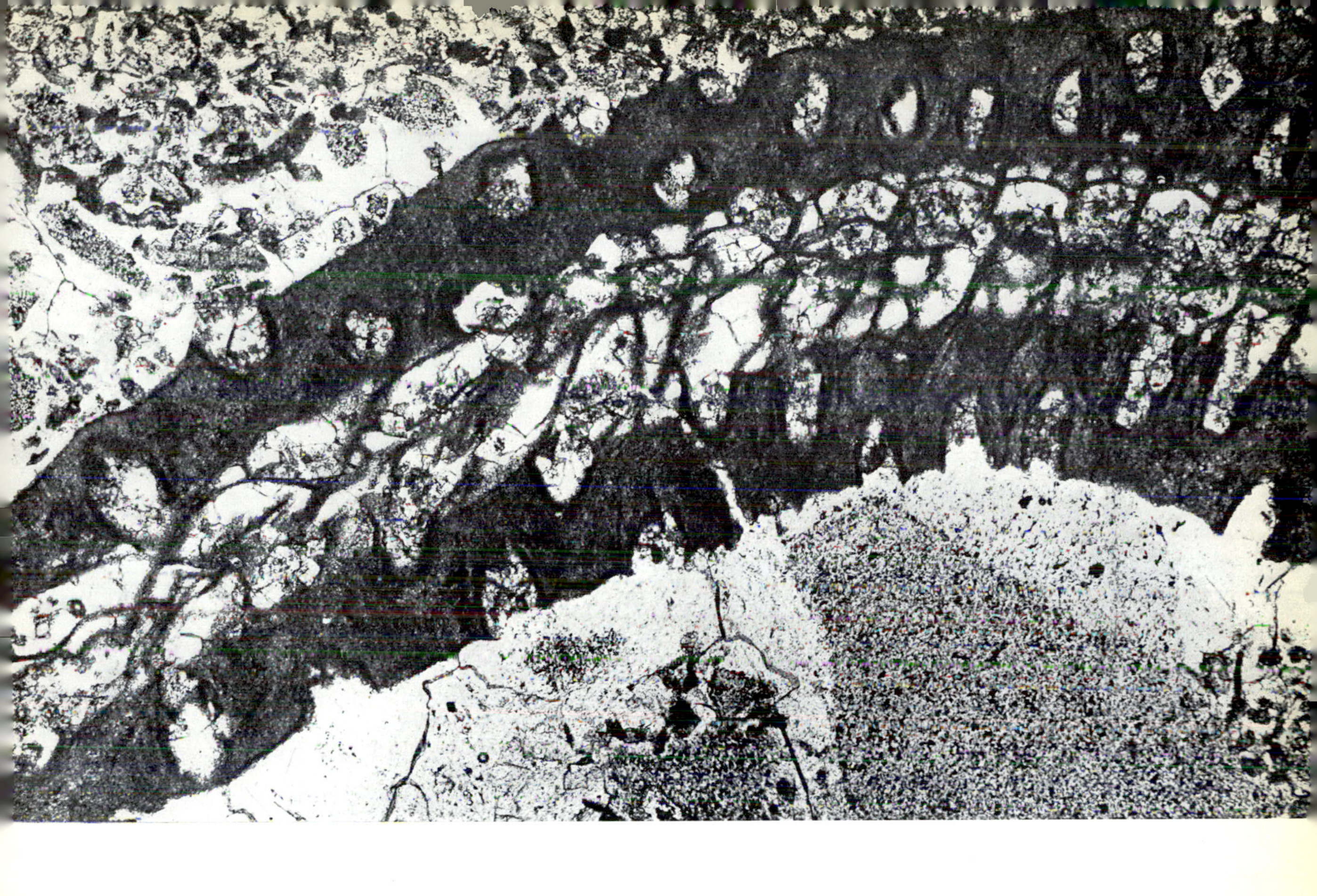

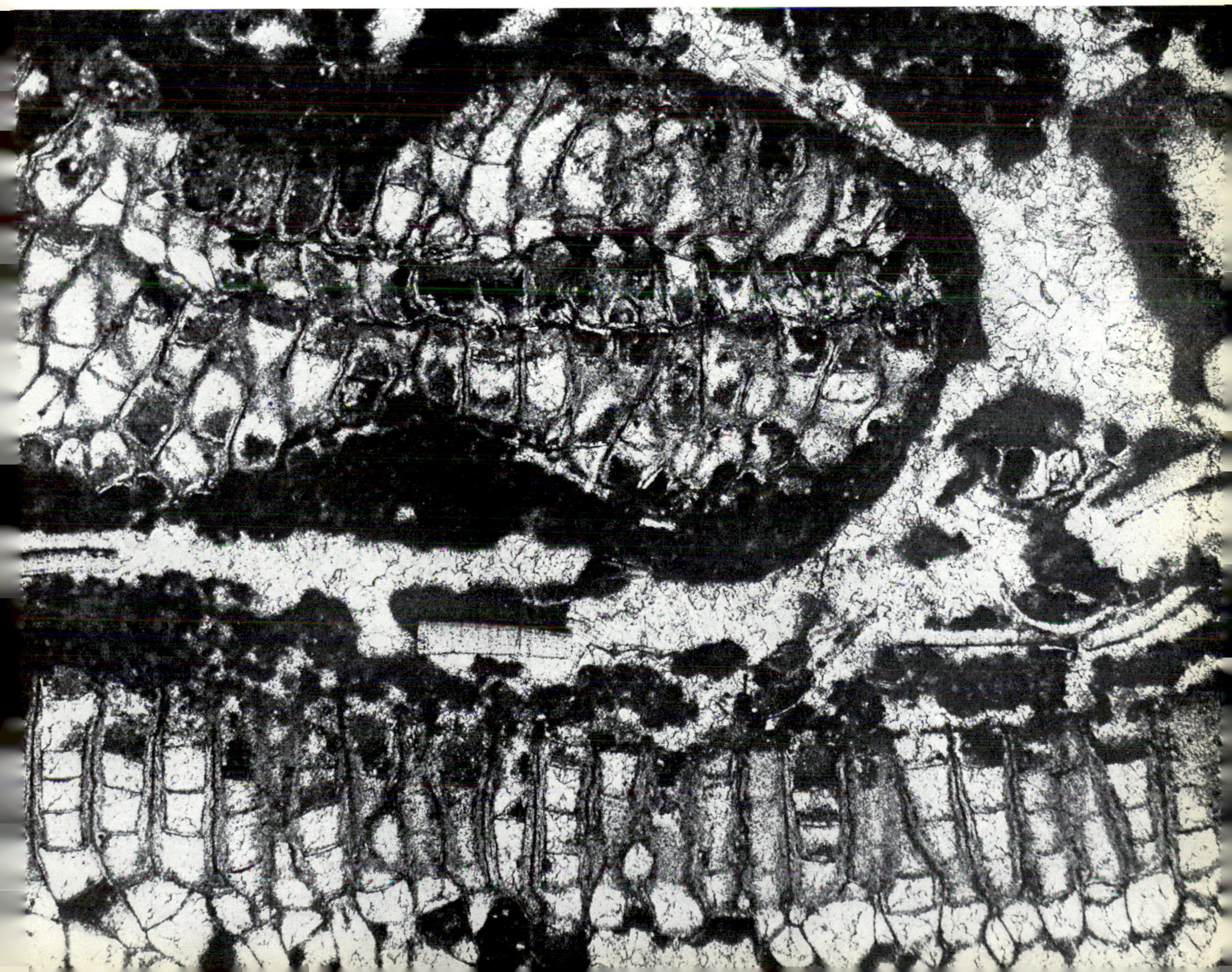

PLATE 29. BRACHIOPODS

1/10 mm
x 20
x 40
x 60
x 80
x100

Fig. 1. Parts of brachiopod shells. One shell (a) exhibits well-developed fibrous internal plates (b). Shell (c) at upper left shows a prismatic layer. Matrix of lime mud contains numerous small dolomite rhombs. Pentamerus Limestone, Zone 7b, Upper Llandoverian, Lower Silurian, quarry $2^1/_2$ kilometers west of Vik, Ringerike area, Busherud, northwest of Oslo, Norway. IU 8099–805, ×20.

Fig. 2. Longitudinal vertical section of both valves of brachiopod. Note internal structures along hingeline at bottom. Geopetal fabric within shell. Matrix of fine pellets and fossil fragments, principally echinodermal plates. Lower Carboniferous, near Moorcock, about 7 kilometers northwest of Hawes, West Riding, Yorkshire, England. IU 8099–63B, ×20.

Fig. 3. Fragments of brachiopod valves, principally from along hingeline, exhibiting internal plates (a) and prismatic shells. Fine-grained calcareous mud matrix containing a few small echinodermal plates and fine brachiopod debris. Pentamerus Limestone, Zone 9a, Upper Wenlockian, Middle Silurian, road section between Vik and Sundvollen, Ringerike area, Busherud, northwest of Oslo, Norway. IU 8099–810, ×10.

Fig. 4. Fibrous punctate shell in matrix of pellets and echinodermal debris. Gerster Formation, Upper Permian, west side of central Butte Mountains, sec. 25, T. 20 N., R. 59 E., White Pine County, Nevada, U.S.A. IU 8099–567, ×20.

Fig. 5. Pseudopunctate shell showing clear pillars in fibrous shell structure. Matrix of bryozoan and echinodermal debris in clear calcite cement. Jeffersonville Limestone, Middle Devonian, active quarry, sec. 34, T. 7 N., R. 8 E., North Vernon, Jennings County, Indiana, U.S.A. IU 8099–80B, ×40.

Fig. 6. *Lingula borealis* BITTNER. Cross section of phosphatic shell displaying layering. Punctae not visible. Dinwoodie Formation, Lower Triassic, Pole Canyon, sec. 13, T. 1 S., R. 4 W., Madison County, Montana, U.S.A. IU 8099–1075, ×40.

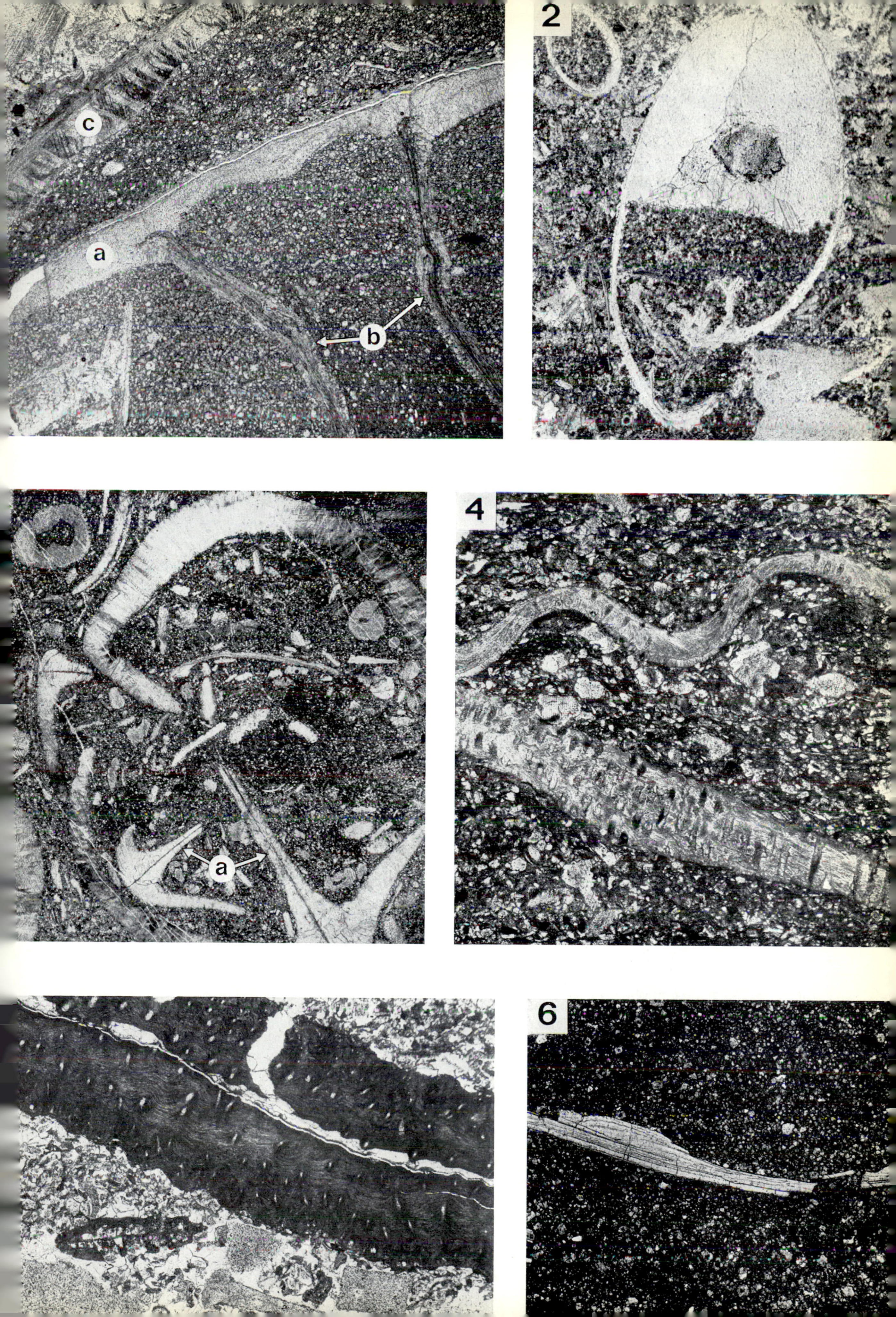

c
a
b
2
a
4
6

1/10 mm
x 20
x 40
x 60
x 80
x 100

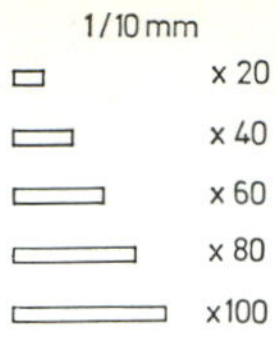

Fig. 1. Fibrous shell showing longitudinal heart-shaped cross section of single valve, parallel to plane of commissure, sulcate (a) at anterior edge. Pelletal matrix. Note geopetal fabric within shells. Finely porous echinodermal grains and quartz sand are also conspicuous. Poorly preserved shell texture in brachiopod at upper left. Circular shell at right, possibly molluscan (gastropod?), has been dissolved and subsequently infilled by calcite spar. A few grains show oolitic coatings. Bethel Formation, Chester Series, Upper Mississippian, Hardin County, Kentucky, U.S.A. IU 8099–P210, ×40.

Fig. 2. Fibrous brachiopod fragment at left exhibiting ornamentation (a), probably off-center base of spine. Note cross sections of fibrous brachiopod spines (b), echinoderm plate and fenestrate bryozoan (c) at right. Lower Carboniferous, Lathkill Dale, near Monyash, Derbyshire, England. IU 8099–39B, ×40.

Fig. 3. Fibrous shell extending across center of photograph. Matrix of quartz sand, porous echinodermal plates, and calcite cement. Some dark grains of pre-existing lime mud. As fig. 1, IU 8099–P209, ×40.

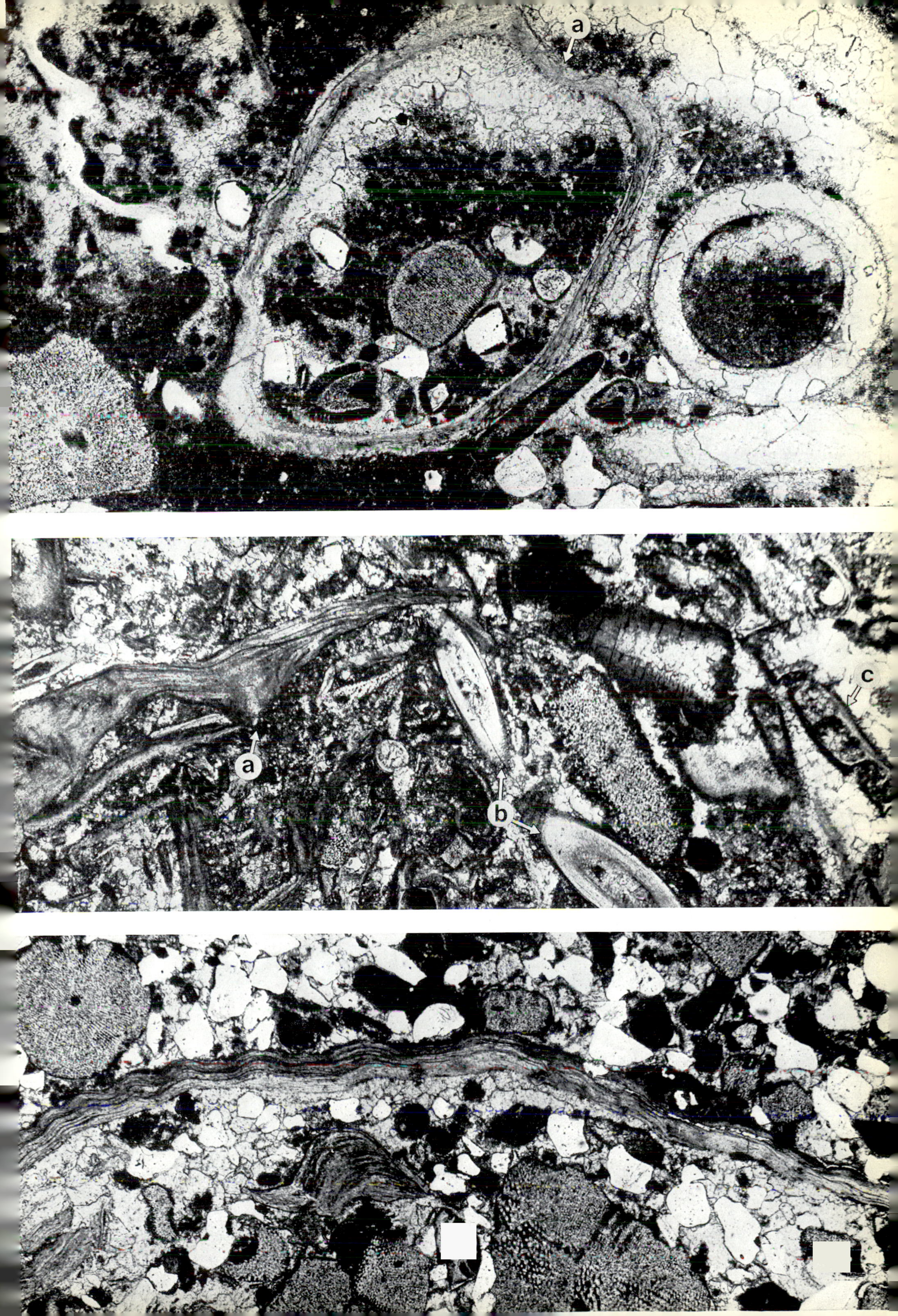
a
a
b
c

1/10 mm
x 20
x 40
x 60
x 80
x100

Fig. 1. Bryozoan-brachiopod-echinodermal limestone. Conspicuous transverse sections of hollow brachiopod spines and fragmentary fenestrate bryozoan cross sections in a matrix of pellets and sparry calcite cement. Geopetal fabric is present in the spine interiors. Lower Carboniferous, Lathkill Dale, near Monyash, Derbyshire, England. IU 8099–39, ×40.

Fig. 2. Longitudinal section of a crushed brachiopod spine at upper center. Cross section of fenestrate bryozoan and fibrous brachiopod shells are shown at the upper left and lower left, respectively. Matrix is apparently a mixture of pelletal mud and calcite spar. Inconspicuous echinoderm plates (a). As fig. 1.

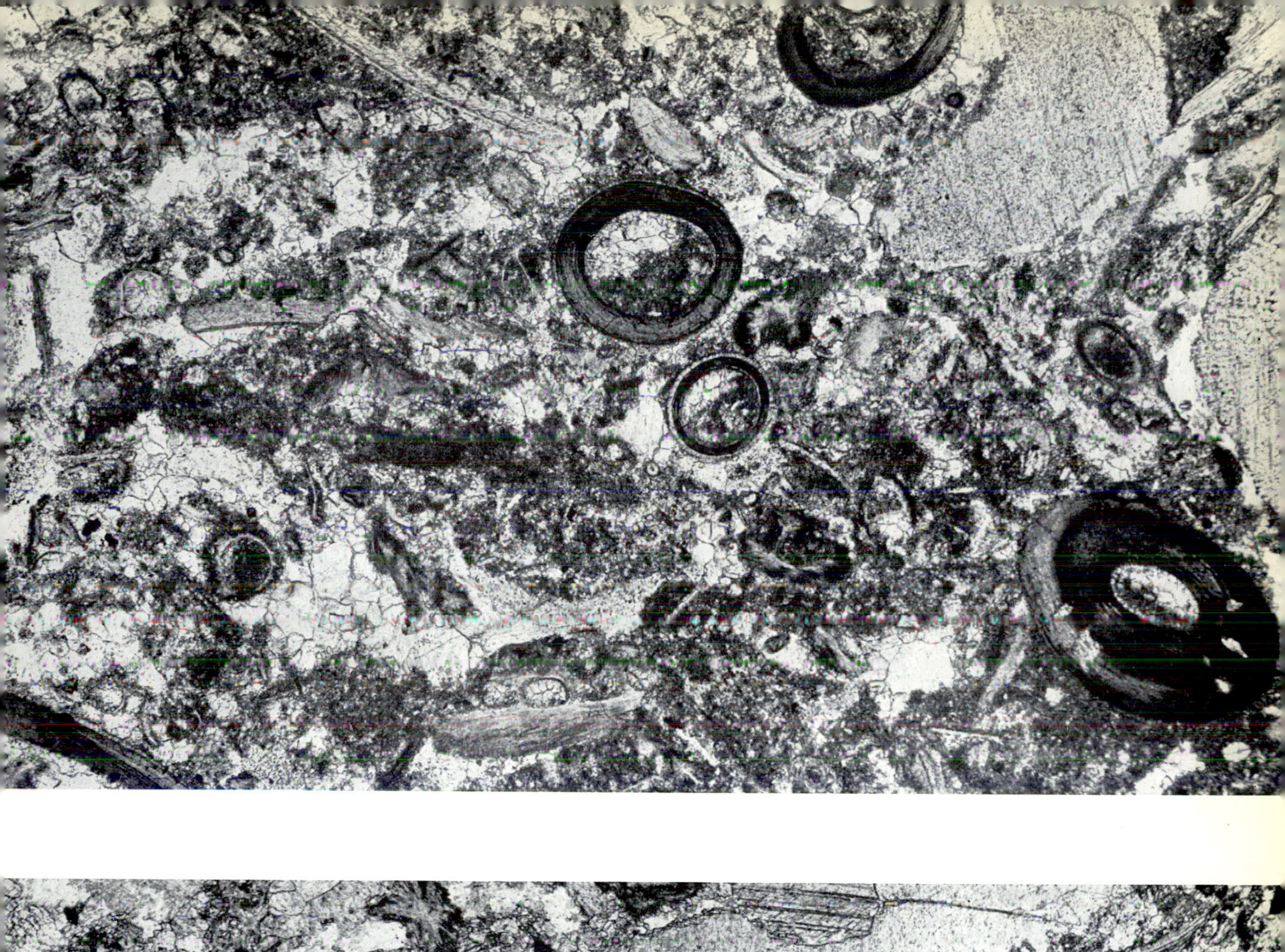

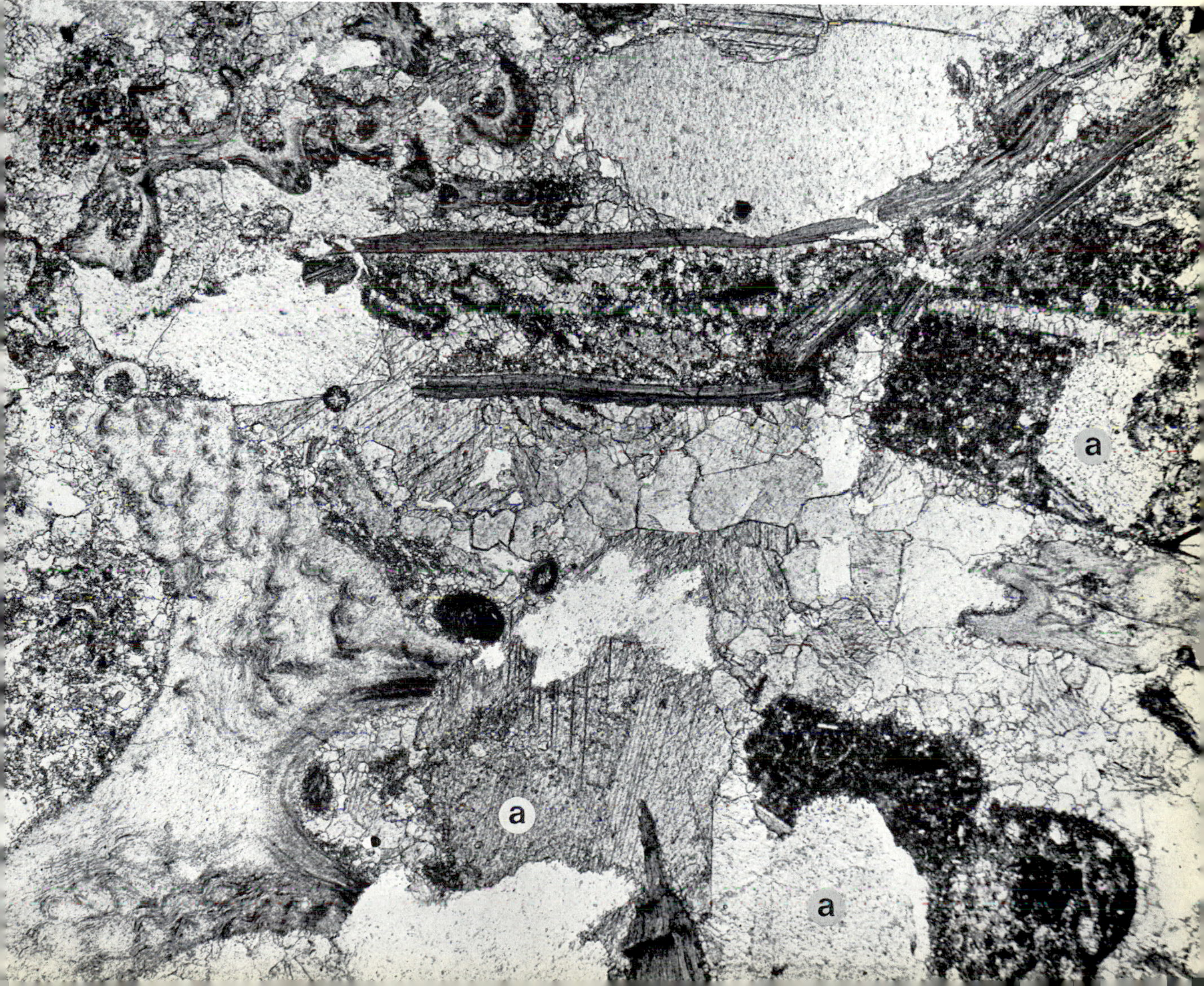
a
a
a

PLATE 32. SMALL CONOIDAL SHELLS OF UNCERTAIN AFFINITIES

1/10 mm
x 20
x 40
x 60
x 80
x100

Fig. 1. A coquina exhibiting longitudinal and transverse sections of *Salterella*, a hyolithid, possibly related to mollusks. Note solution packing of shells as indicated by stylolitic boundaries. Shell structure poorly exhibited possibly because of recrystallization. Buelna Formation, Lower Cambrian, Provcedora Hills, Caborca area, Sonora, Mexico. IU 8099–623, ×20.

Fig. 2. Longitudinal and transverse sections of tentaculitids and other unidentified debris. Note transverse ribbing in longitudinal sections. Micritic matrix. Manlius Limestone, Lower Devonian, Helderberg Mountains, 20 miles southwest of Albany, Rensselaer County, New York, U.S.A. IU 8099–837, ×20.

Fig. 3. Mostly transverse sections, some of which exhibit external ornamentation (transverse ribs). Note one shell within another; compare with fig. 6. Micrite matrix. As fig. 2.

Fig. 4. Note transverse ornamentation and faint laminated shell microstructure in tentaculitids. Shell at upper left (a) shows a few internal septa (transverse partitions). Micrite matrix. Hungry Hollow Formation, Middle Devonian, 2 miles east of Arkona, West Williams Township, Middlesex County, Ontario, Canada. IU 8099–834, ×20.

Fig. 5. Coquina of tentaculitids having poorly preserved shell microstructure. Genundewa Limestone, Upper Devonian, 18 Mile Creek, south of Buffalo, Erie County, New York, U.S.A. IU 8099–836P, ×20.

Fig. 6. Longitudinal and transverse cross sections of tentaculitids forming a coquina. Note faintly laminated wall and transverse ribbing in longitudinal sections. Longitudinal section shows three shells stacked one within the other. Micrite matrix. Geopetal fabrics are in different orientations and suggest some post-depositional movement of shells prior to lithification. As fig. 4.

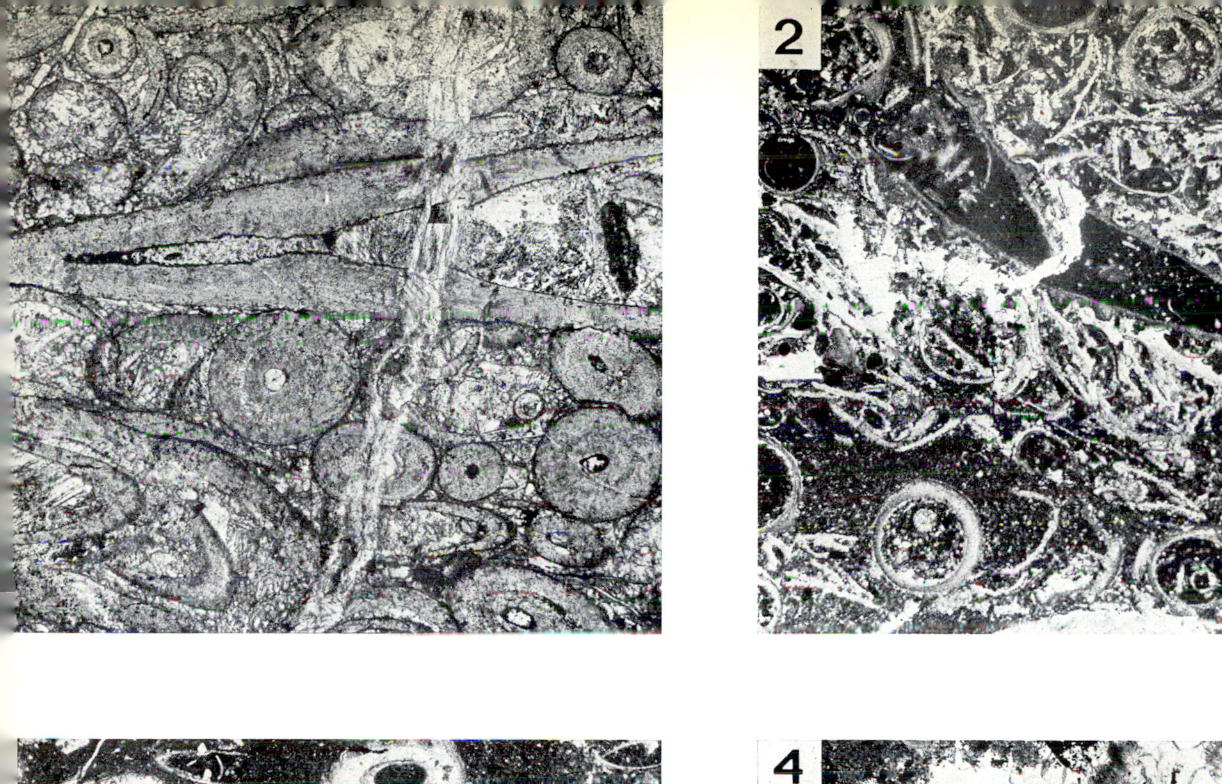
2
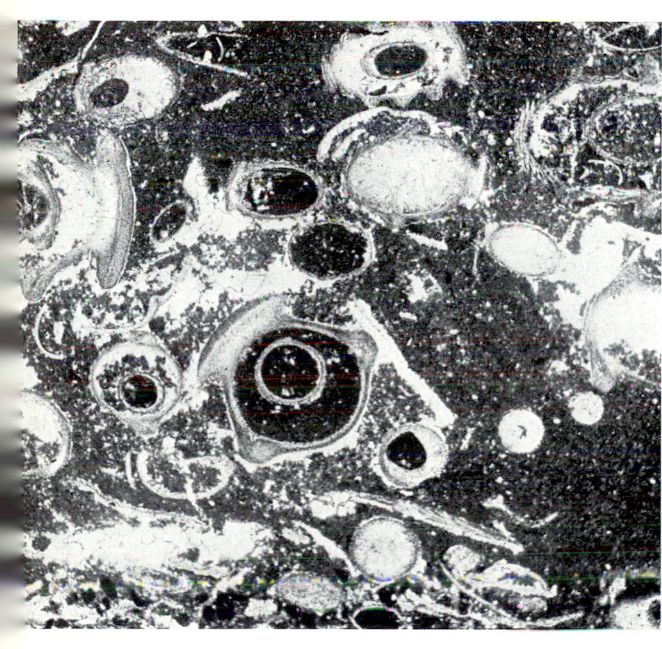
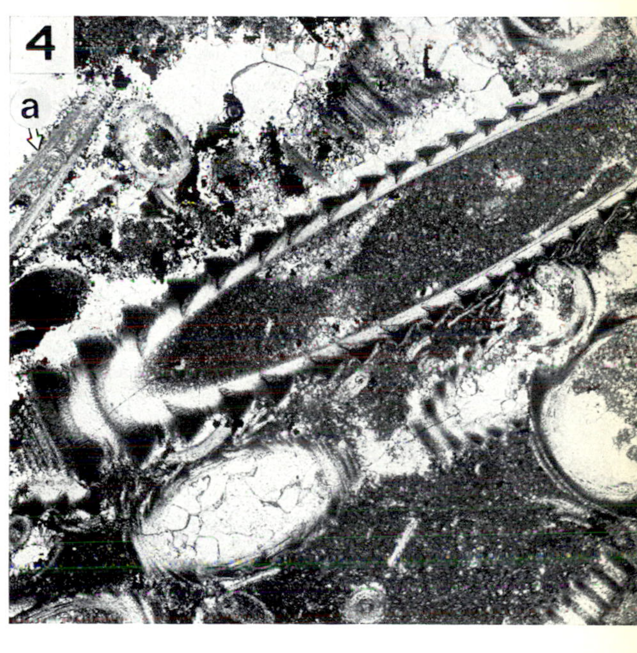
4
a
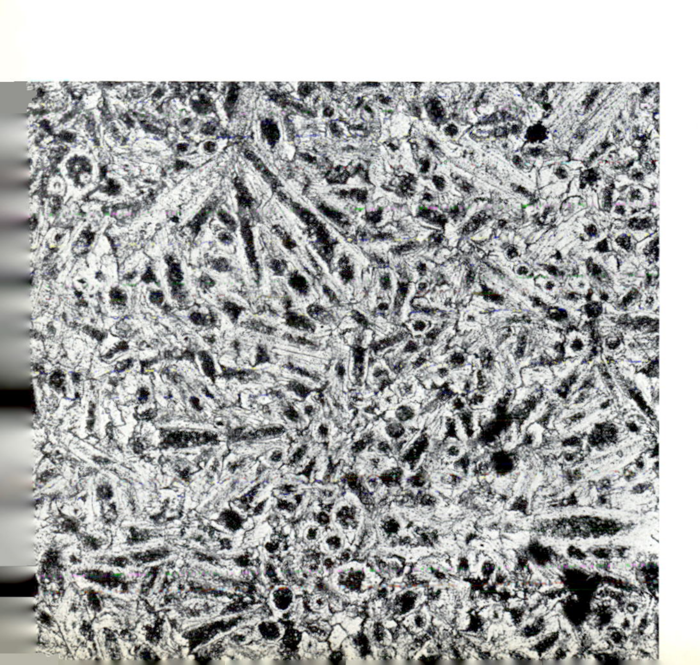
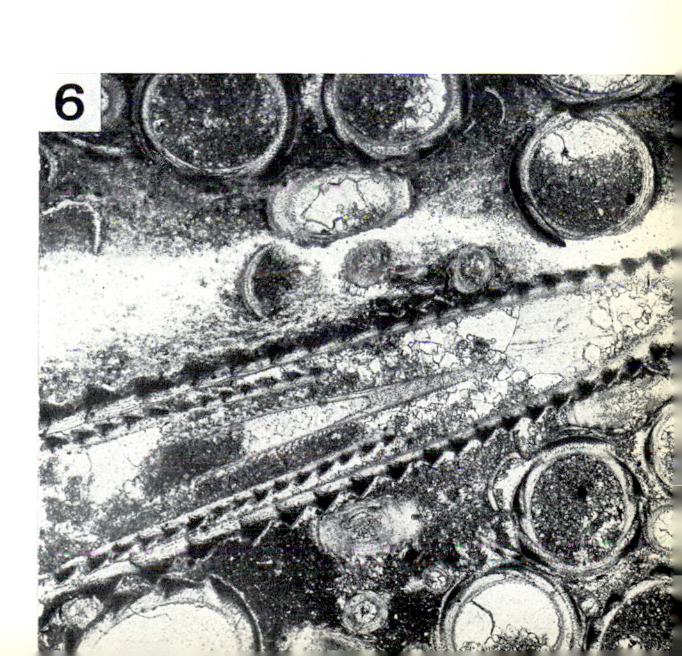
6

1/10 mm
x 20
x 40
x 60
x 80
x100

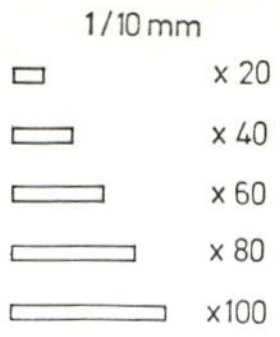

Fig. 1. Articulation of pelecypod valves at hingeline. Sparry calcite infilling becomes coarser away from shell boundaries. Silty calcareous muddy matrix. Fox Hills Formation, Upper Cretaceous, NW$\frac{1}{4}$ sec. 16, T. 12 N., R. 25 E., Armstrong County, South Dakota, U.S.A. IU 8099–721, ×10.

Fig. 2. Fragments of molluscan shells exhibiting crossed-lamellar microstructure under cross nicols. Silty calcareous muddy matrix. Mesozoic ?, Hsü Chia Ho (River), Szechuan Province, China. IU 8099–727, ×20.

Fig. 3. Cross section of single valve of pelecypod in muddy limestone containing quartzose silt. As fig. 1.

Fig. 4. Enlargement of fig. 6 showing shell microstructure, principally crossed-lamellar molluscan microstructure. As fig. 1, ×20.

Fig. 5. Cross section of pelecypod valves at mid-shell parallel to hingeline. Matrix is muddy quartzose siltsone. As fig. 1.

Fig. 6. Cross section of pelecypod valves at mid-shell perpendicular to hingeline. Note that valve shapes are mirror images of one another; this is generally not true of brachiopods (Pl. 29–2). See fig. 4 for enlargement. As fig. 1.

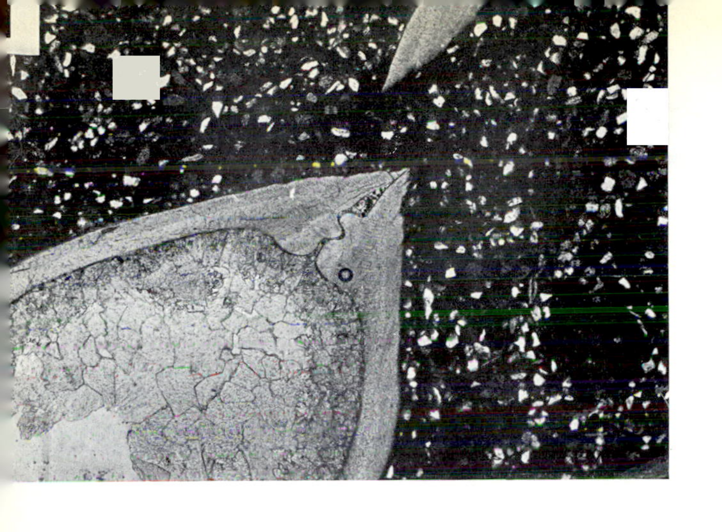

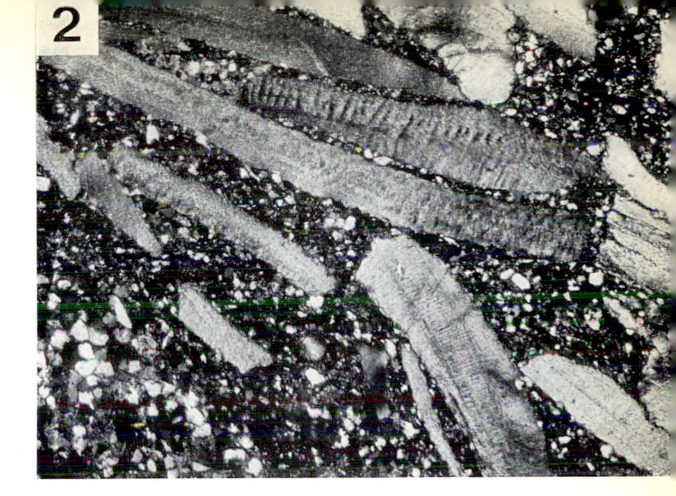
2

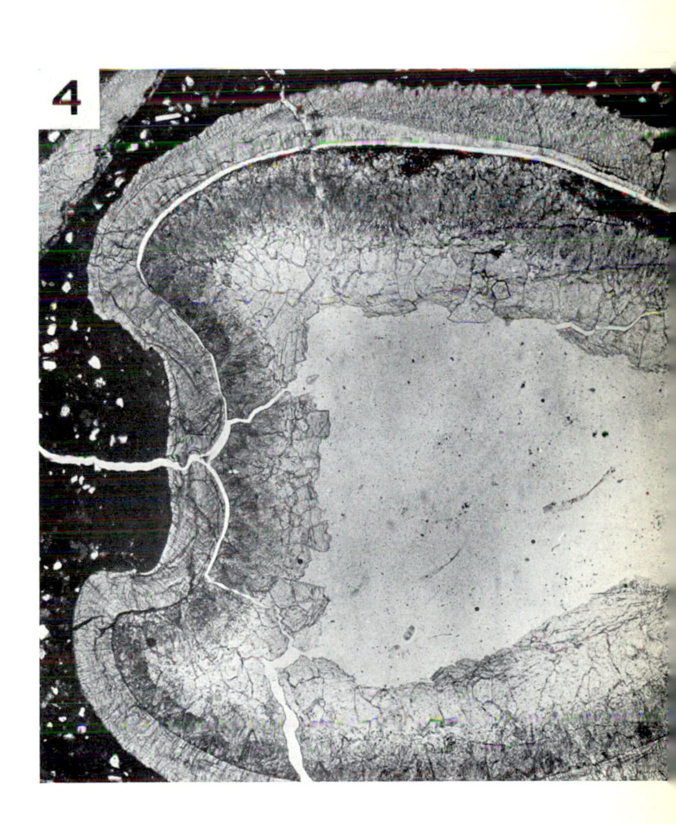
4

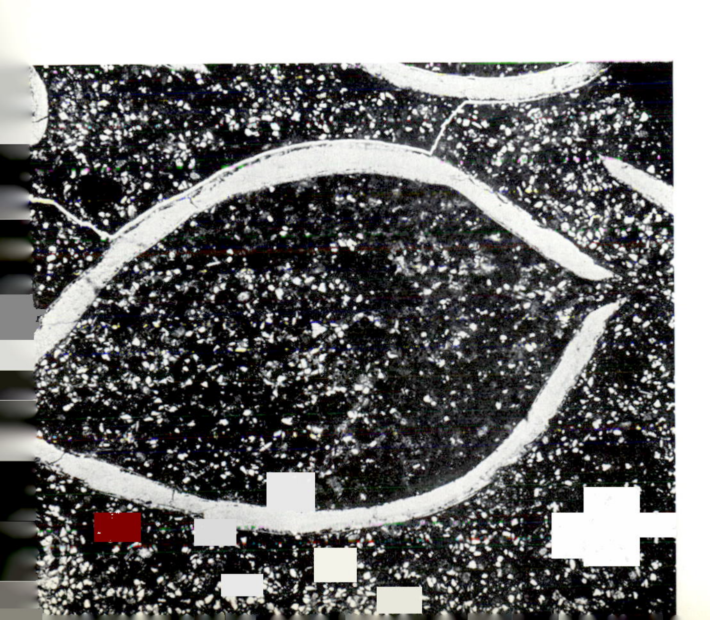

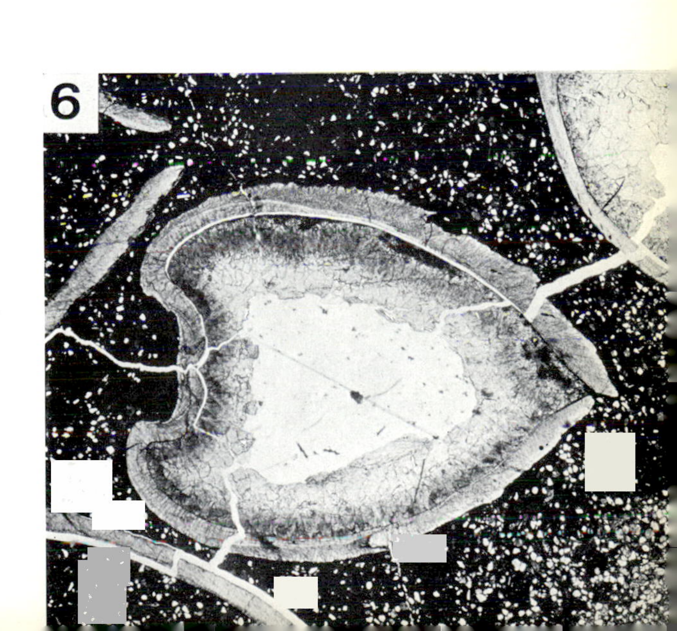
6

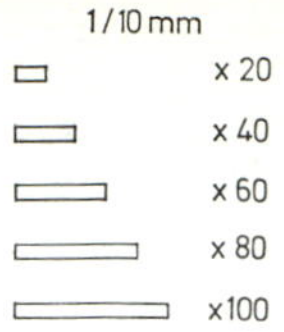

1/10 mm

x 20

x 40

x 60

x 80

x100

Inoceramus coquina exhibiting prismatic shell structure shown in plane and polarized light. Individual prisms act as single crystals; when disarticulated they can form a sand (Inoceramite, Pl. 43–4 and Pl. 52–1) in which each grain has unit extinction. In contrast to unit extinction in echinoderm plates, these prisms do not have a minutely porous structure. Fairport Chalk Member, Carlile Shale, Upper Cretaceous, SW$\frac{1}{4}$ SE$\frac{1}{4}$ sec. 10, T. 15 S., R. 20 W., Ellis County, Kansas, U.S.A. IU 8099–101, $\times 30$.

Fig. 1. *Inoceramus* prisms in section cut perpendicular to shell surface. Note branching of prisms in some shells. Plane light.

Fig. 2. As fig. 1, polarized light.

Fig. 3. *Inoceramus* prisms in section cut parallel to shell surface. Plane light.

Fig. 4. As fig. 3, polarized light.

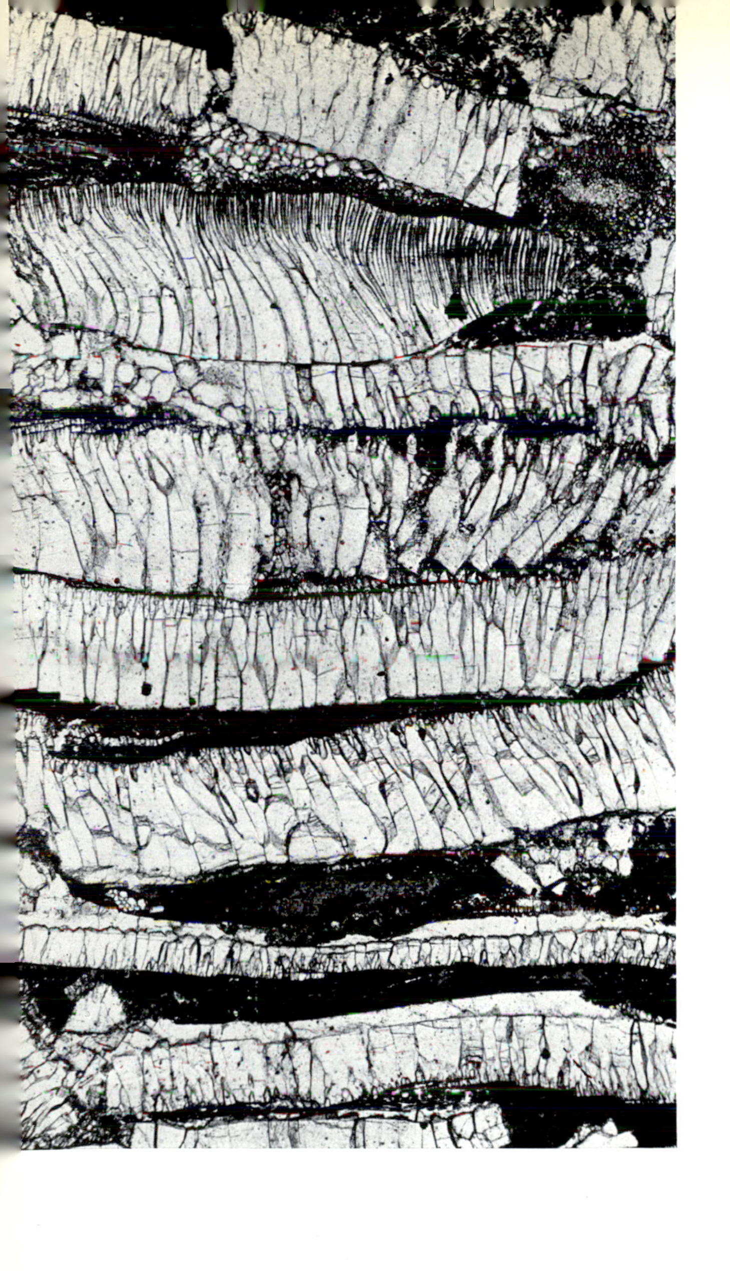

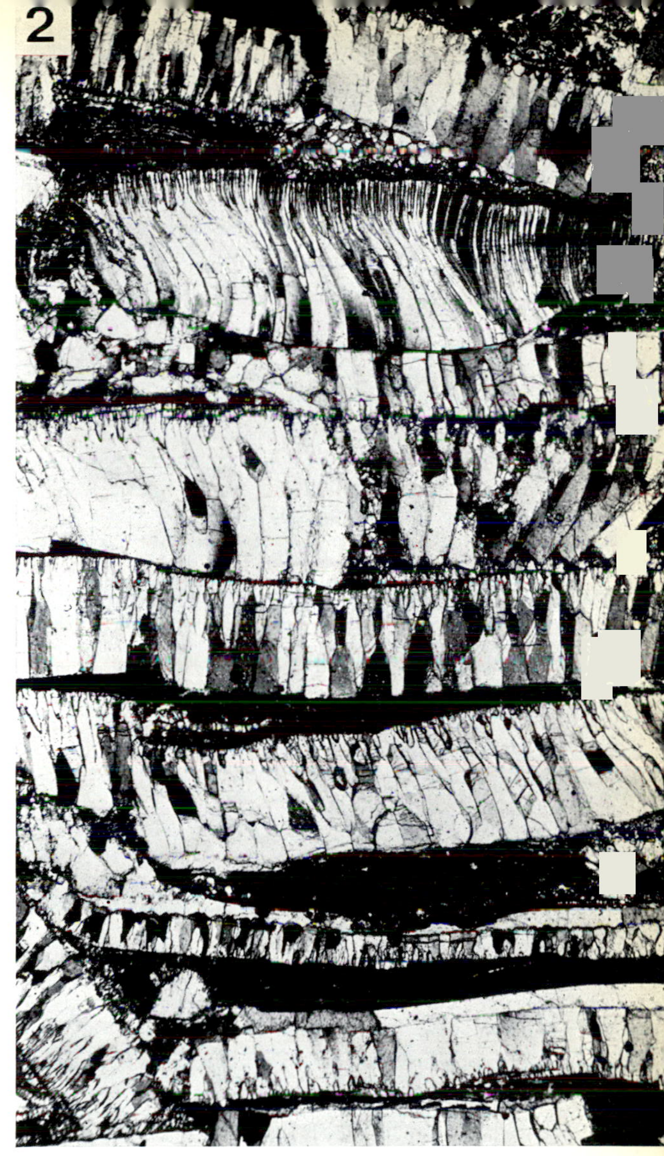
2

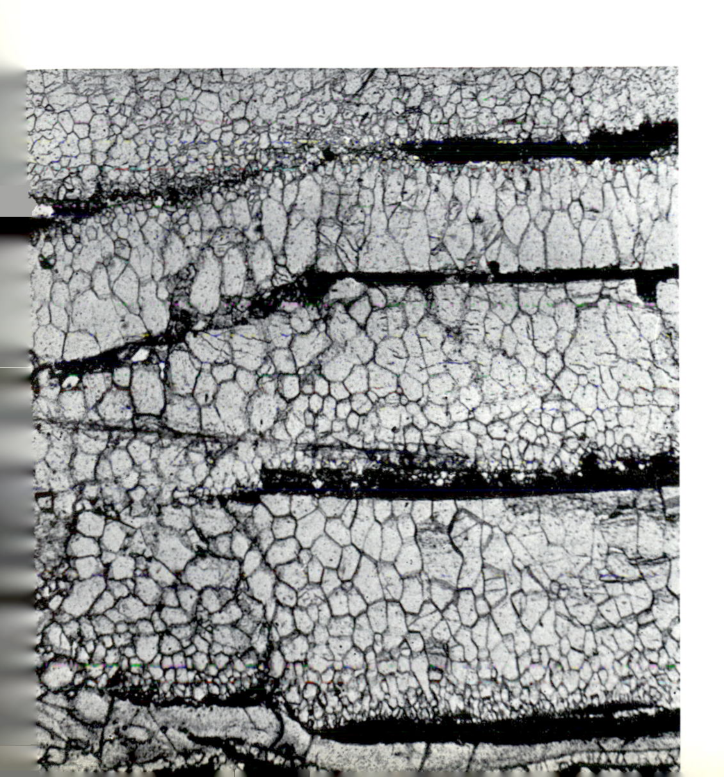

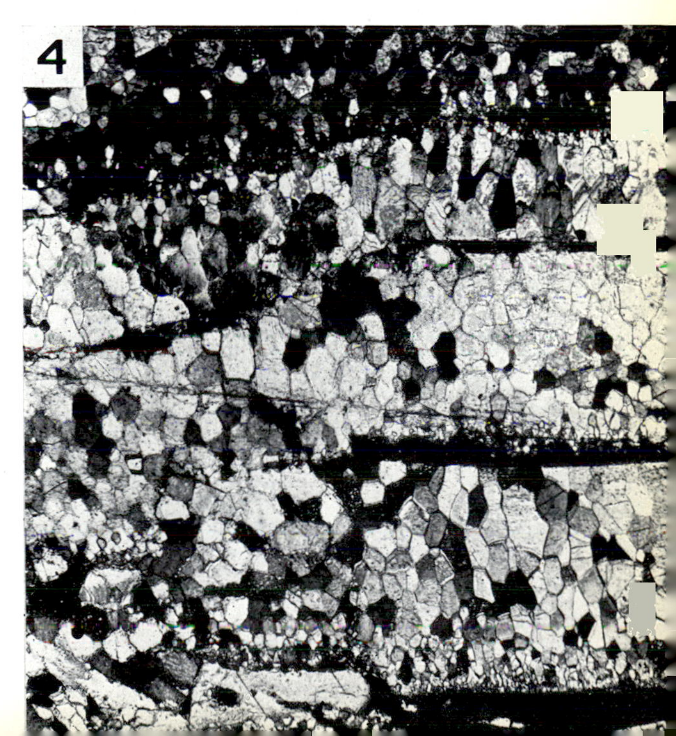
4

1/10 mm
x 20
x 40
x 60
x 80
x100

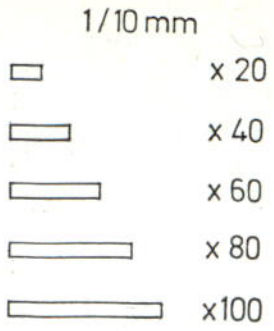

Fig. 1. Pycnodontid (pelecypod) skeletal microstructure. The "vesicular" tissue resembles cross sections of prisms (compare with Pl. 34–1), but the "vesicles" do not act optically as single crystals of calcite under cross polarizers. The pycnodontids are encrusting relatives of oysters. Denton Clay Member, Denison Formation, Lower Cretaceous, one mile south of north boundary of Cobb Park, Fort Worth, Tarrant County, Texas, U.S.A. IU 8099–102B, ×20.

Fig. 2. See fig. 1. Note large patch of finely fibrous or laminated skeletal microstructure. As fig. 1, 8099–102, ×20.

Fig. 3. See fig. 1. Vesicular tissue of pycnodontid lies between fibrous or laminated shell layers. Clear sparry calcite infilling. Fort Hays Limestone Member, Niobrara Formation, Upper Cretaceous, Kansas, U.S.A. IU 8099–745, ×20.

Fig. 4. Large pycnodontid pelecypod in matrix of recrystallized mud and other smaller pelecypod debris. Texture of pelecypod microstructural layers is similar to that of a bryozoan encrusting a brachiopod as illustrated on Pl. 25–3. Details of wall microstructure and internal structures in zooecial tubes differentiate bryozoans from pycnodontid shell structure. As fig. 2.

Fig. 5. Crossed nicols of prismatic layer of pelecypod shell. Individual prisms act as single crystals of calcite under cross polarizers. Rak Carbonate, Upper Cretaceous, Well D11–102, depth 8664 feet, Latitude 29° 07′ 40″ North, Longitude 21° 29′ 17″ East, Libya. IU 8099–153, ×40.

Fig. 6. Slightly oblique section through *Inoceramus* prismatic shell layer (bottom). Pycnodontid pelecypod encrusts prismatic layer (top). As fig. 3.

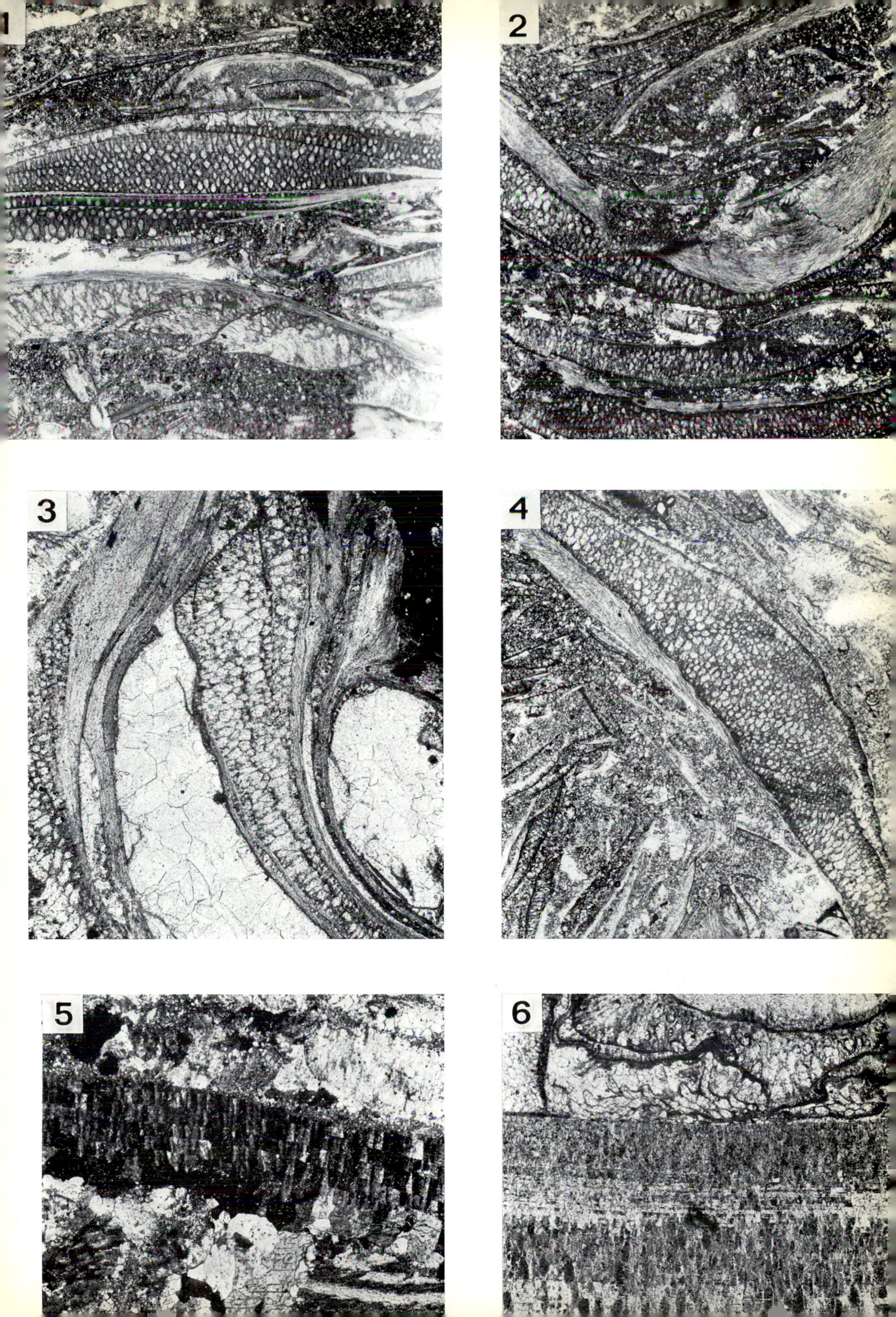
1
2
3
4
5
6

1/10 mm

x 20

x 40

x 60

x 80

x100

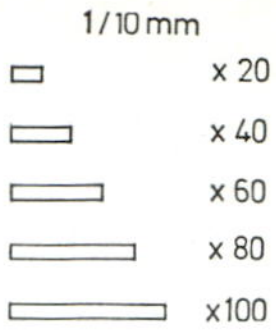

Fig. 1. Fragment of rudistid in echinodermal limestone. Original rudistid microstructure largely destroyed either by recrystallization or by partial leaching and infilling with dark mud (left center). Rectangular or rhombic grid is typical or thick rudistid shell layers. Rak Carbonate, Upper Cretaceous, Well D11–102, depth 8664 feet, Latitude 29° 7′ 40″ North, Longitude 21° 29′ 17″ East, Libya. IU 8099–153, ×20.

Fig. 2. Rudist shell structure partially recrystallized (? from aragonite) and partially leached and infilled by dark lime mud. However, shell wall still clearly displays rudistid microstructure (general rectangular grid). Ghosts of some original finely prismatic microstructure visible at lower left. Upper Cretaceous, road cut on north-facing slope of Rio Jueys Valley, 0.25 miles south of La Zanja trail intersection, Barrio Rio Jueys, Municipio de Salinas, Puerto Rico. IU 8099–759, ×20.

Fig. 3. Enlargement of rudist shell structure. Calcite within each unit or "prism" of rectangular or rhombic grid is polycrystalline possibly because of recrystallization. The "walls" of "prisms" or units appear finely fibrous perpendicular to "walls". Note minutely fibrous textures (relic ?) near dark coating of rudistid wall. Shivta Formation, Turonian, Upper Cretaceous, Negev, Israel. IU 8099–643, ×40.

1
2

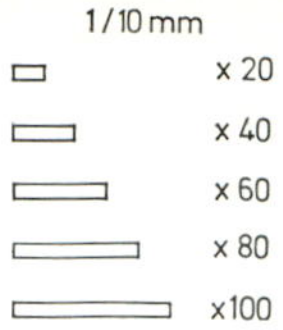

Fig. 1. Thin-walled rudistid microstructure. Note layering; rectangular grid poorly developed. Shell wall has been fractured and bored and later infilled by lime mud. Stylolitic solution boundary separates rudistid shell and mud matrix and molluscan debris at right. Melones Limestone, Upper Cretaceous, Dominican Republic. IU 8099–731, ×20.

Fig. 2. Prismatic pelecypod shell disaggregated at right into individual prisms. Caught in the act! Compare with Pl. 34. Berriedale Limestone, Lower Permian, Mount Nassau, near Hobart, Tasmania, Australia. IU 8099–893, ×20.

Fig. 3. Slender laths of typical molluscan crossed-lamellar structure. Infra-Trappan Dudukur Beds, Cretaceous, Dudukur, 12 miles west of Rajahmundry, Godavari District, Andhra Pradesh, India. IU 8099–927, ×40.

Fig. 4. Pelecypod having circular boring (a) infilled with the pelletal mud as in the matrix. Note at left that euhedral dolomite rhombs replace shell at boundary of shell and matrix. Sylhet Limestone Stage, Eocene, south of Cherrapunji, south of Shillong Plateau, United Khasi-Jaintia Hills District, Assam, India. IU 8099–929, ×40.

Fig. 5. Molluscan debris, poorly preserved in mud matrix, has faint prisms perpendicular to the shell wall. Shells are broken and shell walls poorly defined. Phosphatic Triassic conodont (?, a) in center. Lower Triassic, Bogdo Mountain, Baskunchak, Lower Povlozhe, U.S.S.R. IU 8099–682, ×40.

Fig. 6. Molluscan crossed-lamellar microstructure. Cross hatching that illustrates the opposite orientations of the second-order lamels or laths is rare in thin section (see MacKay, 1952, Pl. 1, fig. 3; Pl. 2., fig. 2). Crossed lamellar structure most commonly is represented by the edges of first-order lamels (laths). Oppositely oriented laths appear as dark and light stripes in thin section. See fig. 3. Pelletal mud and calcite cement matrix. Upper Lutetian, Middle Eocene Côte des Basques, south edge of Biarritz, Basses Pyrénées, France. IU 8099–14, ×20.

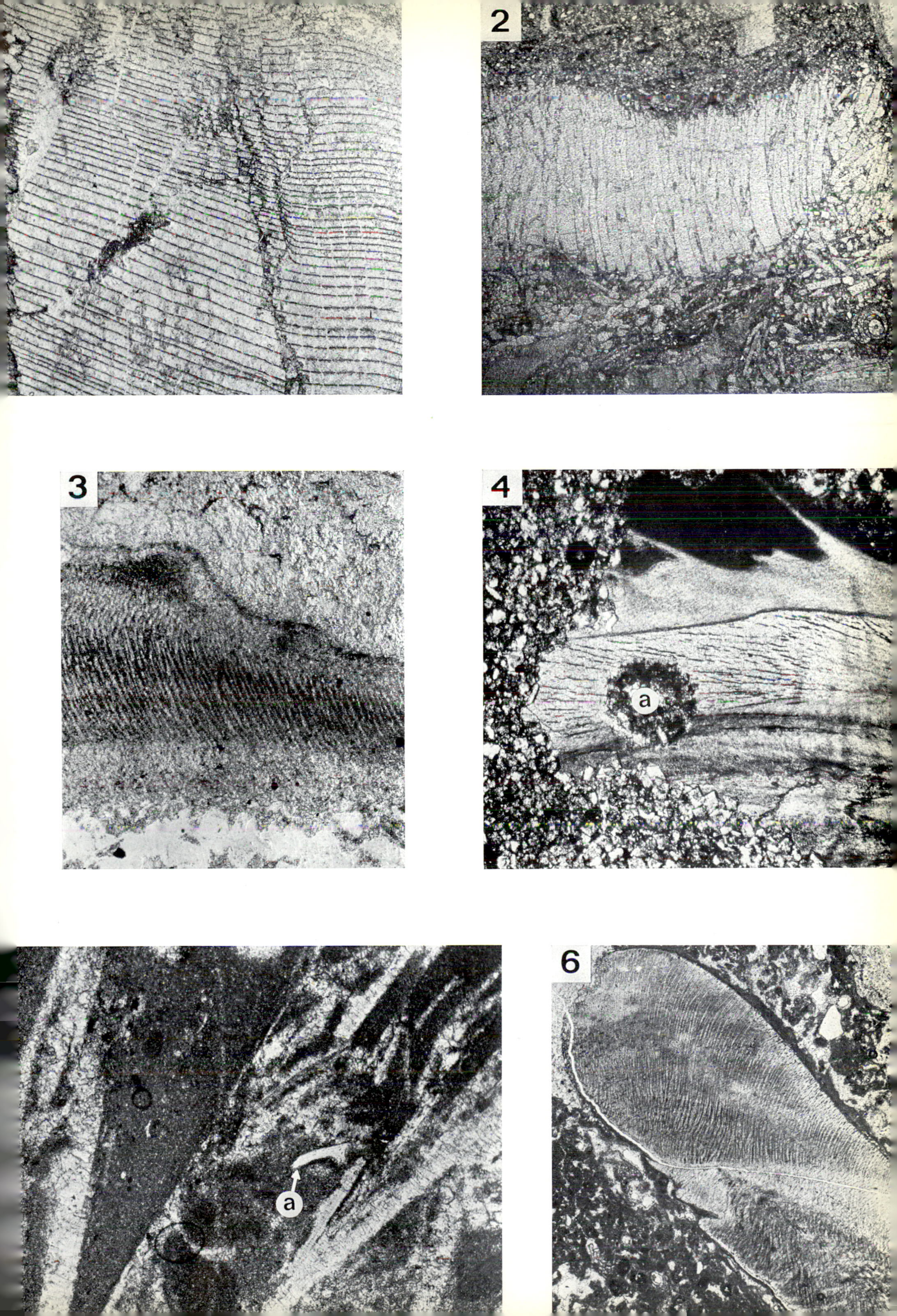
2
3
4
a
a
6

1/10 mm
x 20
x 40
x 60
x 80
x100

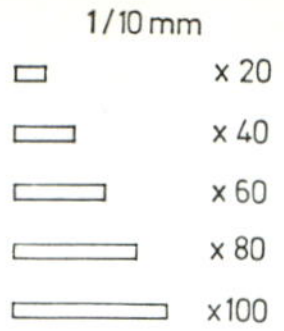

Fig. 1. Sections through fragments and large shells of pelecypods in mud matrix. Shell microstructure has thick fibrous layer and thin prismatic layer. Lack of prismatic layer on both sides of shell suggests that layer is secreted by organism and does not represent later incipient cementation. Upper Triassic, Trestenic, Dobrogea, Romania. IU 8099–380, ×30.

Fig. 2. Molluscan fragmental limestone. Microstructure partly altered but prismatic layer visible in some shell fragments. Note rudistid fragment at upper center. Matrix is mixture of pelletal mud and calcite cement. Melones Limestone, Upper Cretaceous, Dominican Republic. IU 8099–731, ×40.

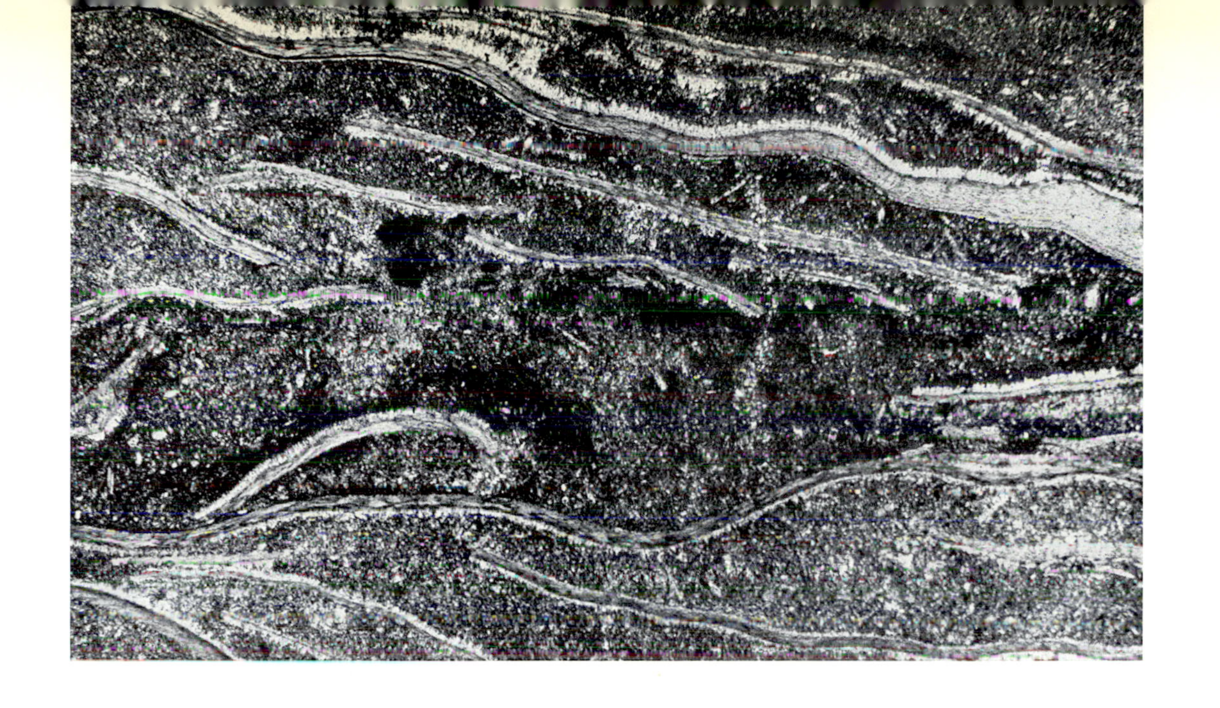

1/10 mm
x 20
x 40
x 60
x 80
x100

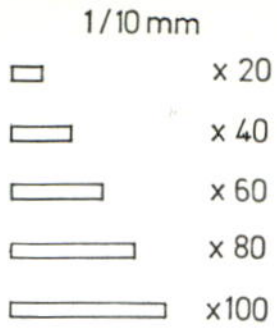

Fig. 1. Section perpendicular to axis of coiling of a gastropod. Fine lamellae are typical of molluscan crossed-lamellar structure. Dark mud matrix. Recent, Chireerete Island, Bikini Atoll, Marshall Islands, Pacific Ocean. IU 8099–476, ×20.

Fig. 2. Section perpendicular to axis of coiling of gastropod. External projections indicate ornamented shell. Shell microstructure destroyed and shell occupied by clear calcite cement. Other fragments of altered molluscan debris in dark pelletal (visible within gastropod) matrix. Cretaceous, Camposauro Mountain, 3 kilometers southwest of Vitulano, Benevento, Italy. IU 8099–376, ×20.

Fig. 3. Longitudinal section, parallel to axis of coiling of high-spired gastropod. Shell microstructure altered and shell infilled with dark pelletal mud. Inner surface of coil (a) has micritic rim. Cretaceous, 2 kilometers southwest of Baia e Latina, Caserta, Italy. IU 8099–368, ×20.

Fig. 4. Gastropod section in mud matrix. Original shell structure completely destroyed. Jaisalmer Limestone, Callovian-Oxfordian, Upper Jurassic, Jaisalmer, Rajasthan, India. IU 8099–928, ×20.

Fig. 5. Transverse section of a gastropod having darker infill than surrounding matrix which suggests redeposition from another environment. Note that dark infilling of shell enhances distinctive outline of coil. As fig. 4.

Fig. 6. Longitudinal section of high-spired gastropod. Interpretation as fig. 5. As fig. 4.

If either mud filling within coils or micritic rims were absent, it would be impossible to assign any sparry calcite area to gastropods in figs. 2 through 6 of this plate.

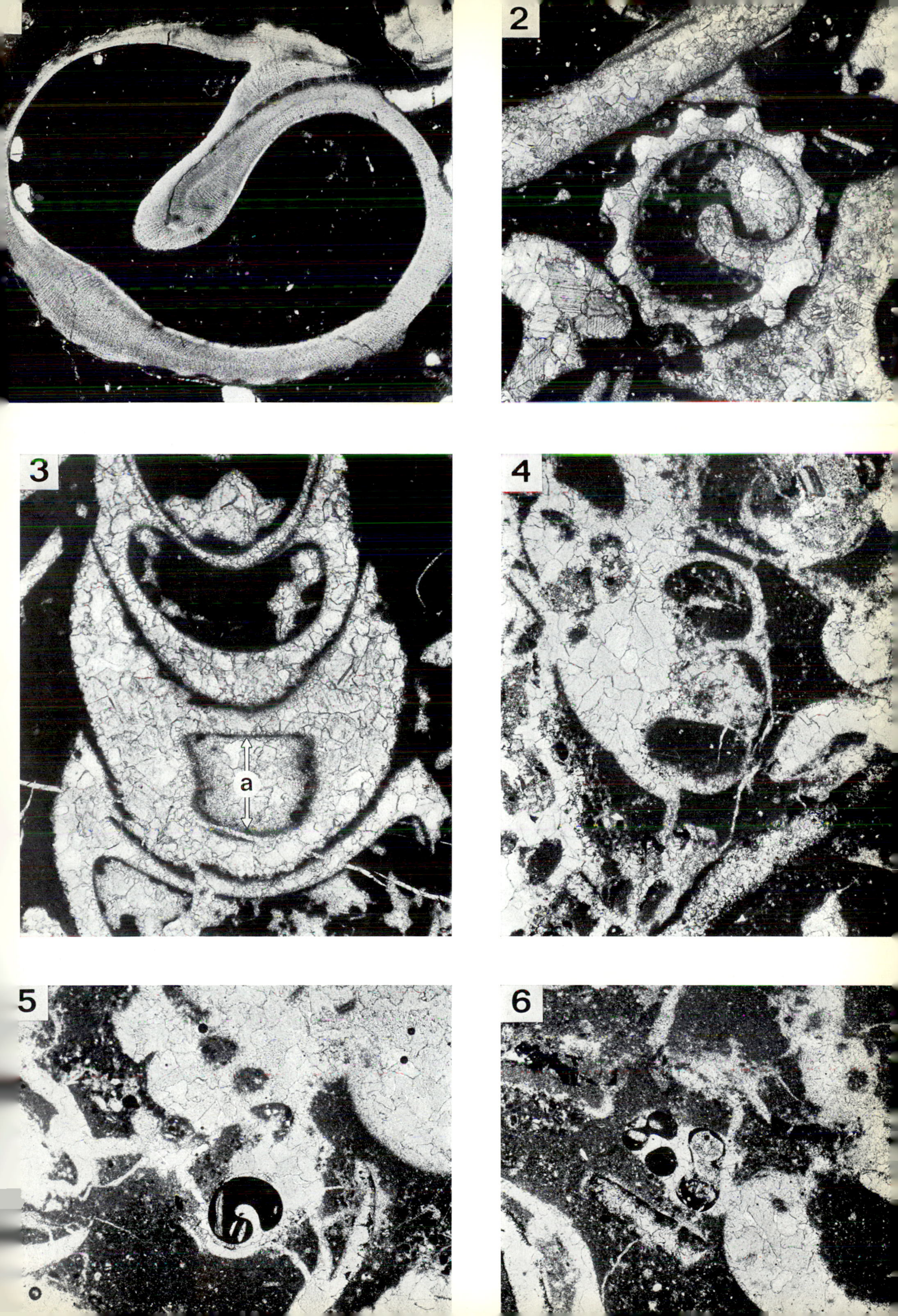
2
3
a
4
5
6

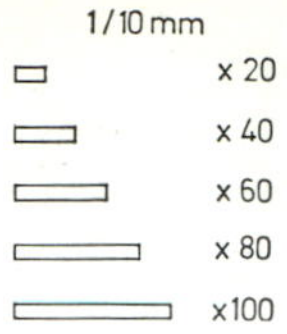

1/10 mm
x 20
x 40
x 60
x 80
x 100

Fig. 1. Recrystallized high-spired shell (? *Nerinella* sp.) exhibiting internal ribbing. Micrite matrix. Beersheva Formation, Upper Jurassic, Makhtesh Ramon, Israel. IU 8099–642, ×20.

Fig. 2. Well-washed oolite with molluscan debris. High-spired gastropods in longitudinal (a) and transverse (b) sections, recrystallized pelecypod debris (c), oolites (d) containing quartz centers (e), and sparry calcite cement (f). Molluscan debris exhibits micritic coats. Corallian, Oxfordian, Upper Jurassic, Osmington Mills, Dorset, England. IU 8099–57B, ×30.

Fig. 3. Extraordinary transverse section of an externally ornamented gastropod exhibits a portion of the ghosts of the originally fine crossed-lamellar structure in the presently coarsely recrystallized shell wall. Ghost structure indicates that shell was recrystallized rather than leached and infilled by calcite. Matrix consists of well-sorted angular silt-size quartz grains and carbonate cement. Upper Cretaceous, Unkwelane Hill, south bank of Umfolozi River, south of Mtubatuba, Zululand, Natal, Republic of South Africa. IU 8099–848, ×40.

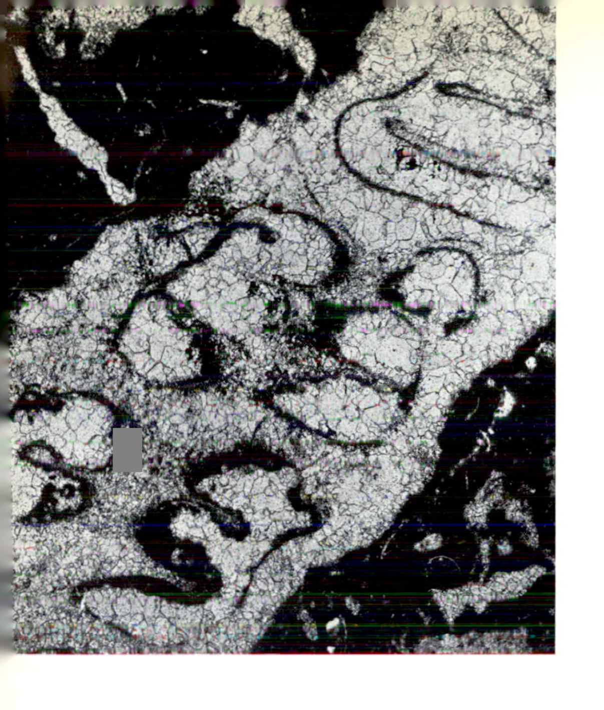

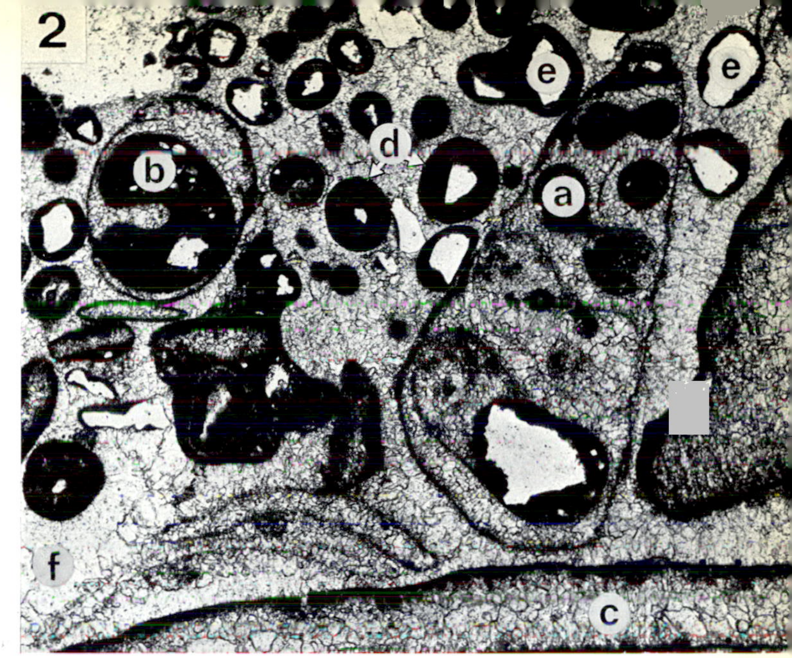
2
e
e
b
d
a
f
c

1/10 mm
x 20
x 40
x 60
x 80
x100

Fig. 1. Portion of gastropod shell showing crossed-lamellar microstructure. Enlargement of Pl. 43–4. Kiowa Shale, Lower Cretaceous, Champion Draw south of Belvidere, S$\frac{1}{2}$ sec. 9, T. 30 S., R. 16 W., Kiowa County, Kansas, U.S.A. IU 8099–110, ×40.

Fig. 2. High-spired gastropod filled with mud and quartz silt and redeposited in a well-washed limestone. Bethel Formation, Chester Series, Upper Mississippian, Hardin County, Kentucky, U.S.A. IU 8099–P211, ×40.

Fig. 3. Small medium-spired gastropod (left) and large gastropod (right) cut through body chamber (last shell whorl). Shells rather thin-walled; smaller one with ornamentation on last whorl. Shell microstructure destroyed. Matrix dark fine-grained silty micrite. Echinoderm plate at center within large gastropod. Upper Pequop Formation, Upper Permian, North Cherry Creek Mountains, sec. 12, T. 29 N., R. 63 E., Elko County, Nevada, U.S.A. IU 8099–570, ×40.

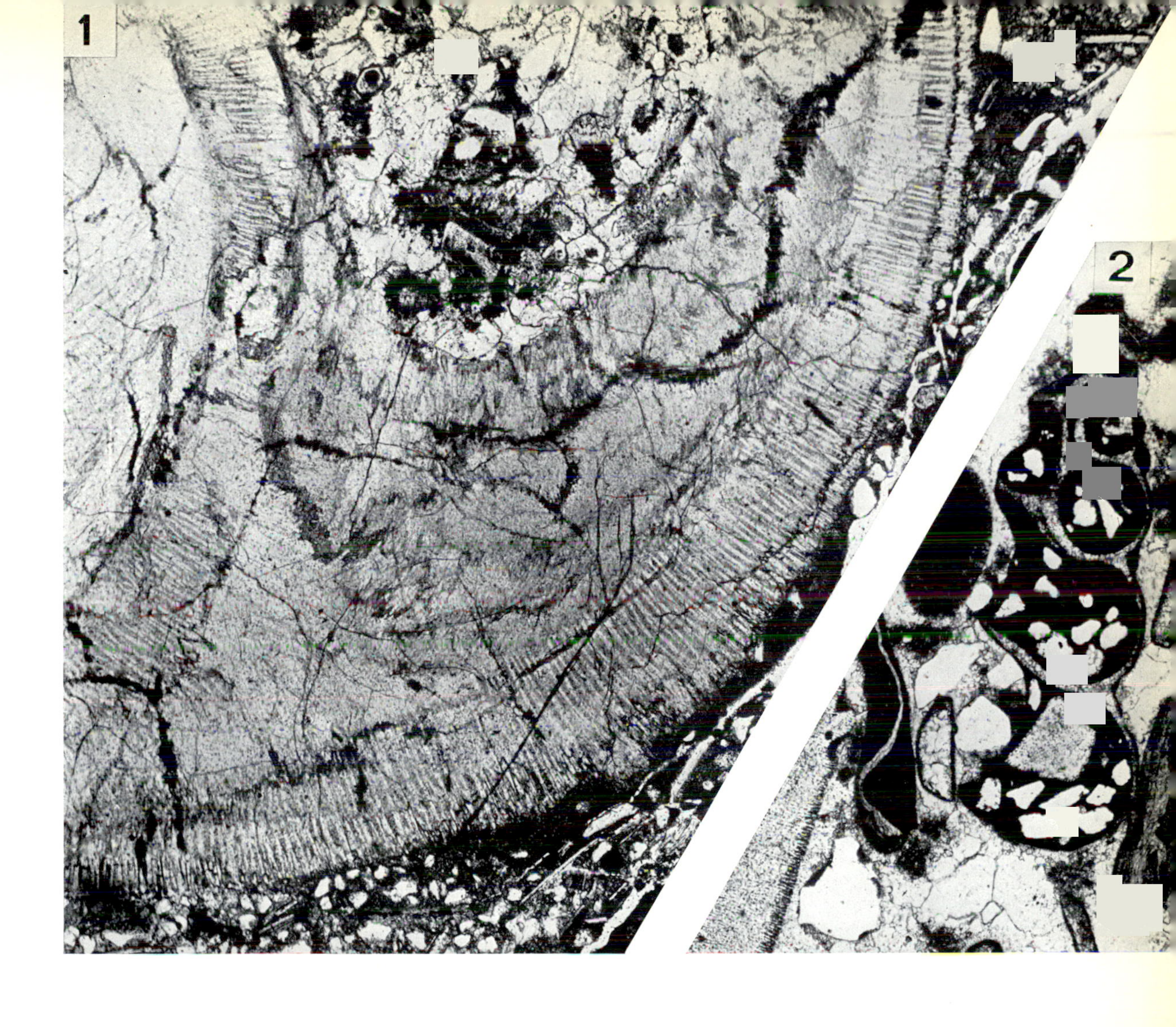
1
2

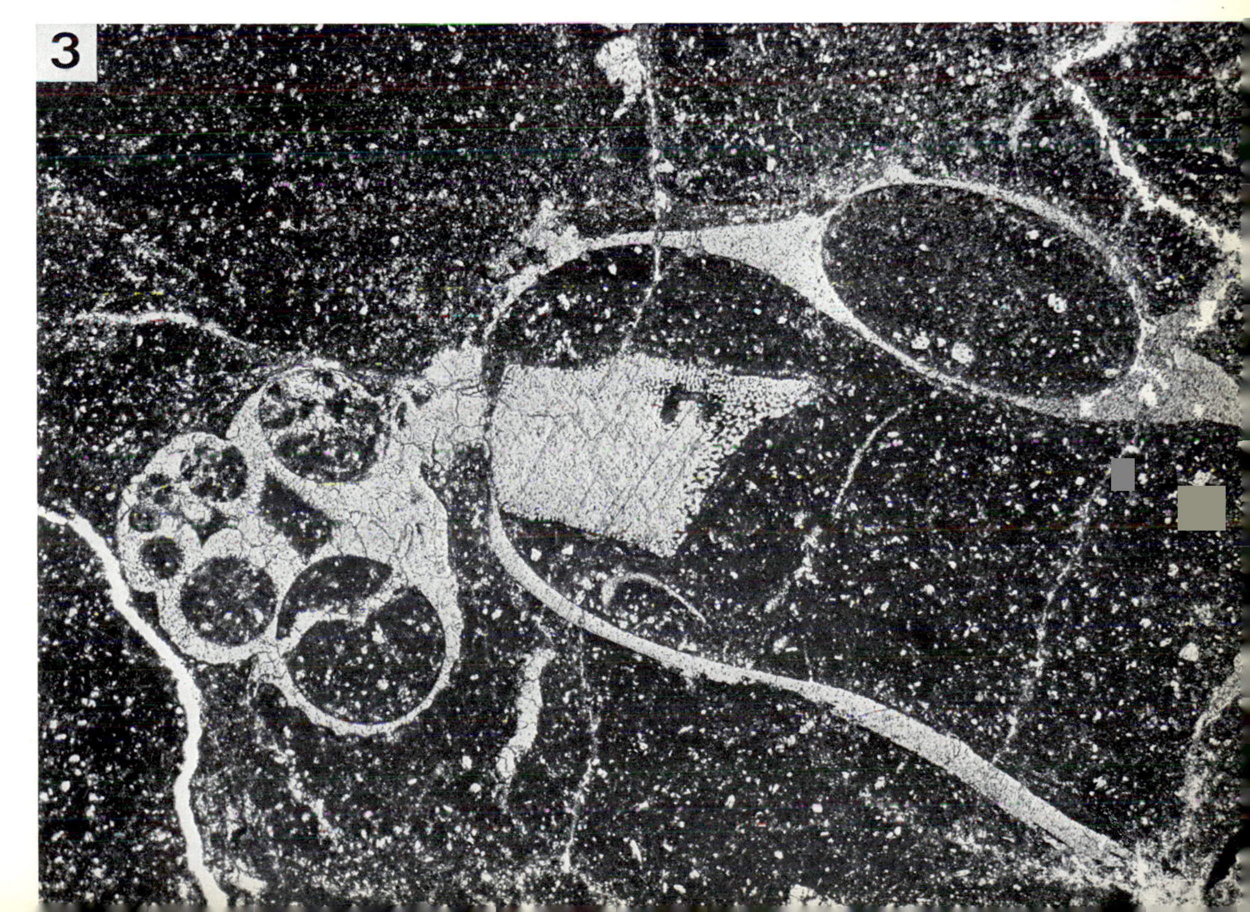
3

PLATE 42. GASTROPODS

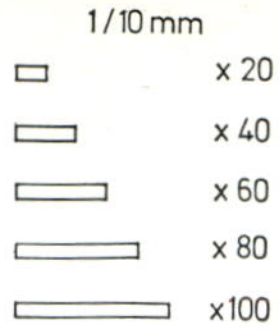
1/10 mm
x 20
x 40
x 60
x 80
x100

Prominent fragments of gastropod shells exhibiting typical crossed-lamellar microstructure best shown in the larger shell fragments. Matrix consists of quartz silt, calcite spar, and mud. Note shell ornamentation at upper right. Cretaceous, near Tiruchirapalli (Trichinopoly), Tiruchirapalli District, Madras, India. IU 8099–930, ×50.

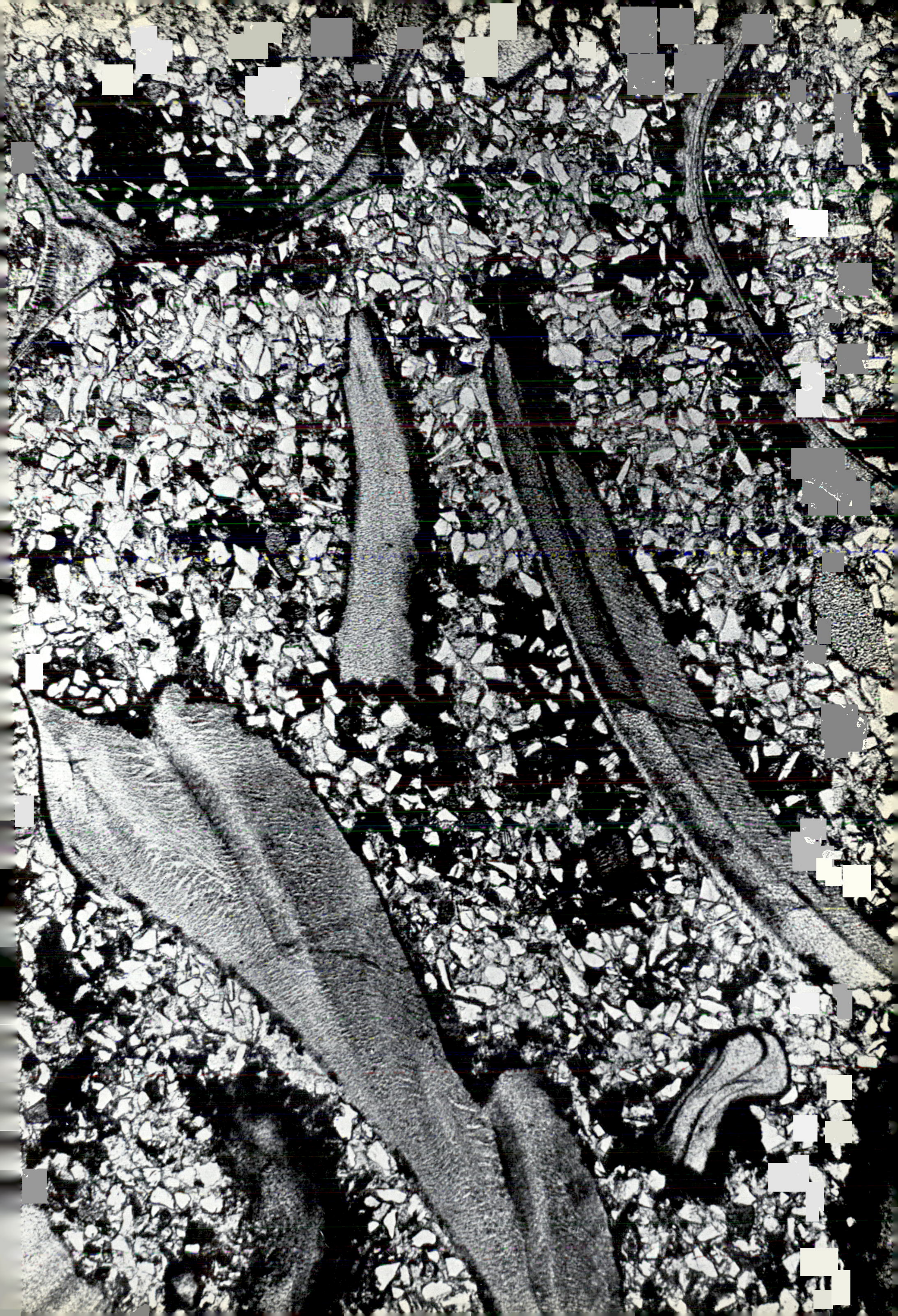

1/10 mm
x 20
x 40
x 60
x 80
x100

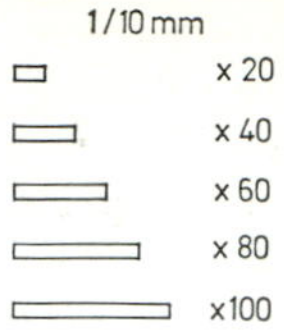

Fig. 1. Cross nicols. Cavities formed by leached molluscan debris and lined by calcite rim cement. Matrix is calcareous quartz silt. Pleistocene, near Villa Cisneros, Spanish Sahara. IU 8099–495B, ×20.

Fig. 2. Fossil debris with micritic envelopes. Note rim cement both on inside and outside of shell walls. Some fragments are only partially filled with calcite cement, and cement does not fill all the interstices between grains. No features are present to indicate that this is molluscan debris; similar preservation is observed among some Paleozoic leafy calcareous algae. In post-Paleozoic rocks one surmises that originally aragonitic molluscan debris was the common source of such altered and broken shell fragments. Examination of hand specimens frequently resolves such problems. Molluscan limestone, Eocene, Kamish-Burun, Kerchensk Region, Crimea, U.S.S.R. IU 8099–686, ×20.

Fig. 3. A singularly obscure picture illustrating micritic envelopes on shell surfaces and "ghost" remnants of former shell microstructure. Cretaceous, Luanda, Portuguese Angola. IU 8099–693, ×20.

Fig. 4. Transverse section of gastropod showing microstructural layers in shell resulting from different orientations of the structural units (fine laths) of crossed-lamellar shell microstructure. See Pl. 41–1 for enlargement of shell microstructure. Gastropod is infilled with well-sorted close-packed *Inoceramus* prisms and some quartz silt and phosphatic fish plates. Kiowa Shale, Lower Cretaceous, Champion Draw south of Belvidere, S$\frac{1}{2}$ sec. 9, T. 30 S., R. 16 W., Kiowa County, Kansas, U.S.A. IU 8099–110, ×20.

Fig. 5. Longitudinal section of high-spired gastropod in calcareous siltstone. Thin-shelled pelecypod at right. Upper Cretaceous, Unkwelane Hill, south bank of Umfolozi River, south of Mtubatuba, Zululand, Natal, Republic of South Africa. IU 8099–848, ×20.

Fig. 6. Longitudinal section of high-spired gastropod in calcareous siltstone. As fig. 5.

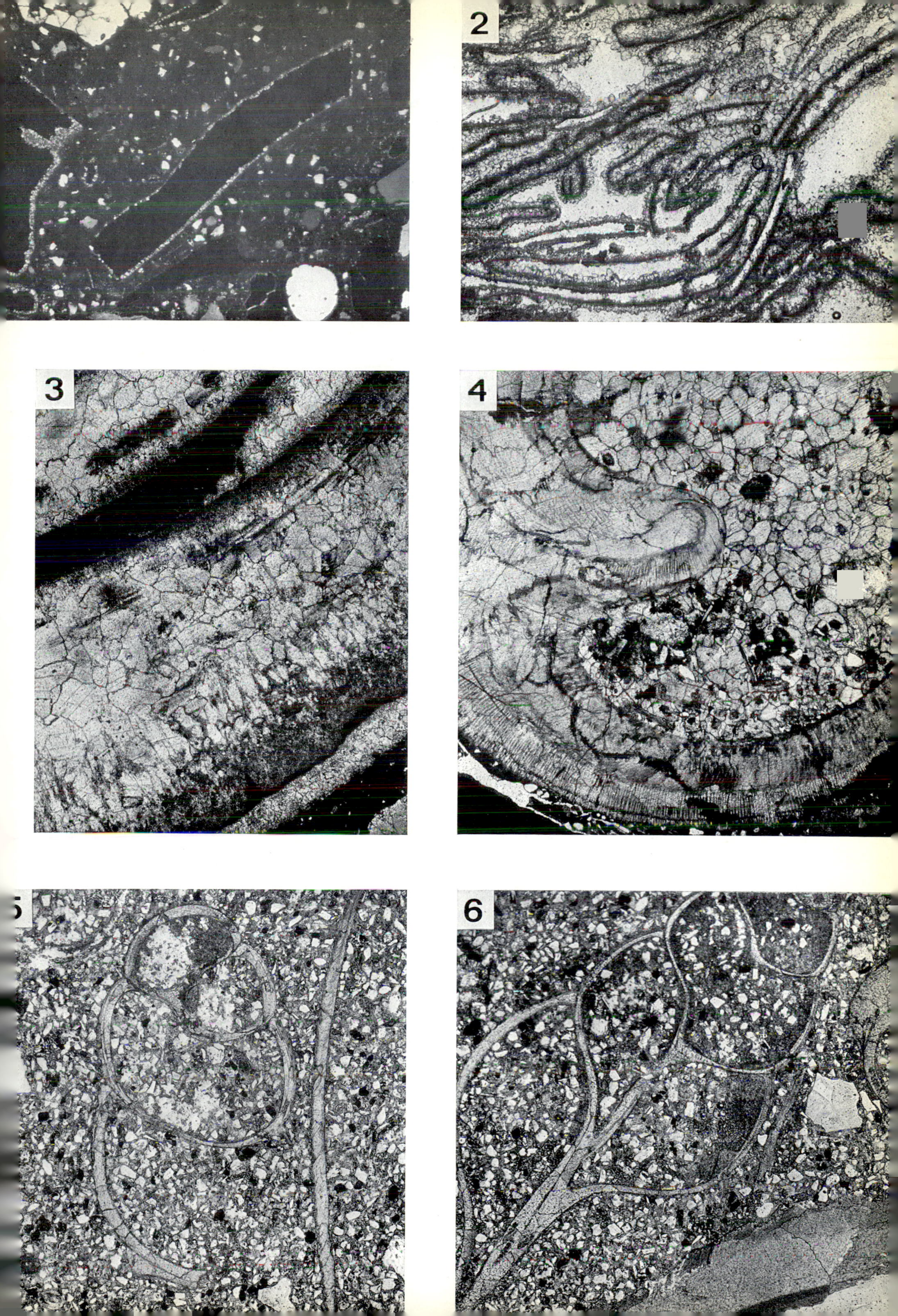
2
3
4
6

PLATE 44. CEPHALOPODS

1/10 mm
x 20
x 40
x 60
x 80
x100

Cross section of planispirally coiled cephalopod. Shell wall has been altered (recrystallized or replaced). Septa not evident in figure. Small specimen at lower left (see Pl. 45–3, 4 for other views). Matrix principally mud. Thin-walled ostracodes and broken cephalopod fragments are common in matrix. Planispirally coiled gastropods usually do not exhibit so uniform a depression on both sides of the axis of coiling. The depression (umbilicus) is commonly on one side only in gastropods. Middle Triassic, Prida Formation, Fossil Hill, secs. 19 and 30, T. 28 N., R. 35 E., Pershing County, Nevada, U.S.A. IU 8099–708, ×50.

1/10 mm
x 20
x 40
x 60
x 80
x100

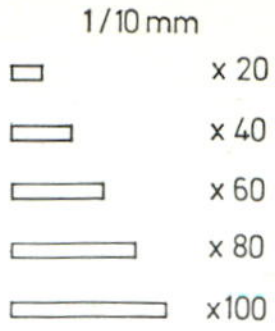

Fig. 1. Straight (orthoconic) cephalopod broken in middle but showing internal septa and thin altered shell walls. Matrix of dark carbonate mud partially fills cephalopod. Salem Limestone, Middle Mississippian, SW$\frac{1}{4}$ NW$\frac{1}{4}$ sec. 26, T. 5 N., R. 1 W., Bedford, Lawrence County, Indiana, U.S.A. IU 8099–426, ×20.

Fig. 2. Brachiopod fragments in a carbonate cement. Serrated fragment from ornamented shell. Compare with similar ornamentation on molluscan shell, Pl. 46–1. Mud attached to shell suggests it was transported from original muddy environment to final resting place. Salem Limestone, Middle Mississippian, active quarry, NW$\frac{1}{4}$ SW$\frac{1}{4}$ sec. 34, T. 12 N., R. 2 W., Morgan County, Indiana, U.S.A. IU 8099–81, ×20.

Fig. 3. Transverse section of a planispirally coiled cephalopod. Shell partially crushed and infilled at right by pelletal mud. Portions of much larger cephalopod shells at left show altered wall microstructure and chambers filled principally by calcite cement. Prida Formation, Middle Triassic, secs. 19 and 30, T. 28 N., R. 35 E., Pershing County, Fossil Hill, Nevada, U.S.A. IU 8099–708, ×40.

Fig. 4. Similar to previous figure but another larger cephalopod exhibiting thin internal septa (a). As fig. 3.

Fig. 5. Section through tightly coiled (involute) cephalopod showing cross sections of interior whorls. Wall microstructure altered and exhibits no structural detail. Shell infilled with sparry calcite cement. Note increase in calcite crystal size away from walls. As fig. 1.

Fig. 6. Typical appearance of a gastropod section perpendicular to axis of coiling. Matrix of quartz sand or silt and fine skeletal debris. Shell microstructure partially altered but appears finely prismatic in places, which may represent orientation of original crossed-lamellar laths. Kiowa Shale, Lower Cretaceous, Champion Draw south of Belvidere, S$\frac{1}{2}$ sec. 9, T. 30 S., R. 16 W., Kiowa County, Kansas, U.S.A. IU 8099–110, ×20.

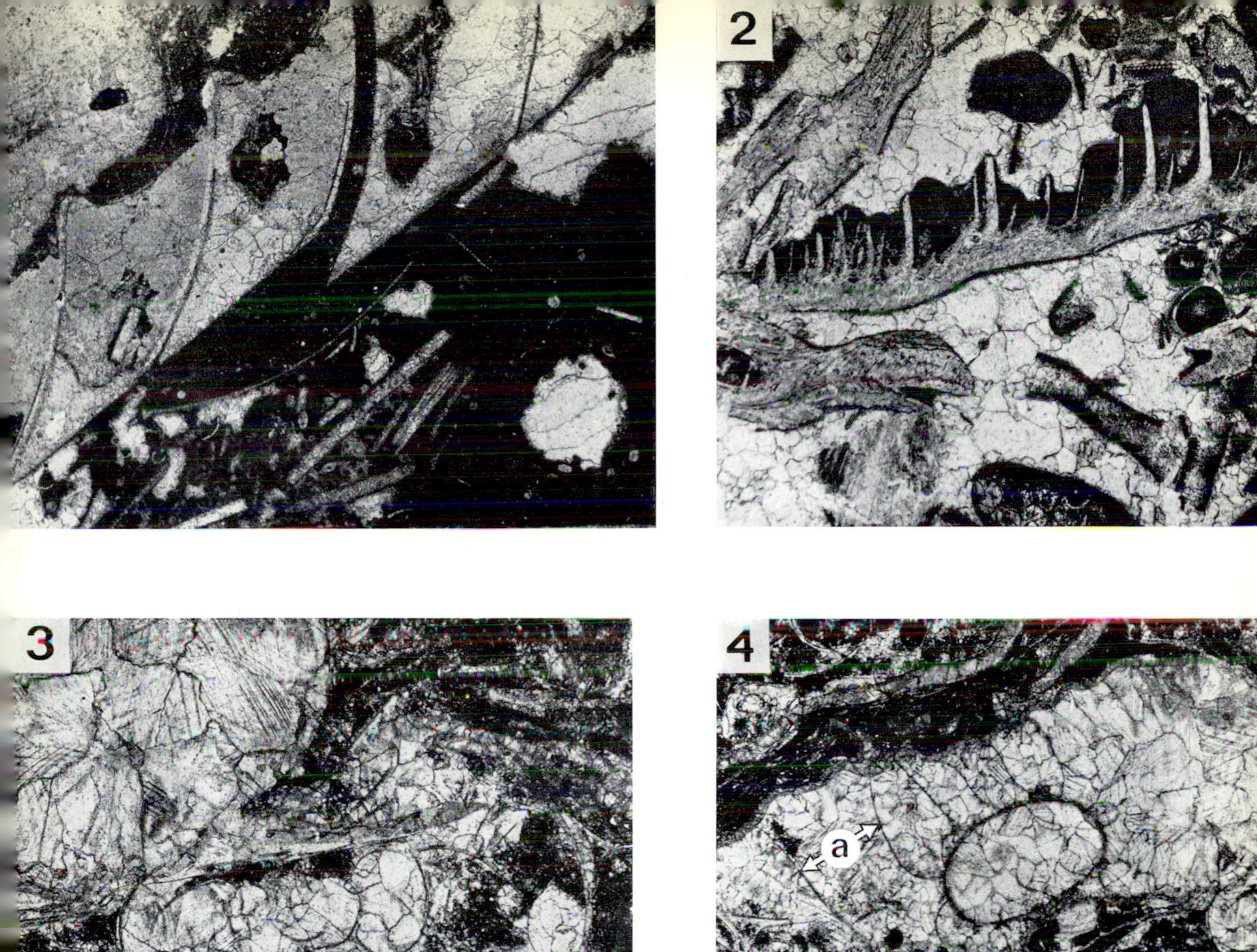
2

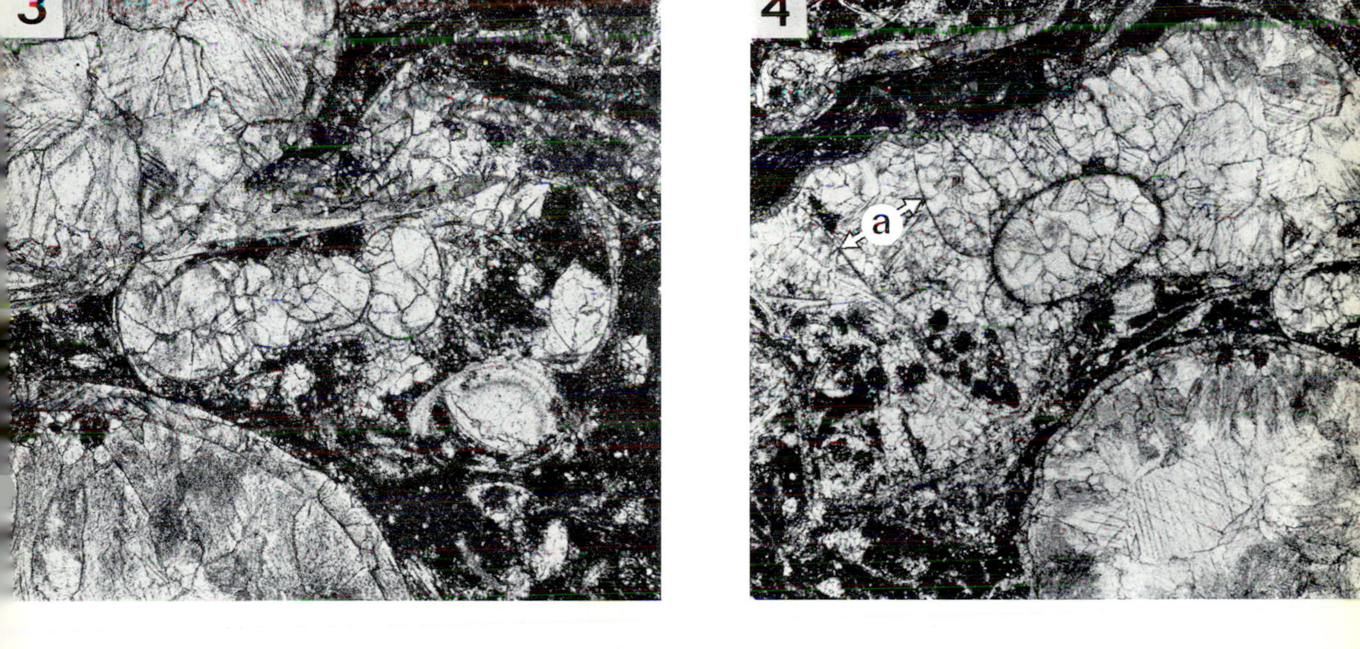
3
4
a

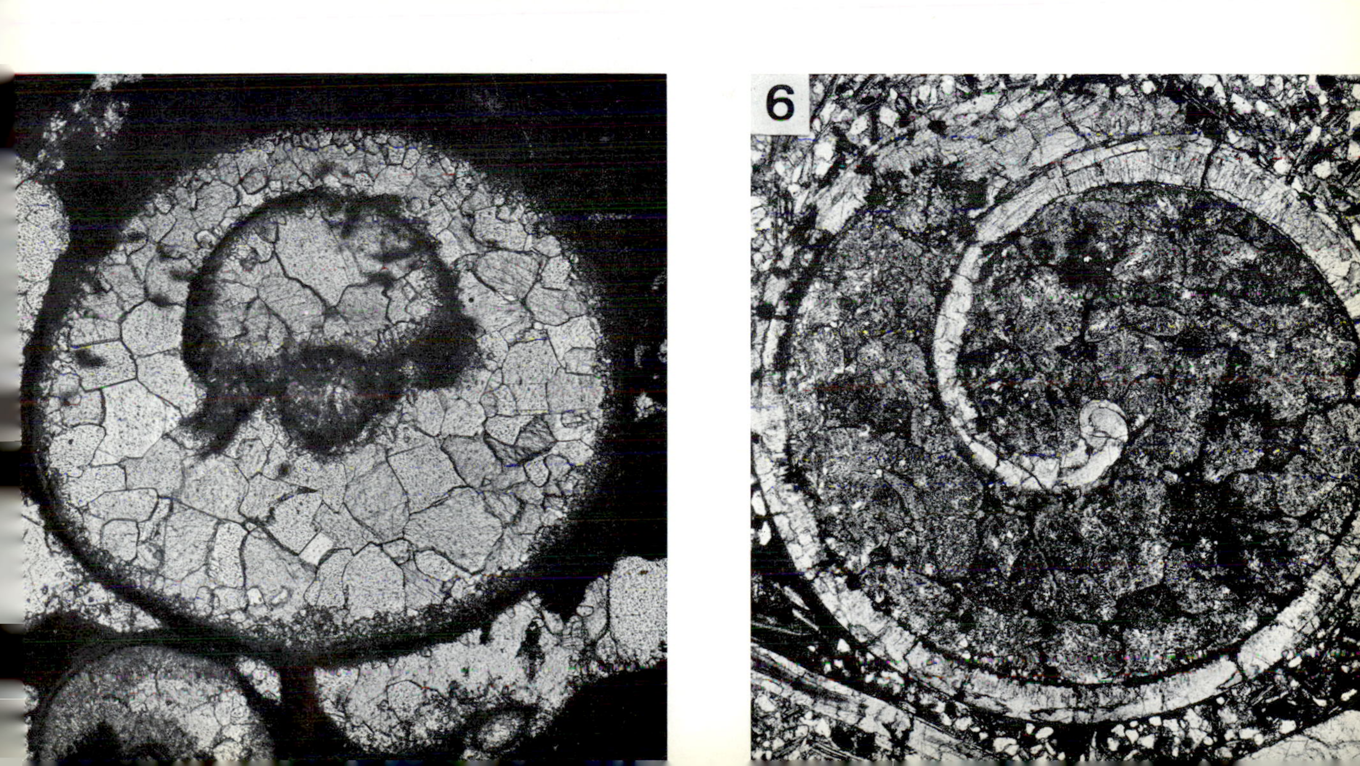
6

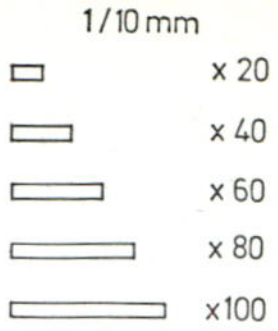

1/10 mm

x 20

x 40

x 60

x 80

x100

Fig. 1. Pelecypod shell showing ornamented outer (upper) surface. Compare with Pl. 45–2. Single ostracode valve at upper left. Dark mud matrix. Miocene-Pliocene ?, from Nahlozi, Hells Gates, Lake St. Lucia, Zululand, Natal, Republic of South Africa. IU 8099–843, ×40.

Fig. 2. Transverse section of belemnite. Note prominent growth bands, probably annual. Lower Jurassic, Comana Valley, Persani Mountains, Romania. IU 8099–389, ×10.

Fig. 3. Transverse section of belemnite exhibiting no growth bands. Jurassic, Poliwna, on the Volga River, U.S.S.R. IU 8099–735, ×20.

Fig. 4. As fig. 2 but under crossed polarizers. Note large radially oriented prisms of calcite.

Fig. 5. As fig. 3 but under crossed polarizers. Note large radially oriented prisms of calcite.

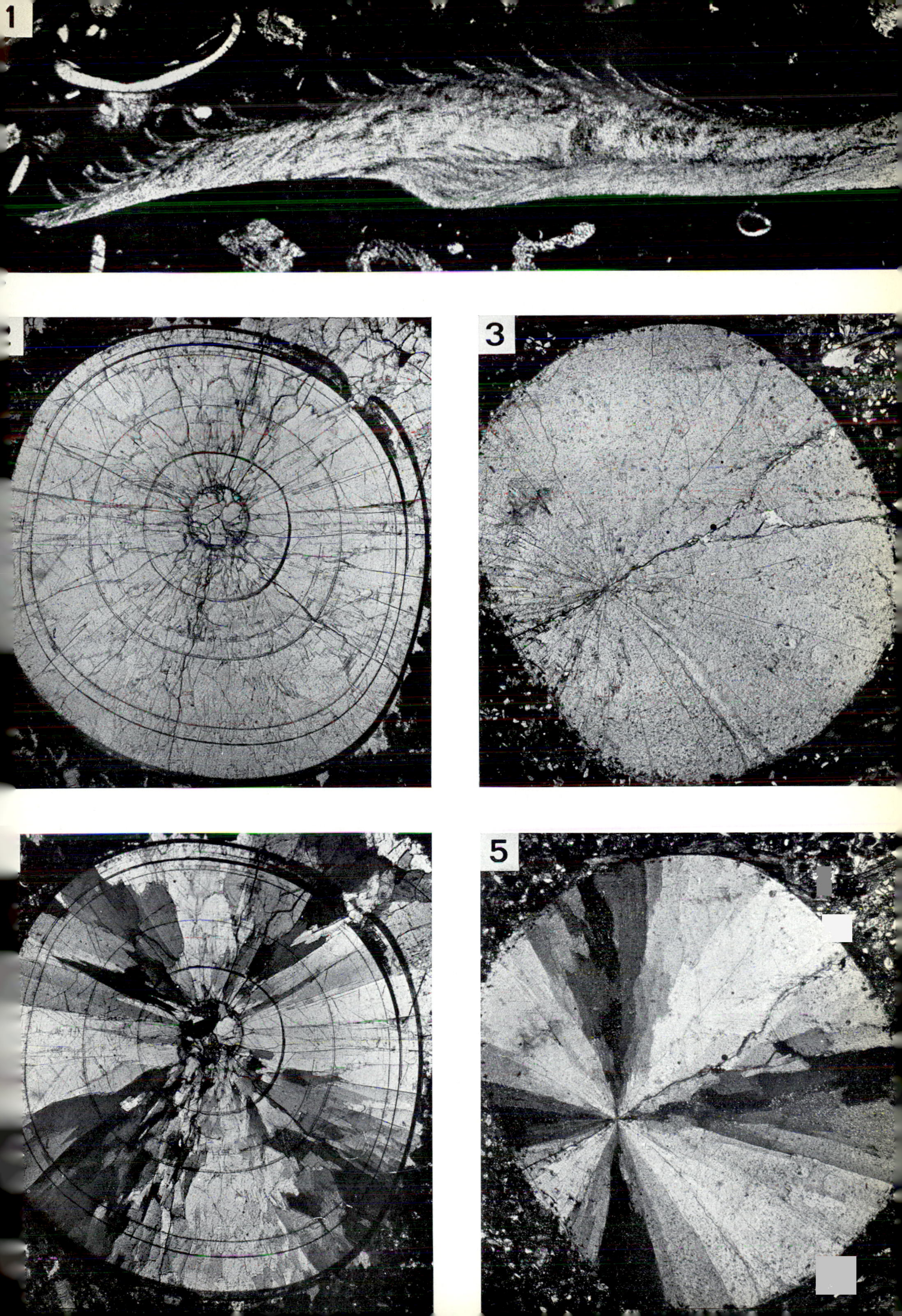
1
3
5

1/10 mm
x 20
x 40
x 60
x 80
x100

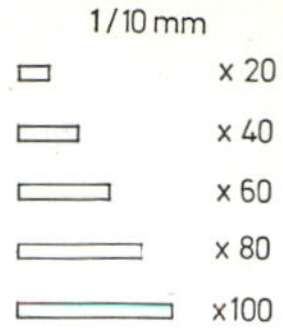

Fig. 1. Trilobitic-echinodermal limestone with mud matrix. "shepherds crook" of large trilobite dominates photomicrograph. Meagher Formation, Middle Cambrian, South Boulder Creek, T. 1 S., R. 3 W., Madison County, Montana, U.S.A. IU 8099–S58, ×30.

Fig. 2. Poorly sorted trilobite debris in micrite matrix. Observe large recurved shell containing faint, very fine prismatic texture. Orthoceratite Limestone, Arenigian, Lower Ordovician, Bornholm, Denmark. IU 8099–813, ×20.

Fig. 3. Trilobite in framework of echinodermal debris. Brassfield Formation, Lower Silurian, C N$\frac{1}{2}$ SW$\frac{1}{4}$ sec. 6, T. 5 N., R. 11 E., Jefferson County, Indiana, U.S.A. IU 8099–290, ×20.

Fig. 4. Open framework of trilobite debris in muddy matrix. The closed shell fragments represent sections parallel to arched cephala (heads) and pygidia (tails). As fig. 2.

Fig. 5. Well-sorted, skeletal limestone containing large trilobite fragment (a), echinodermal plates (b), bryozoans (c), grains having micritic coats (d), crystallized molluscan fragment (e), quartz (f), sparry calcite cement (g), and pelletal mud (h). Bethel Formation, Chester Series, Upper Mississippian, Hardin County, Kentucky, U.S.A. IU 8099–P211, ×40.

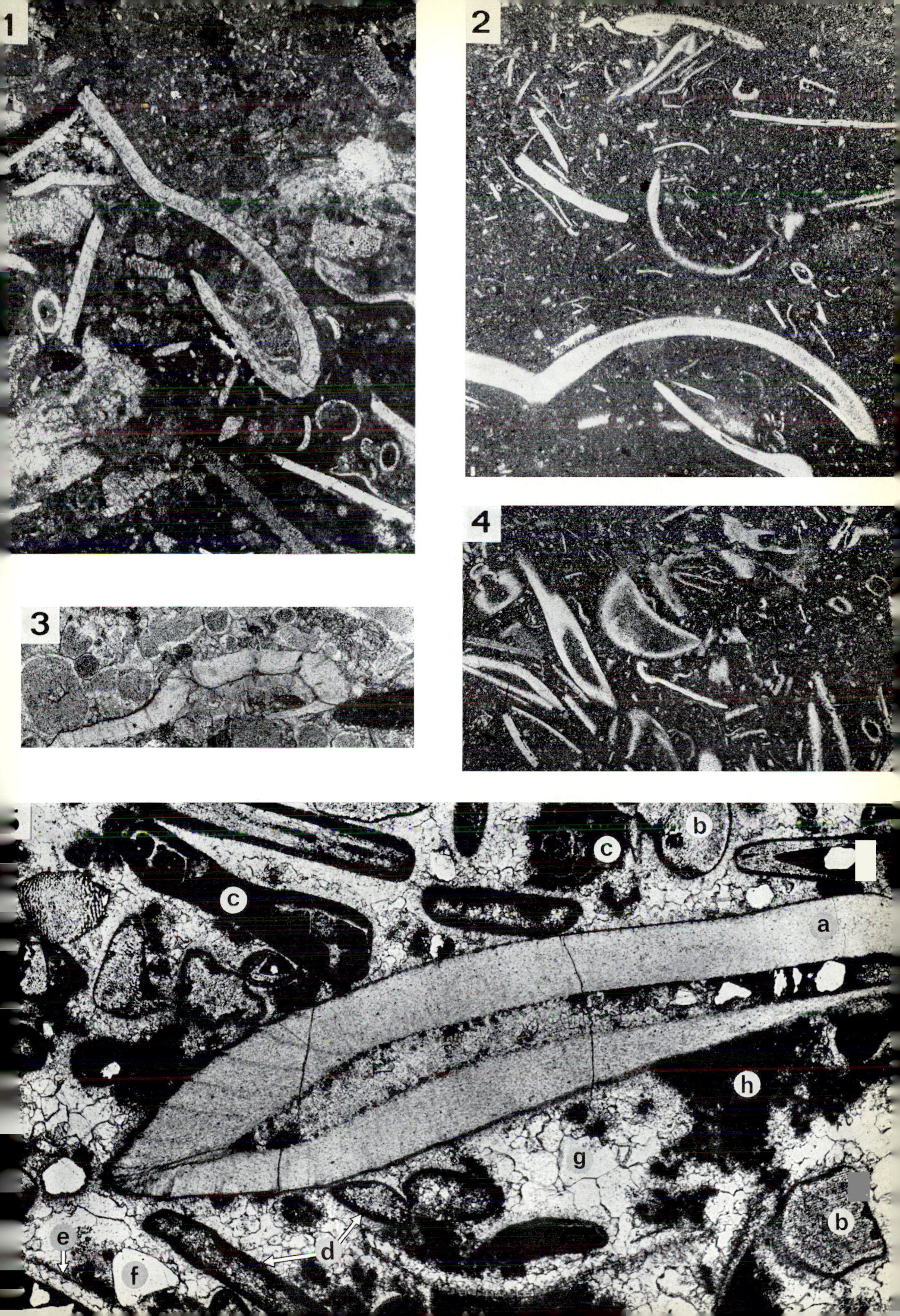
1
2
3
4
a
b
b
c
c
d
e
f
g
h

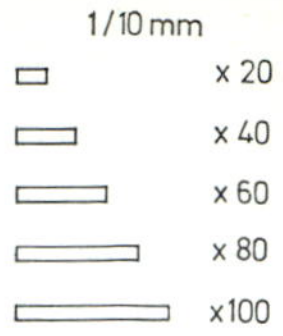

1/10 mm
x 20
x 40
x 60
x 80
x100

Fig. 1. Trilobite debris associated with microspar and some possible coarse recrystallized spar or pyrite. Note characteristic recurved trilobite shells ("shepherd's crooks"), some of which have been crushed so that they appear segmented. Conaspis Zone, Snowy Range Formation, Upper Cambrian, Gullis Creek Section, Snowy Range, Montana, U.S.A. IU 8099–616, ×30.

Fig. 2. Trilobite (a) and echinodermal debris in micrite. Note coarsely porous echinoderm plates in sections transverse (b) and parallel (c) to plate surfaces, detrital quartz (d), and conspicuous secondary dolomite crystal (e). Some coarse recrystallized spar (f). Crepicephalus Zone, Upper Cambrian, Elk Creek Section, Black Hills, Lawrence County, South Dakota. IU 8099–618, ×30.

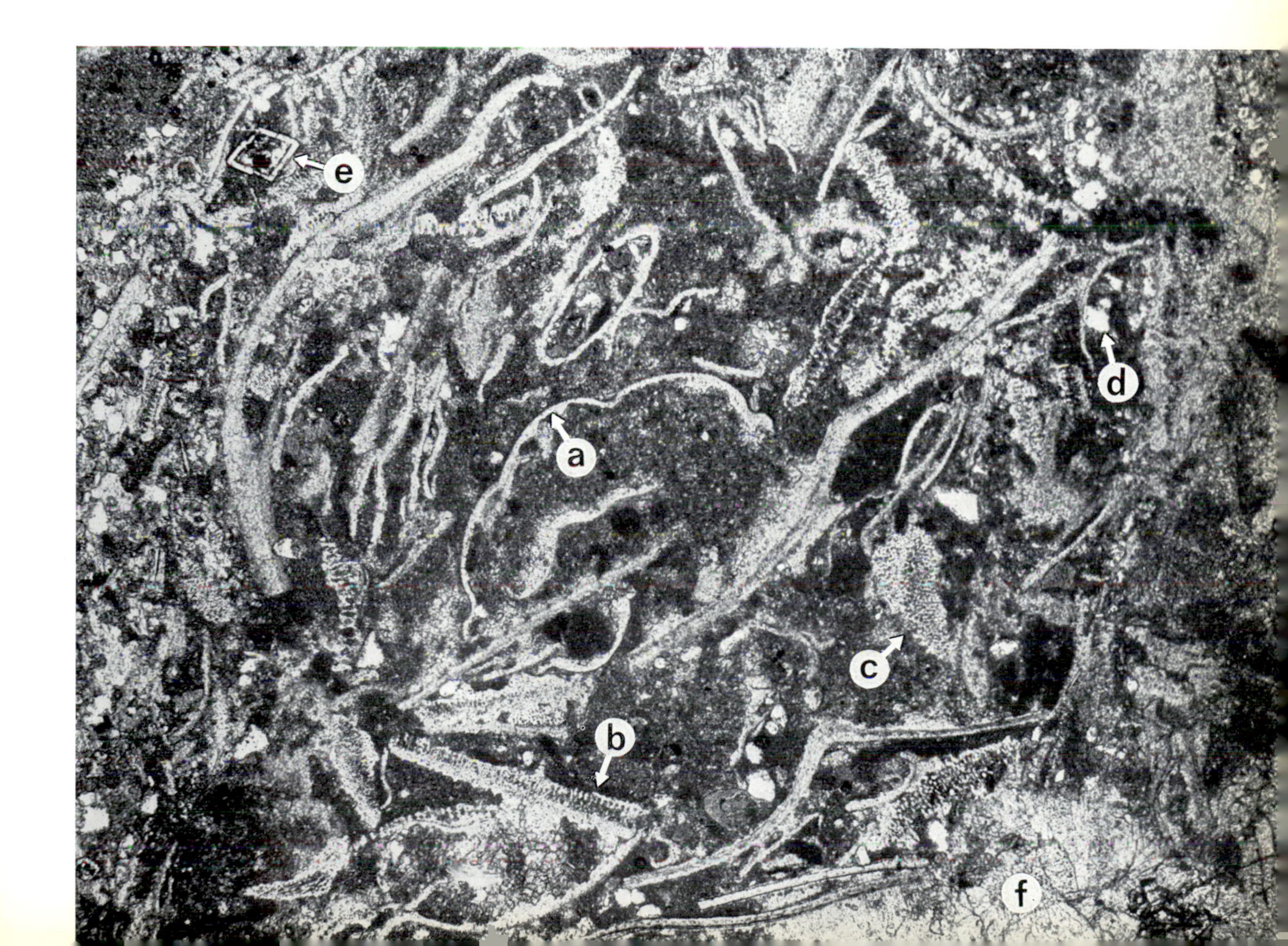
e
d
a
c
b
f

1/10 mm
x 20
x 40
x 60
x 80
x100

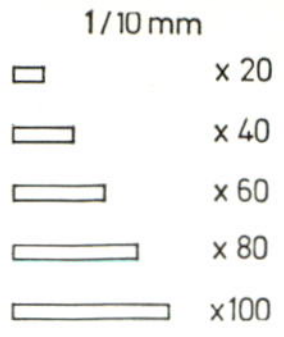

Fig. 1. Cross sections of articulated and disarticulated ostracodes in recrystallized lime mud. Several echinodermal plates. Note characteristic overlap of valves in ostracode at left center. Zone 7b–c, Upper Llandoverian, Lower Silurian, Vik road junction, Ringerike region, Buskerud, northwest of Oslo, Norway. IU 8099–807, ×40.

Fig. 2. Ostracode-filled gastropod. Infilling may be interpreted in three different ways: 1) gastropod is pregnant with ostracodes (that is, ostracodes use gastropod as a brood pouch or a mating ground); 2) gastropod ingested ostracodes and expired; or 3) empty gastropod shell was infilled by ostracodes prior to final deposition in mud matrix. A good example of the method of multiple working hypotheses. Gastropod limestone, Cretaceous, SW$\frac{1}{4}$ sec. 28, T. 1 N., R. 12 W., Granite County, Montana, U.S.A. IU 8099–719, ×40.

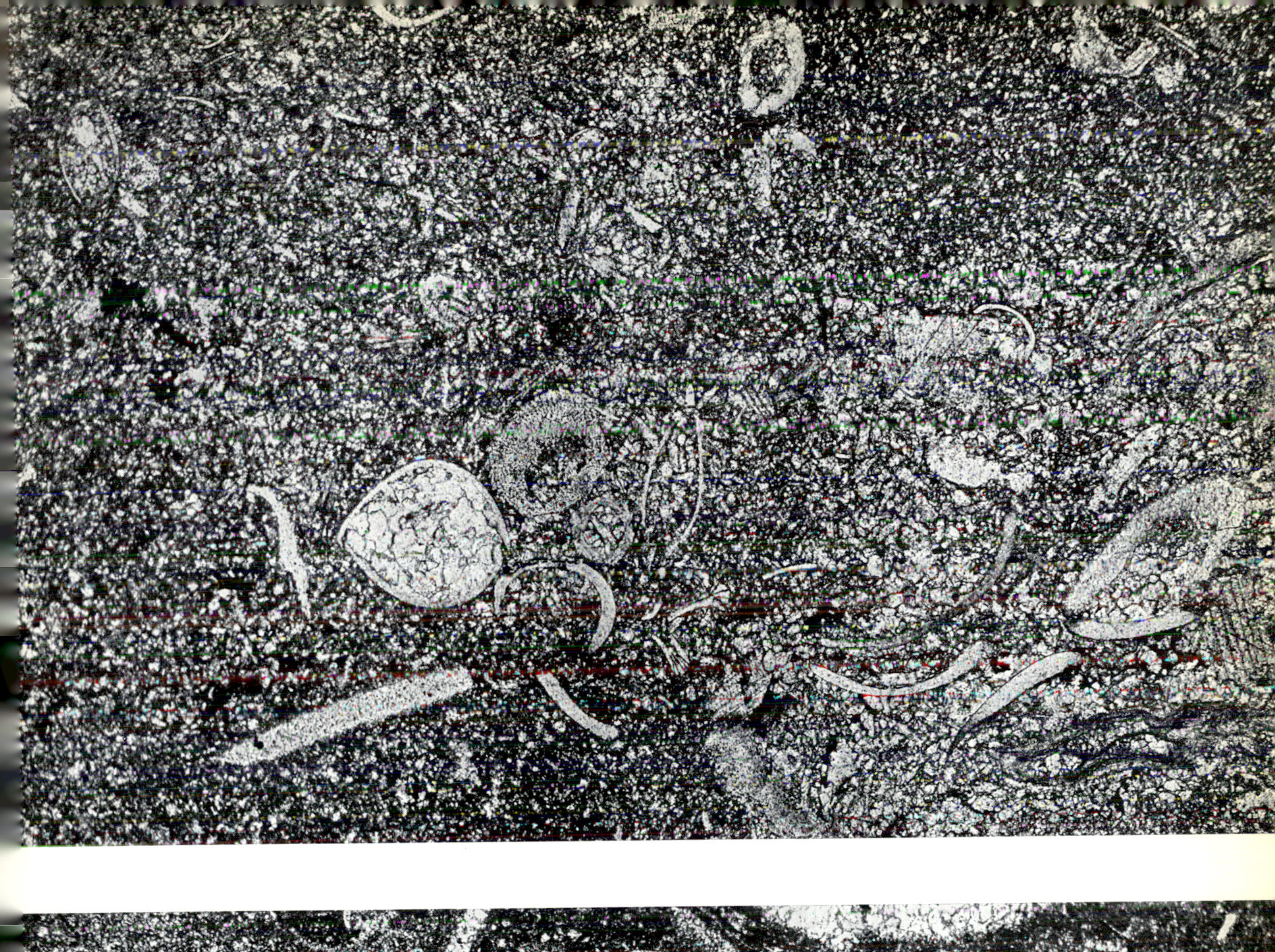

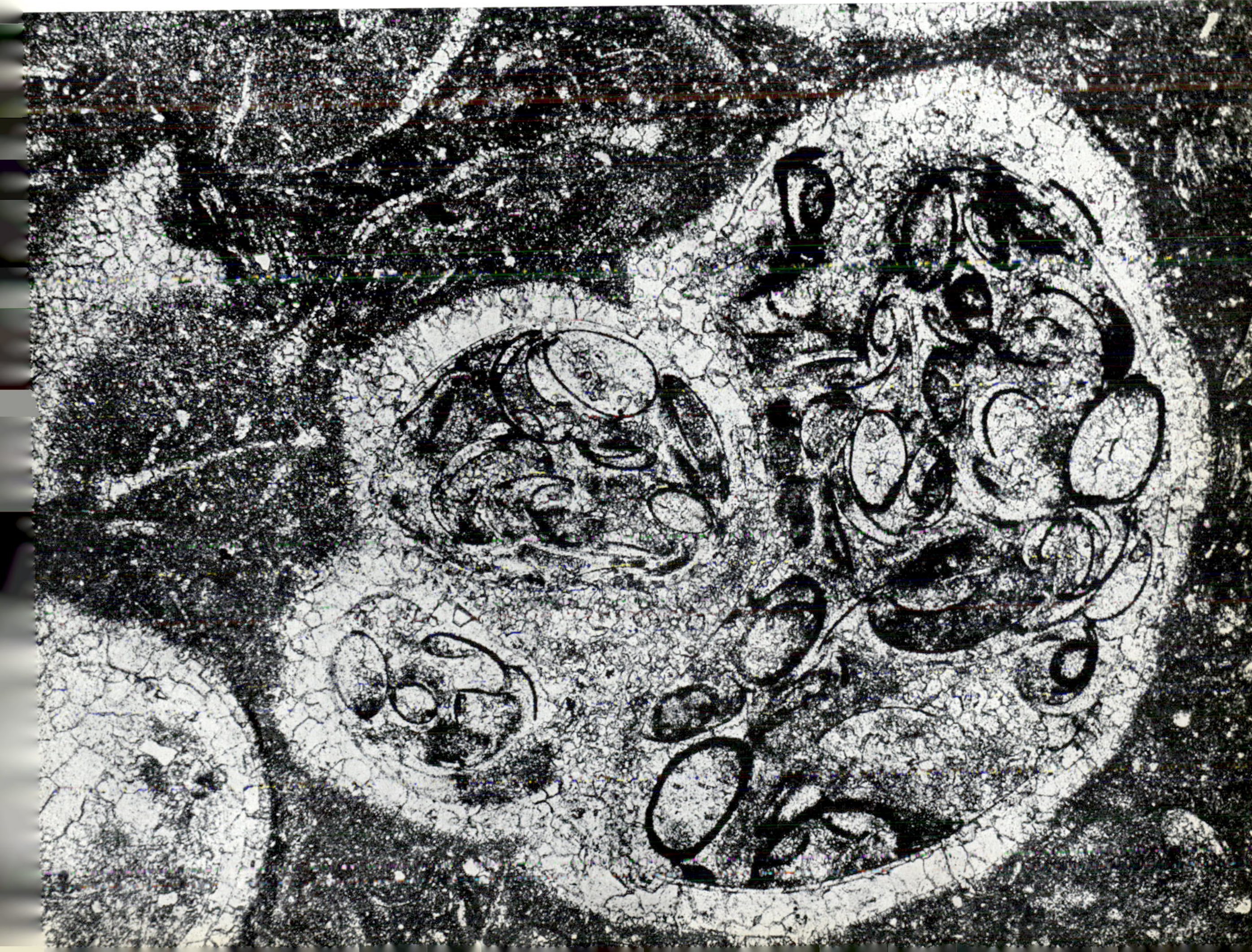

1/10 mm
x 20
x 40
x 60
x 80
x100

Fig. 1. Porous echinodermal plates, including oblique section of echinoid spine (a) at lower right and cross section exhibiting serrate surface of stem plate (b). Quartz grains, brachiopod spines, and sparry calcite cement. Bethel Formation, Chester Series, Upper Mississippian, Hardin County, Kentucky, U.S.A. IU 8099–P211, ×40.

Fig. 2. Oblique section of echinoid spine. Note radiating pore pattern and spar filling of original hollow interior. Compare with Pl. 51–2. Middle Triassic, Morilor Valley, Plopis Mountains, Romania. IU 8099–404, ×40.

Fig. 3. Echinodermal-fenestrate bryozoan limestone, typical of many late Paleozoic assemblages. Echinoderm plates exhibit finely porous structure, accentuated by dark pore fillings (? pyrite). Fenestrate bryozoan skeleton (a) frequently too dark to show fibrous thick, outer wall. Large brachiopod shell with spine base (b). Clear calcite cement. Lower Carboniferous, Lathkill Dale, near Monyash, Derbyshire, England. IU 8099–39B, ×40.

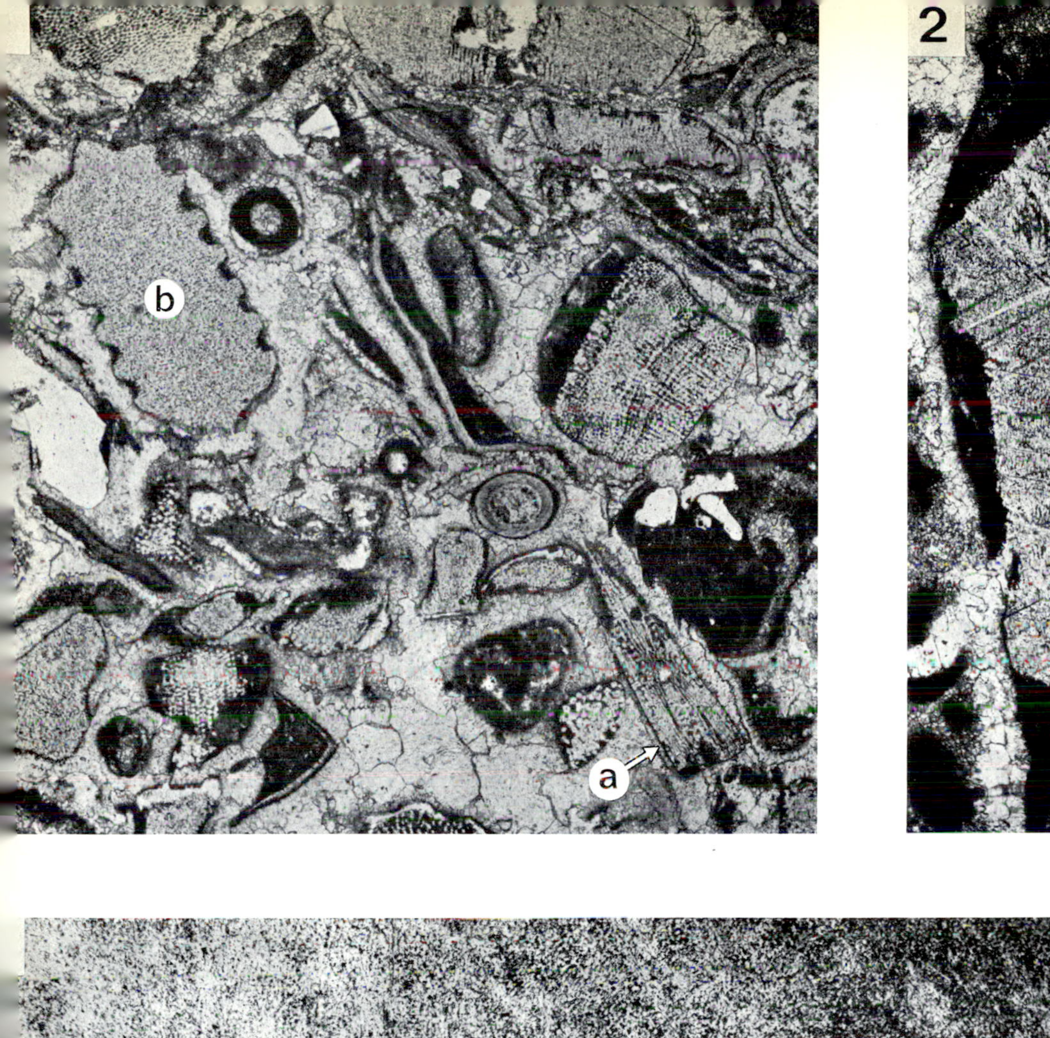
b
a

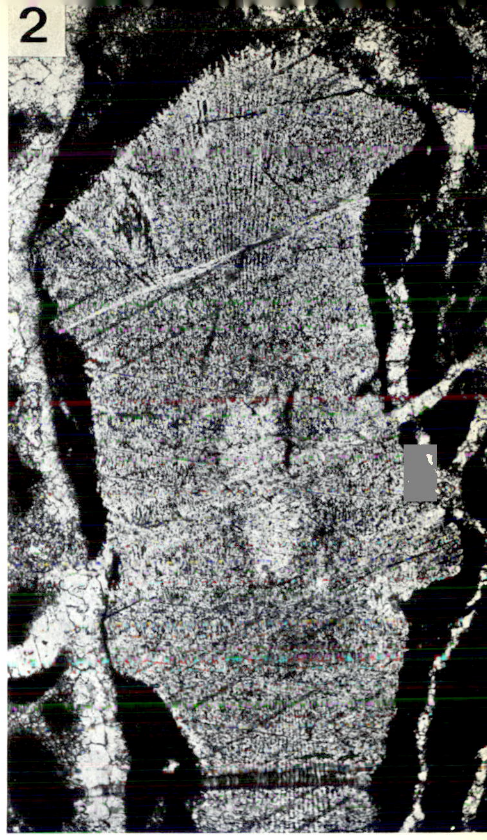
2

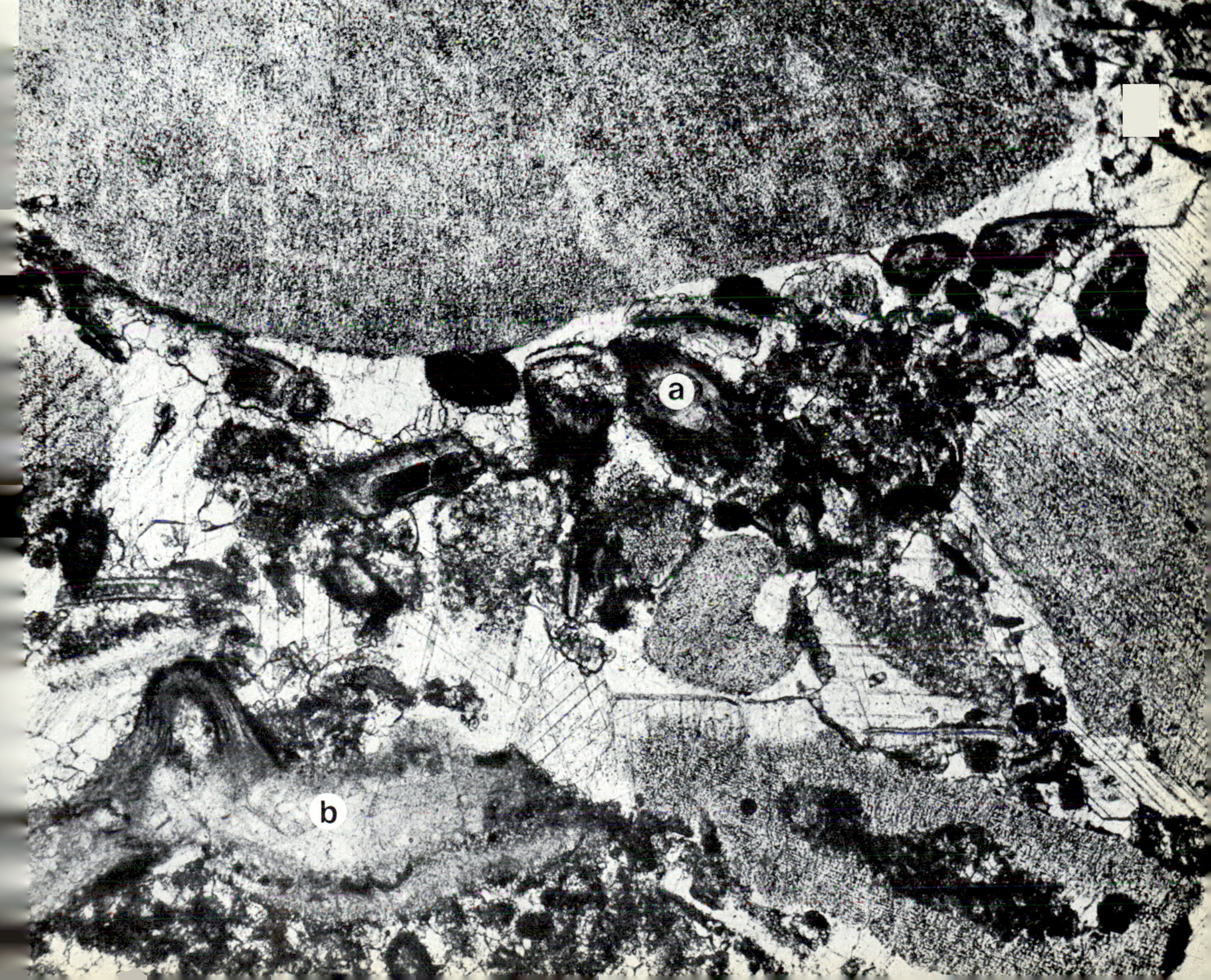
a
b

1/10 mm
x 20
x 40
x 60
x 80
x 100

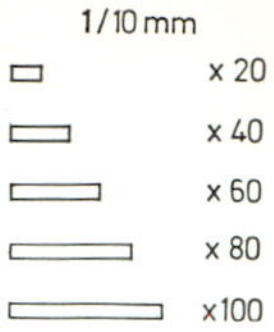

Fig. 1. Longitudinal section of a large echinoid spine in framework of fenestrate bryozoans and echinodermal debris. Sparry calcite cement. Glen Dean Limestone, Chester Series, Mississippian, NW$\frac{1}{4}$ sec. 18, T. 4 S., R. 1 W., Perry County, Indiana, U.S.A. IU 8099–919, ×20.

Fig. 2. Transverse section of echinoid spine. Middle Triassic, Morilor Valley, Plopis Mountains, Romania. IU 8099–404, ×40.

Fig. 3. Transverse section of large encrusted (a), hollow echinoid spine showing radial pattern. Framework includes altered molluscan fragments with micritic coats (b). Pellets conspicuous in interior of spine. Algal (?) incrustation of spine has incorporated various kinds of other skeletal debris. Upper Jurassic, Piatra Cetii, Trascau Mountains, Romania. IU 8099–398, ×20.

Fig. 4. Striking slightly oblique transverse section of echinoid spine filled with mud (lower center). Spine in upper left cut near top and does not show a hollow center. Debris of punctate and impunctate brachiopods. Some coated grains, some pellets and angular, clear quartz. Sparry calcite cement. Bethel Formation, Chester Series, Mississippian, Hardin County, Kentucky, U.S.A. IU 8099–P211, ×50.

Fig. 5. Different cross sections of echinoderm columnals, some having oolitic coatings. A few quartz grains. Pelletal mud matrix on right and calcite cement on left. As fig. 4, IU 8099–P210, ×40.

Fig. 6. Echinoid spine shows radial pattern of pores. Spine has been bored, and borings filled by debris. Carillo Puerto Formation, Pliocene, on Federal Highway 180, 3.9 kilometers southeast of Kantunil, Yucatán, México. IU 8099–125B, ×20.

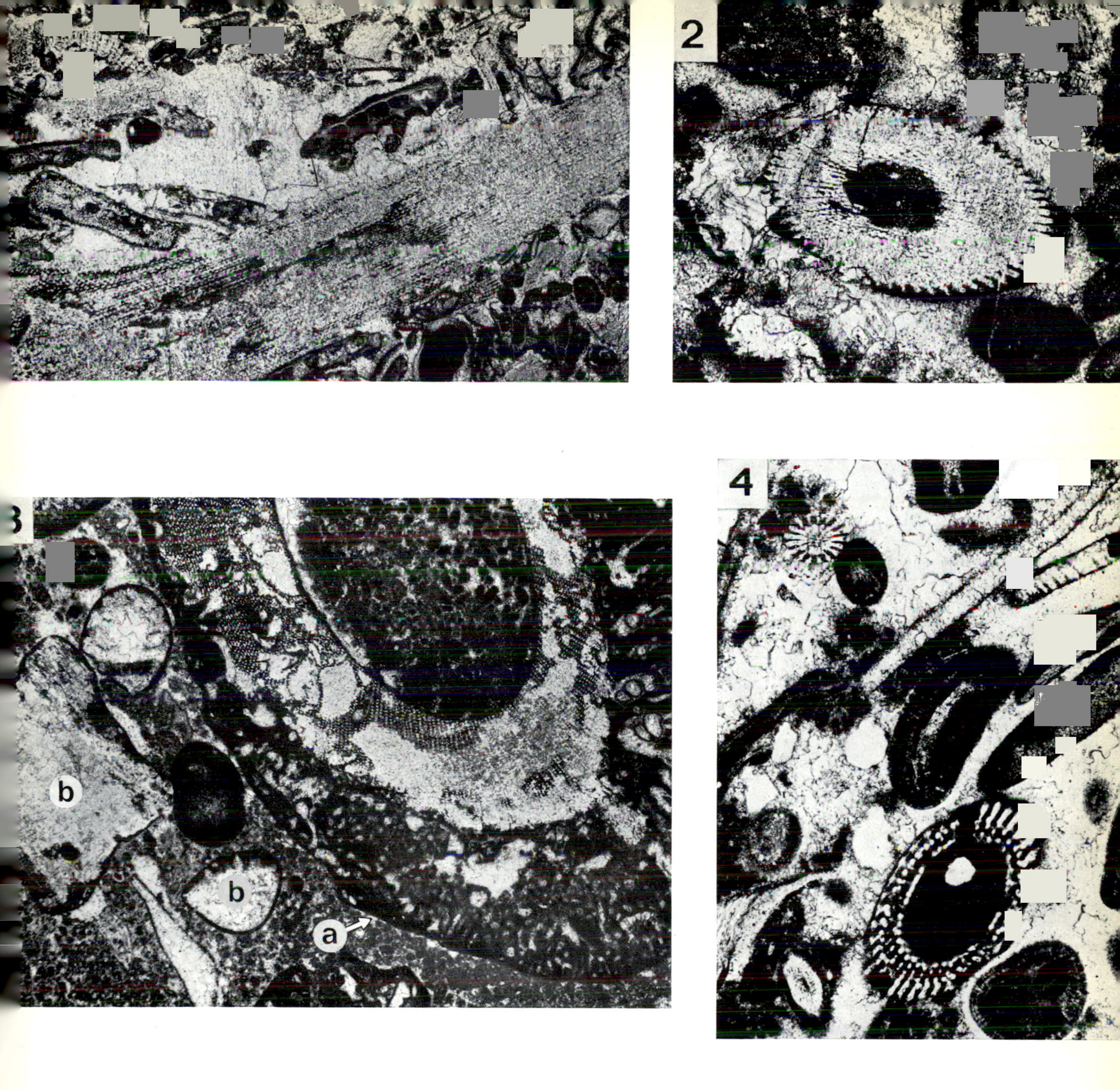
2

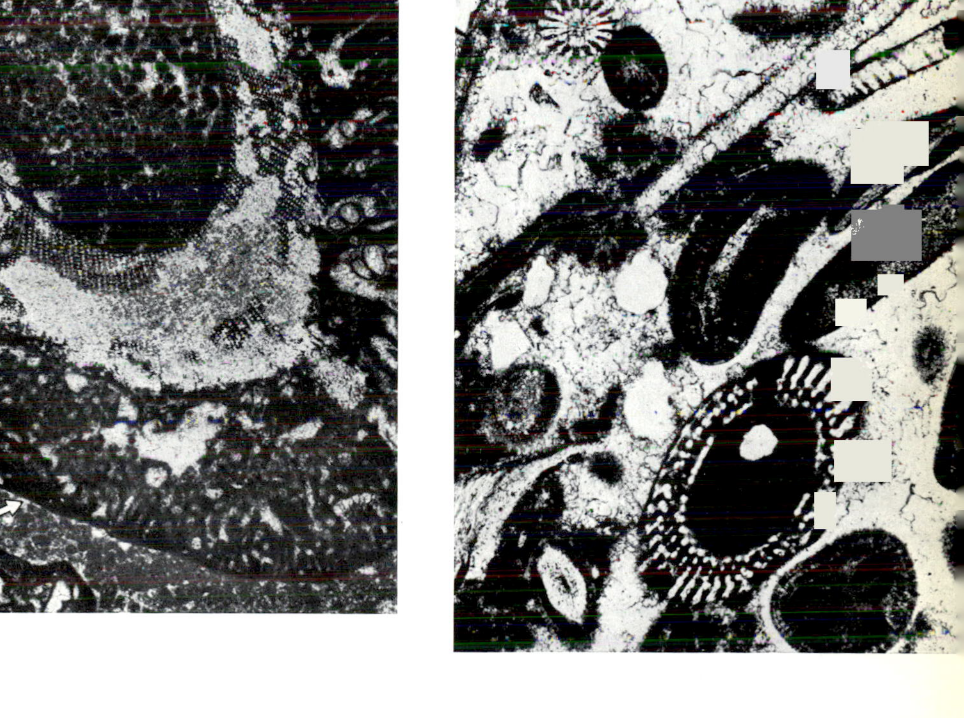
3
4
b
b
a

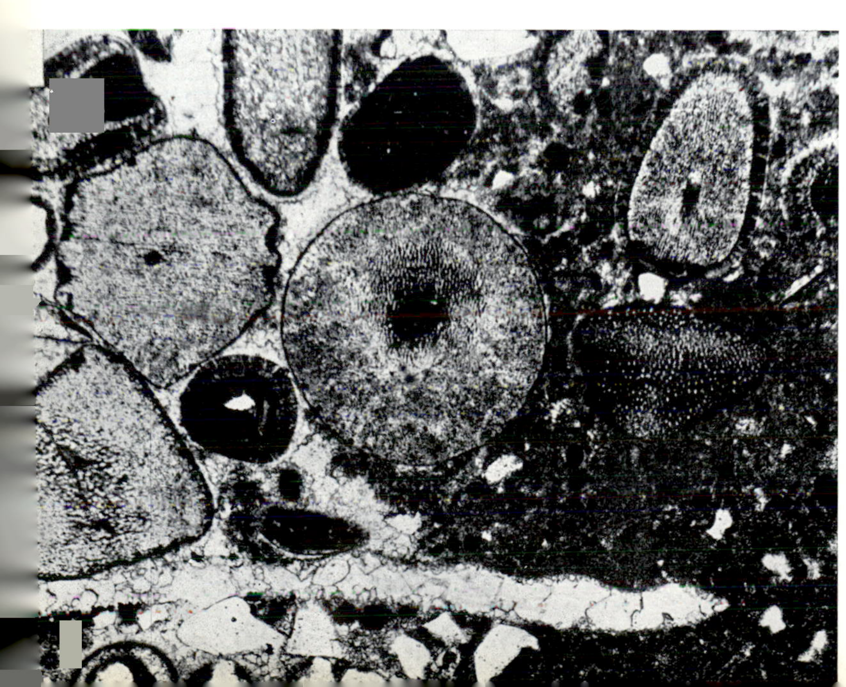

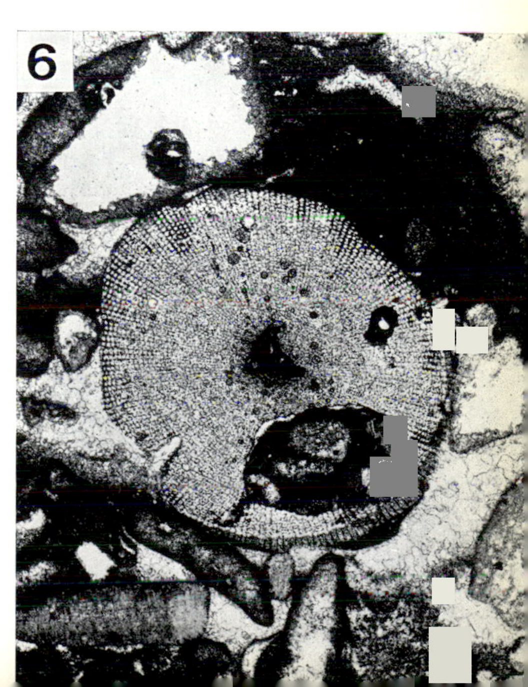
6

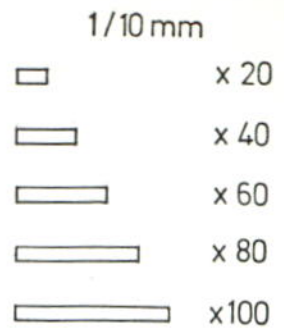

1/10 mm
x 20
x 40
x 60
x 80
x 100

Fig. 1. Cross section of small phosphatic bone. Note open internal meshwork and solid exterior covering. Matrix consists of disarticulated inoceramid prisms. Approximately 5 feet below top of Graneros Shale, Upper Cretaceous, roadcut on east side of U. S. Highway 281, just north of Saline River, NW$\frac{1}{4}$ sec. 35, T. 12 S., R. 14 W., Russell County, Kansas, U.S.A. Hattin Collection, KG–1R, ×80.

Fig. 2. Cross section of fish scale. Note crude layering and homogeneous microstructure. Although not apparent here, distinctive brown color is typical of much phosphatic fossil debris. Matrix consists of globigerinids, disarticulated inoceramid prisms, and dark mud. Approximately 1.5 feet above base of Lincoln Limestone Member, Greenhorn Formation, Upper Cretaceous, cut on west side of county road approximately 3 miles south of Simpson, SE$\frac{1}{4}$ SE$\frac{1}{4}$ sec. 24, T. 8 S., R. 6 W., Mitchell County, Kansas, U.S.A. Hattin Collection, KHG–4L lower, ×40.

Fig. 3. Dense dark-colored (brown in technicolor productions) phosphatic vertebrate grain, probably a tooth. Graneros Shale, Upper Cretaceous, cut bank on a small tributary to Wolf Creek approximately $7^1/_2$ miles north-northwest of Holyrood, NE$\frac{1}{4}$ sec. 5, T. 16 S., R. 10 W., Ellsworth County, Kansas, U.S.A. Hattin Collection, KG–AD=18C, ×20.

Fig. 4. Fish scale cut parallel to scale surface. Contrast with fig. 2. As fig. 2.

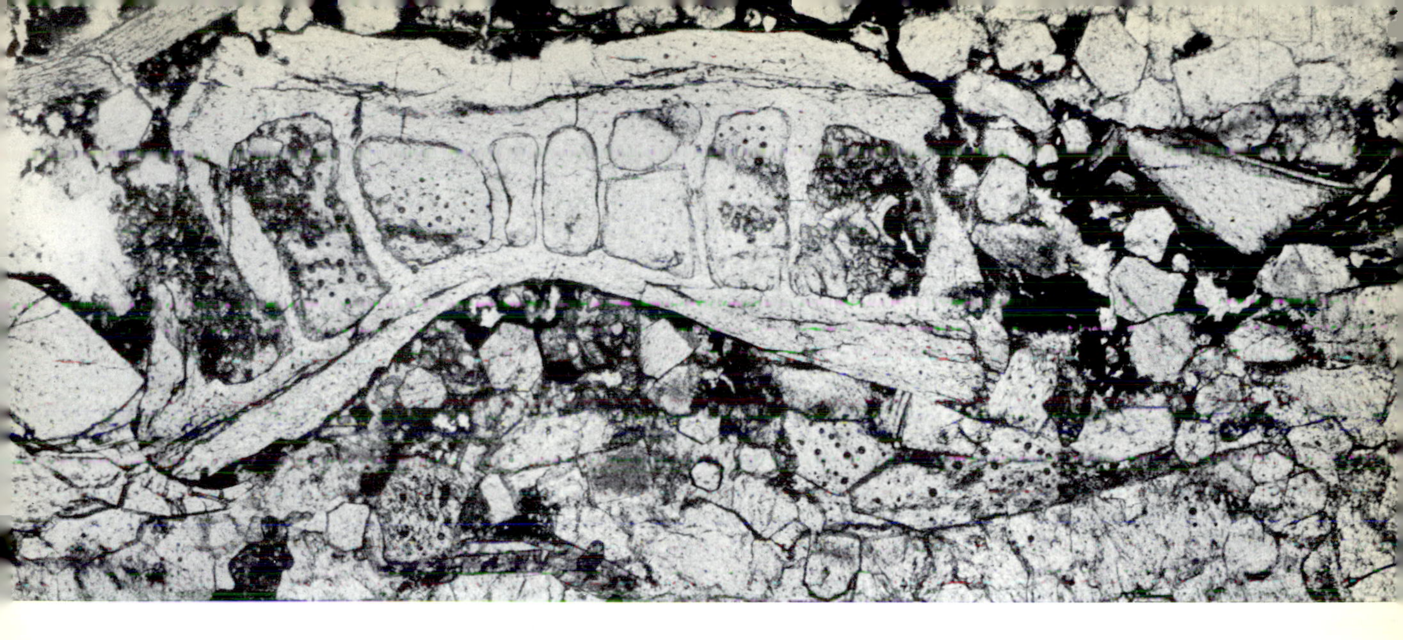

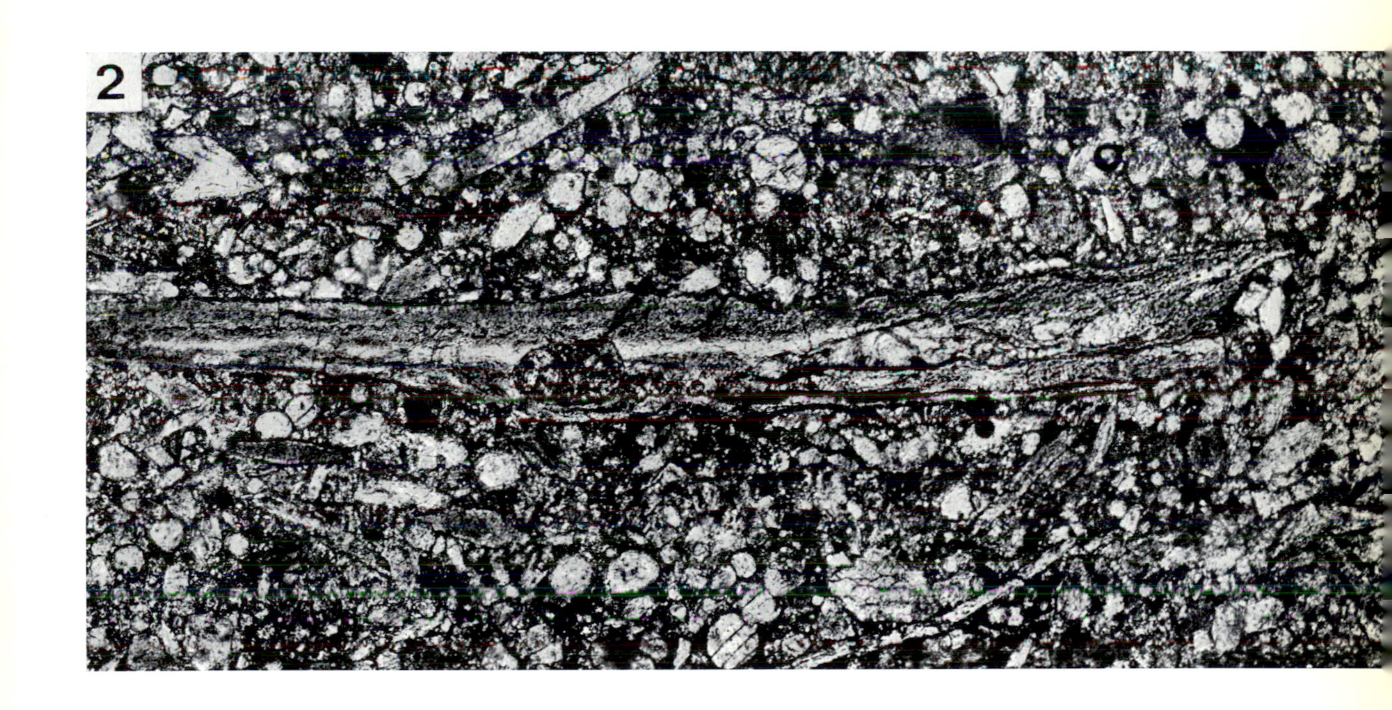
2

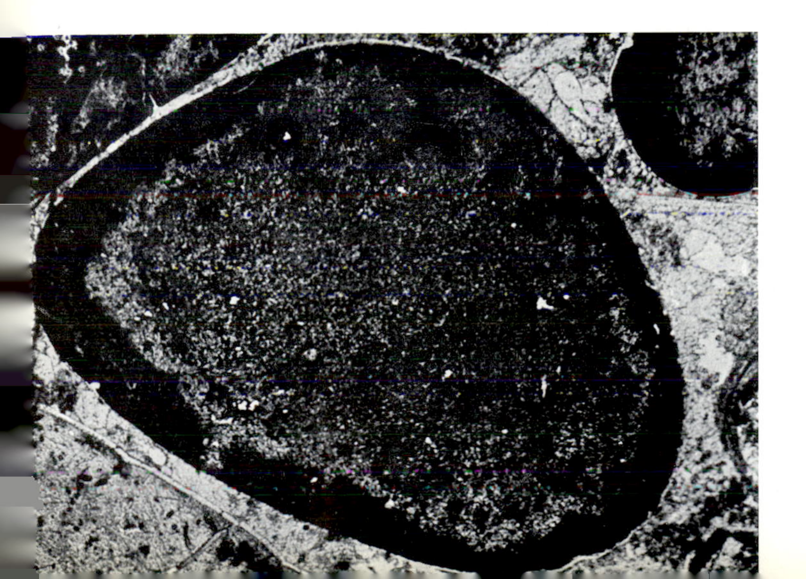

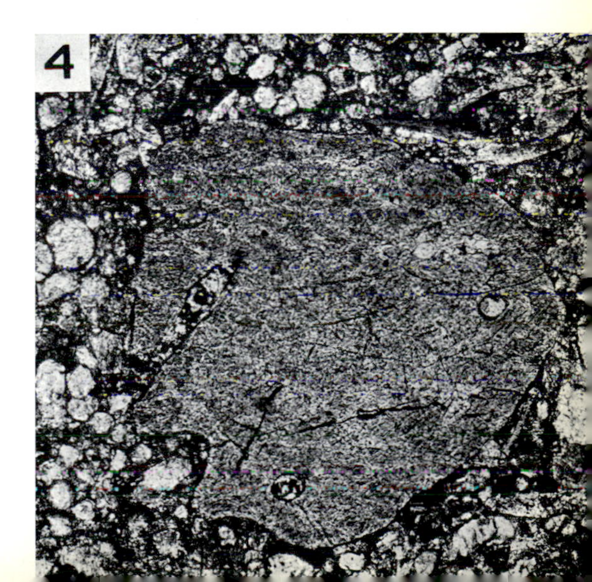
4

PLATE 53. PHOSPHATIC FOSSILS AND WORM (?)

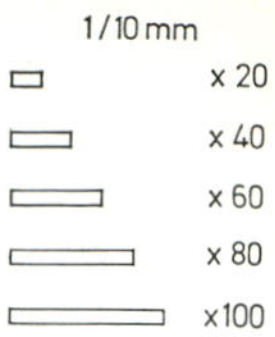

1/10 mm
x 20
x 40
x 60
x 80
x 100

Fig. 1. Fragmental phosphatic debris interpreted as conodonts. Conodont in center exhibits longitudinal section of simple curved cone. In thin section brown color of phosphatic fossils is distinctive. North Vernon Limestone, Middle Devonian, Jennings County, Indiana, U.S.A. IU 8099–525B, × 80.

Fig. 2. Dark fragment at lower center is vertebrate fragment. Phosphatic zones (bone beds) are a persistent feature in Middle Devonian limestones of the Cincinnati Arch region. As fig. 1, IU 8099–525, ×80.

Figs. 3, 4. Variously oriented cross sections of phosphatic conodonts. Layered internal structure visible in portions of some cross sections and not in others. Laminations not different at this scale from those seen in inarticulate brachiopods (Pl. 29–6) or fish scales (Pl. 52–2, 4; fig. 1 above). Jacobs Chapel bed, Lower Mississippian, bed of tributary to Silver Creek at ford, E $\frac{1}{4}$ N $\frac{1}{4}$ lot 288, Clark's Grant, Scott County, Indiana, U.S.A. IU 8099–1076, ×100.

Fig. 5. Coiled calcareous worm tube (?*Spirorbis*) encrusting a skeletal fragment possibly a pelecypod. Shell infilled with sparry calcite. Brachiopod shell debris includes spines exhibited in longitudinal (a) and transverse section (b). Echinodermal plates (c) and some bryozoan debris (d). Lower Carboniferous, Lathkill Dale, near Monyash, Derbyshire, England, IU 8099–39B, ×40.

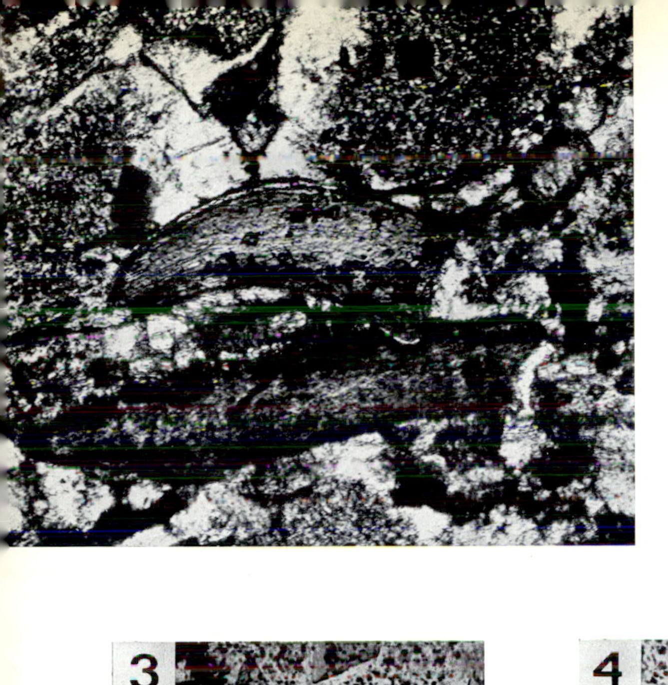

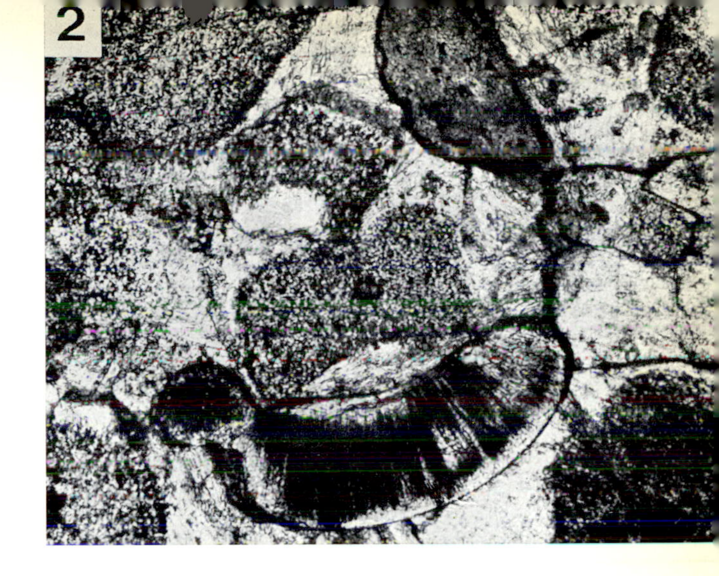
2

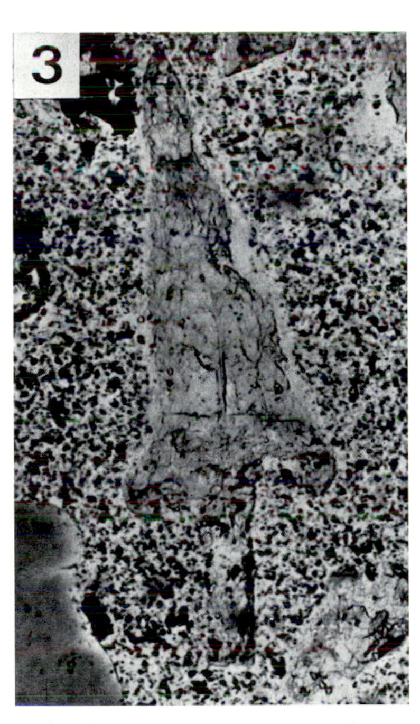
3

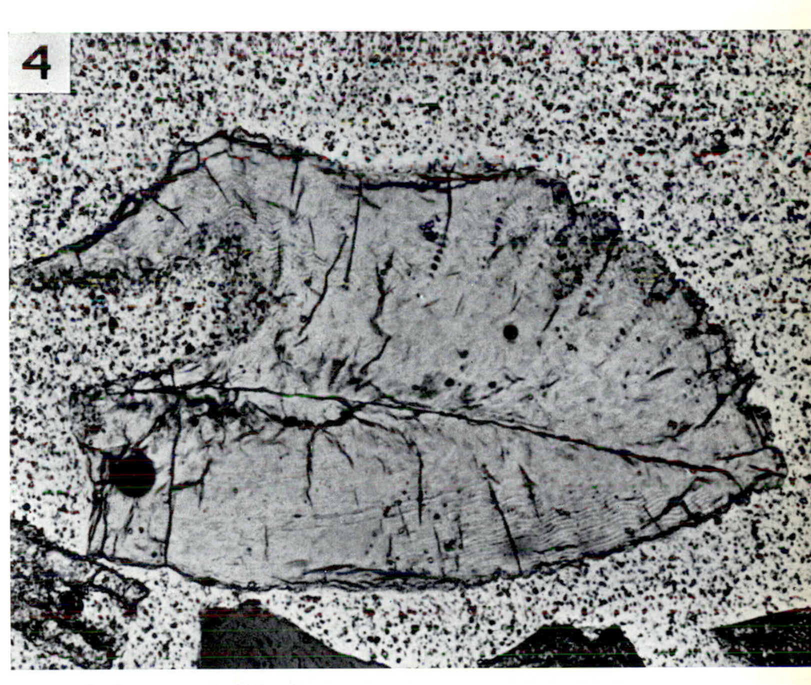
4

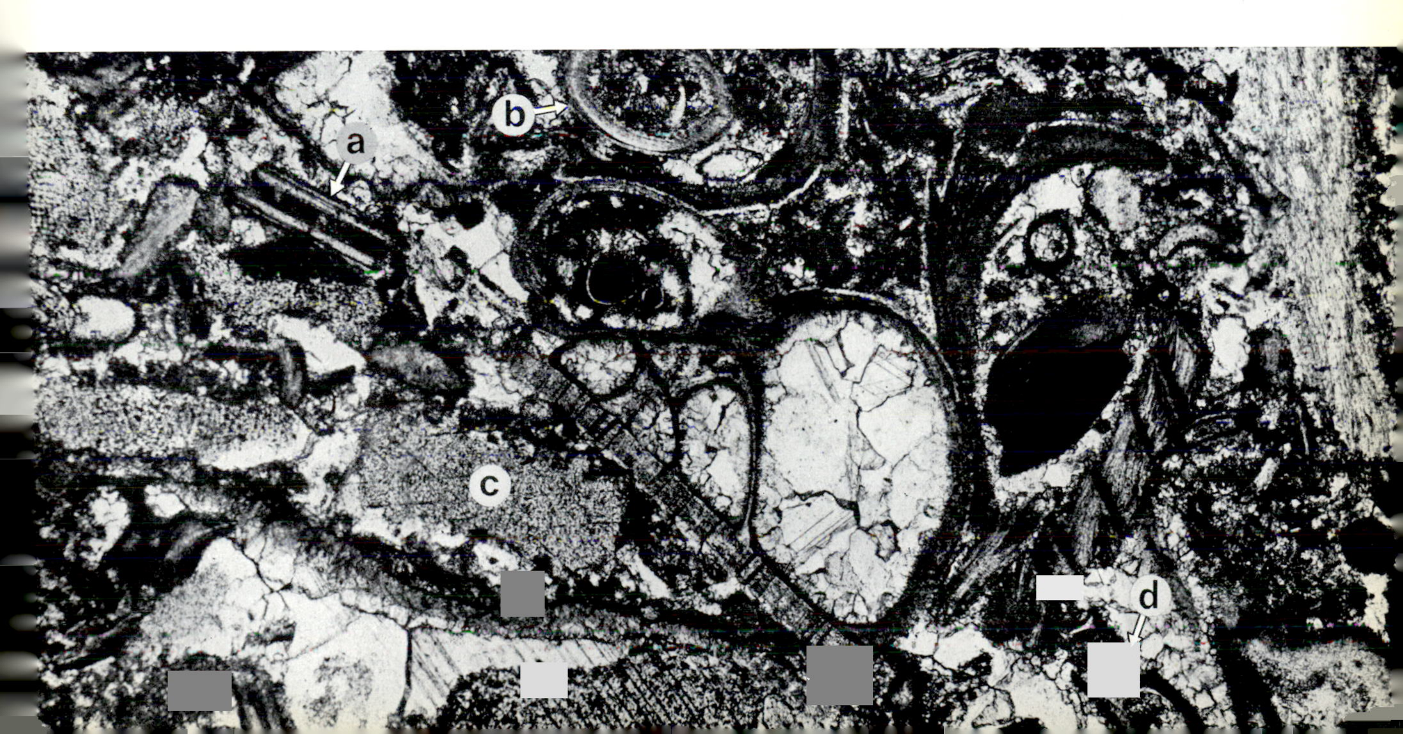
a
b
c
d

1/10 mm
x 20
x 40
x 60
x 80
x100

Fig. 1. *Favreina* cf. *F. salevensis* (PARÉJAS). Very well-sorted loosely packed framework of faecal pellets. Most of these pellets contain an internal pattern of tubules formed by elongate projections from the intestine walls. Tubules are shown in longitudinal and transverse sections. These tubules and original pore space now infilled by sparry calcite. Recent anomuran crustaceans (arthropods) produce similar types of faecal pellets. Neocomian, Lower Cretaceous, Broce, Yugoslavia. Elliott Collection, British Museum (Natural History), Department of Palaeontology Z. 988, ×40.

Fig. 2. *Favreina kurdistanensis* ELLIOTT. Mud matrix obscures recognition of boundaries of faecal pellets similar to those of fig. 1, although some transverse and longitudinal sections of tubules can be seen. Dark circular rims help define outer boundary of pellets. Sarmord Formation, Barremian, Lower Cretaceous, Kara-Zewa, Northern Iraq. Elliott Collection, British Museum (Natural History) Department of Palaeontology, Z. 989, ×40.

Fig. 3. The dense material in this slide photographs poorly. The dark pattern exhibits poorly defined layering as in some stromatolites and suggests pellets that are difficult to recognize because of their dense composition and close packing. The white network, simulating a spicular sponge network, is interpreted as sparry calcite cement, which permits the recognition of the pelletal character of the sample. However, high magnifications of the dense material suggests it could be an altered calcareous skeleton (? calcareous sponge; Mesozoic stromatoporoid or related coelenterate group). מֵילָא, לְפָּחוֹת נִסִינוּ. Saharonim Formation, Triassic, Makhtesh Ramon, Israel. IU 8099–649, ×20.

Fig. 4. A rather typical section of well-sorted silt-sized pellets. As in fig. 3, individual pellets are easily recognized only where calcite cement isolates them. Differential packing of pellets produces weak mottling. Edwards Limestone, Lower Cretaceous, Whitney Dam, Bosque County, Texas, U.S.A. IU 8099–739, ×40.

Fig. 5. Well-sorted silt-sized pelletal limestone with sparry calcite cement plus some finely comminuted fossil debris that is too small for identification at this magnification. Lower Carboniferous, Namur, Belgium. IU 8099–1074, ×20.

Fig. 6. Pelletal limestone having a sparry calcite cement. Note brachiopod showing internal plates. Buchan Caves Limestone, Middle Devonian, right bank of Spring Creek between Royal and Fairy Caves, Victoria, Australia. IU 8099–462, ×40.

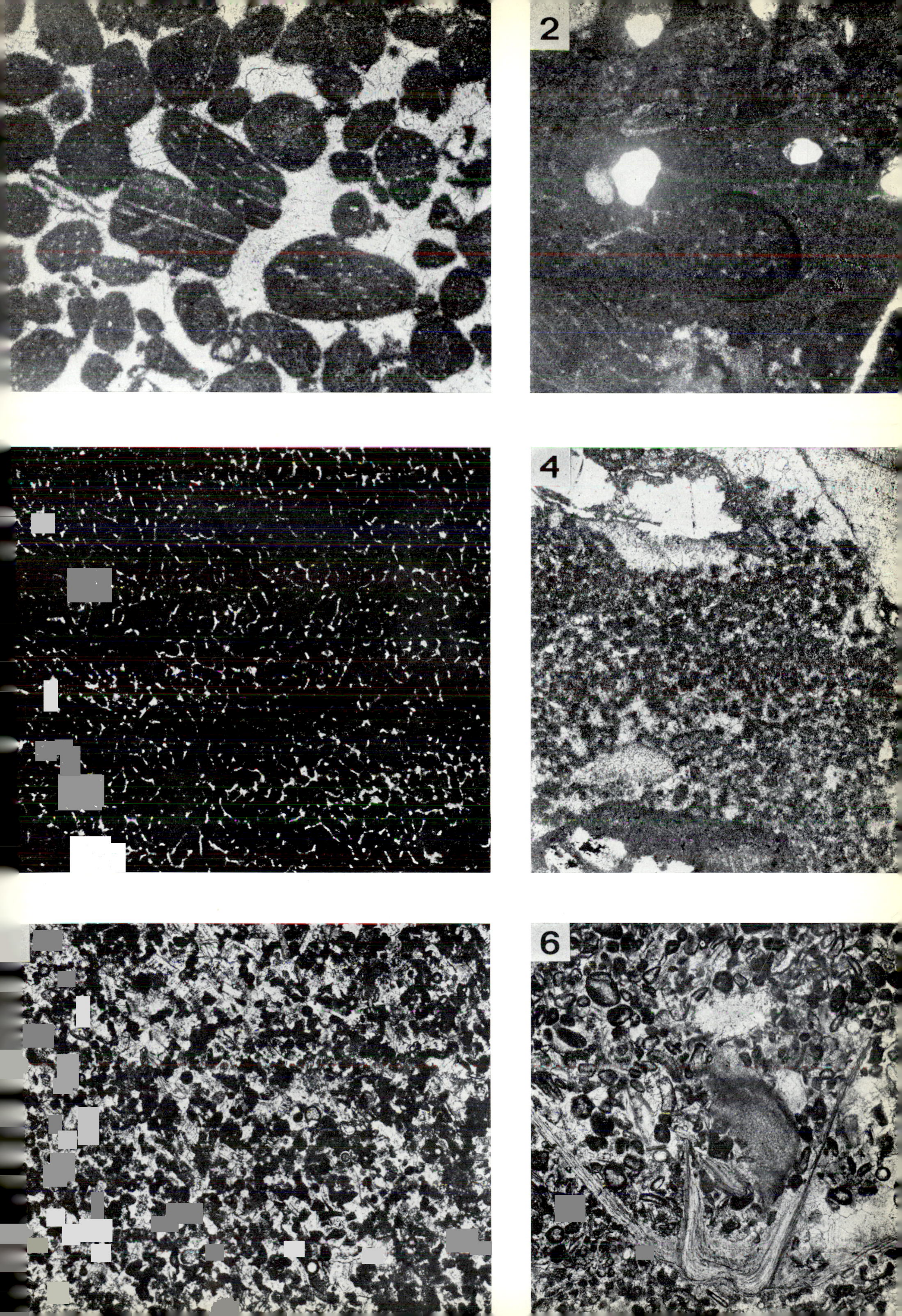

2
4
6

1/10 mm
x 20
x 40
x 60
x 80
x100

Fig. 1. Wood fragment exhibiting dark mud rim, well-defined cell boundaries, and regular cell pattern. Matrix is a complex mixture of mud, mostly calcite, quartz and some dark pellets. Clear sparry calcite cement at lower center. Pleistocene, Ryukyu Limestone, Okino erabu Island, Amami Islands, Kagoshima Prefecture, Japan. IU 8099–1008, ×80.

Fig. 2. Coralline algal fragment in matrix of dark mud and clear calcite grains of indeterminate origin. The alga is probably *Lithophyllum* as it shows central large cells (hypothallus) and outer small cell layers (perithallus). Oigawa Group, Lower Miocene, Shizuoka Prefecture, Japan. IU 8099–965, ×80.

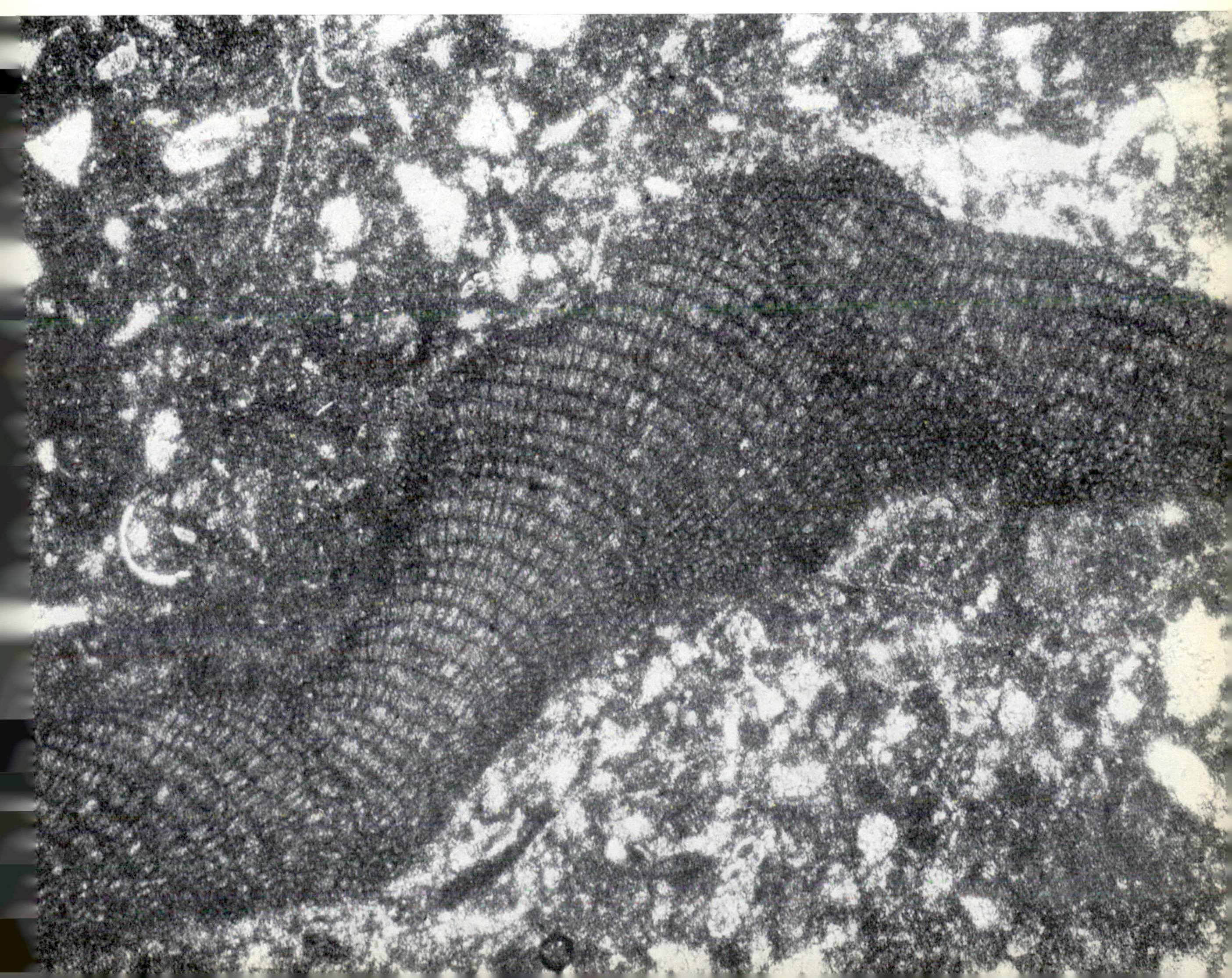

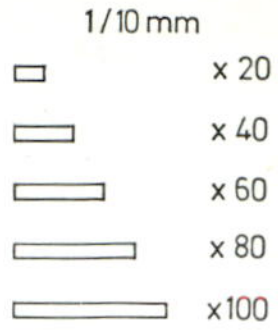

1/10 mm

x 20

x 40

x 60

x 80

x100

Fig. 1. Carbonized wood exhibiting cellular structure. Sandstone matrix. Carbonate cement binds sand and fills cells of wood. Bethel Formation, Chester Series, Mississippian, Hardin County, Kentucky, U.S.A. IU 8099–844, ×80.

Fig. 2. Two large smoothly rounded fragments of articulating coralline algae (*Amphiroa* sp.) in angular loosely packed quartz sand. Note incipient cement around quartz grains. Algae exhibit two cell sizes. Pleistocene (?), exposures on sea floor, $1^1/_2$ miles seaward of Durban Bluff, depth 140 feet, Durban, Natal, Republic of South Africa. IU 8099–842, ×80.

1/10 mm

x 20

x 40

x 60

x 80

x100

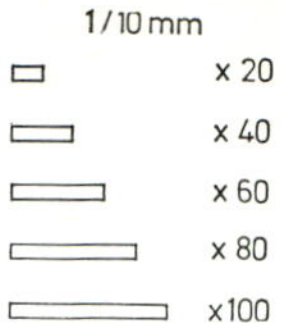

Fig. 1. *Halimeda* displaying thin outer tubes perpendicular to surface and larger randomly oriented internal tubes. In thin section segments of *Halimeda* are characteristically dark and difficult to photograph. Recent, Chireerete Island, Bikini Atoll, Marshall Islands, Pacific Ocean. IU 8099–476, ×80.

Fig. 2. Articulating coralline algal fragment. Note regular ranks of cells. Umatac Formation, Miocene-Oligocene, Guam, Marianas Islands, Pacific Ocean. IU 8099–487, ×80.

Fig. 3. Encrusting coralline algae, possibly *Lithothamnium*, exhibiting sporangia (a). Matred Formation, Eocene, Negev, Israel. IU 8099–645, ×80.

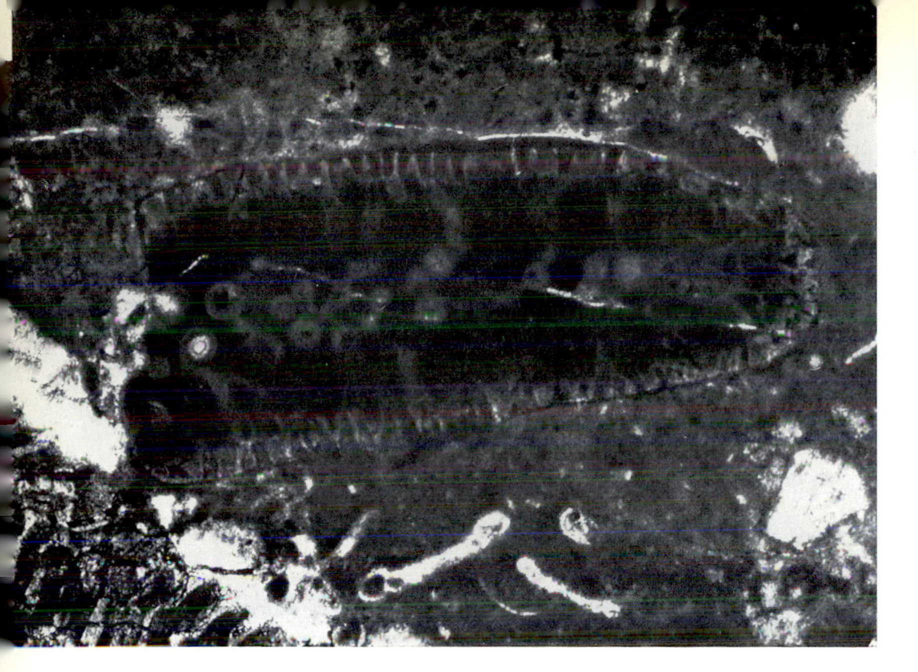

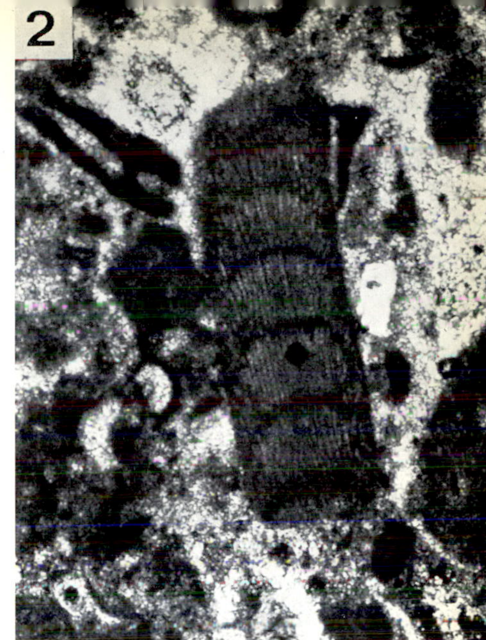
2

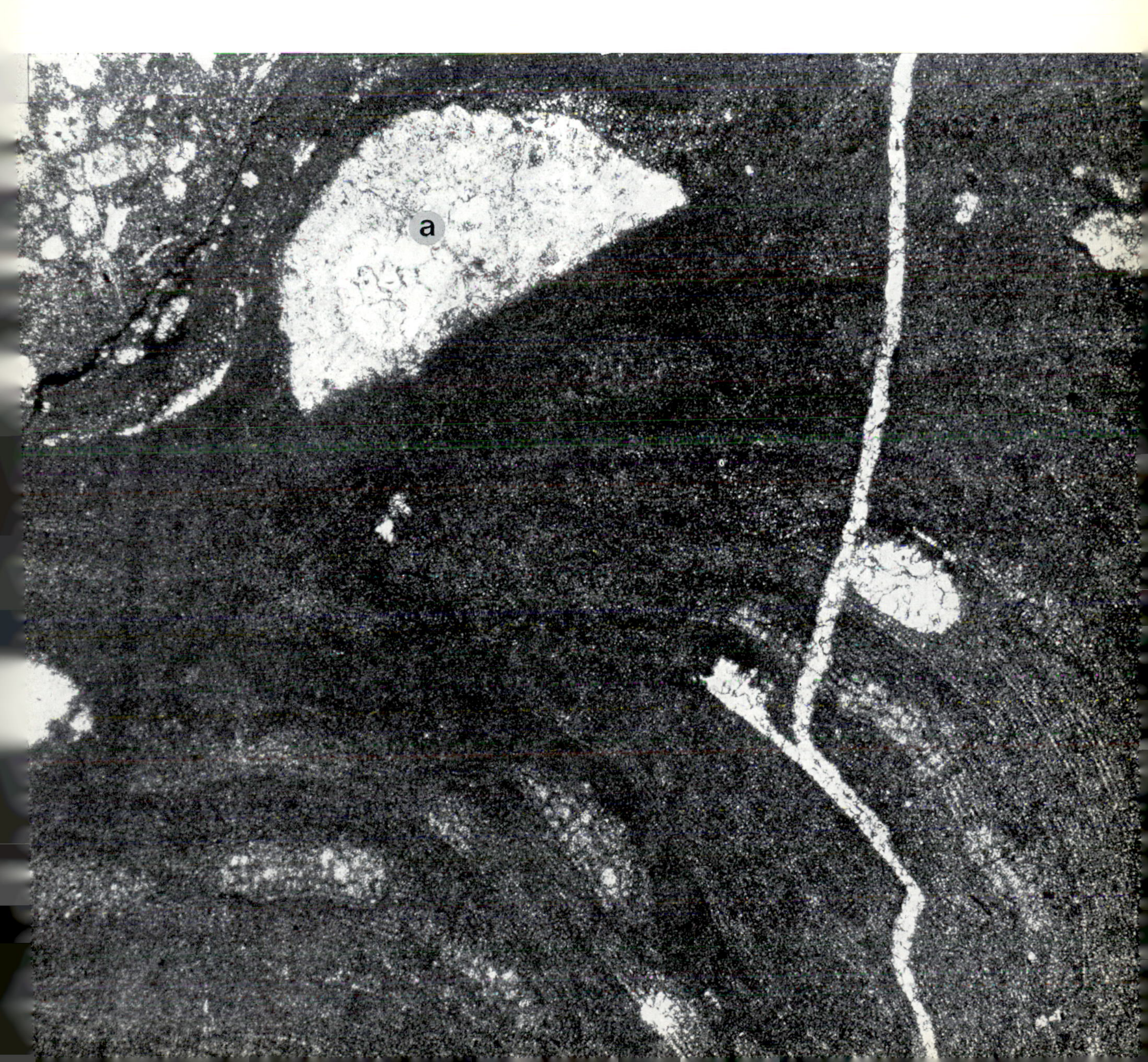
a

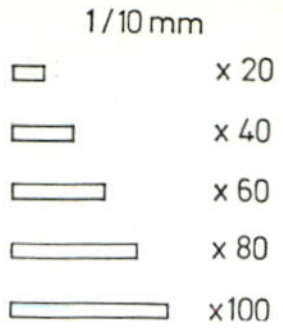

1/10 mm
x 20
x 40
x 60
x 80
x100

Fig. 1. Coralline algae, *Archaeolithothamnium*. Note calcite-filled individual ovate sporangia and layered rectangular cellular microstructure. Darker lines represent interruptions in algal growth. Small foraminifers at upper right. Crystal River Formation, Eocene, abandoned quarry, NE $\frac{1}{4}$ SW $\frac{1}{4}$ sec. 6, T. 19 S., R. 18 E., Citrus County, Florida, U.S.A. IU 8099–283, ×80.

Fig. 2. Algal limestone featuring crustose, forms probably *Lithophyllum*, with very thin perithallus (a) and thicker hypothallus (b). Oligocene, subsurface well south of München, Bavaria, Germany. IU 8099–448, ×80.

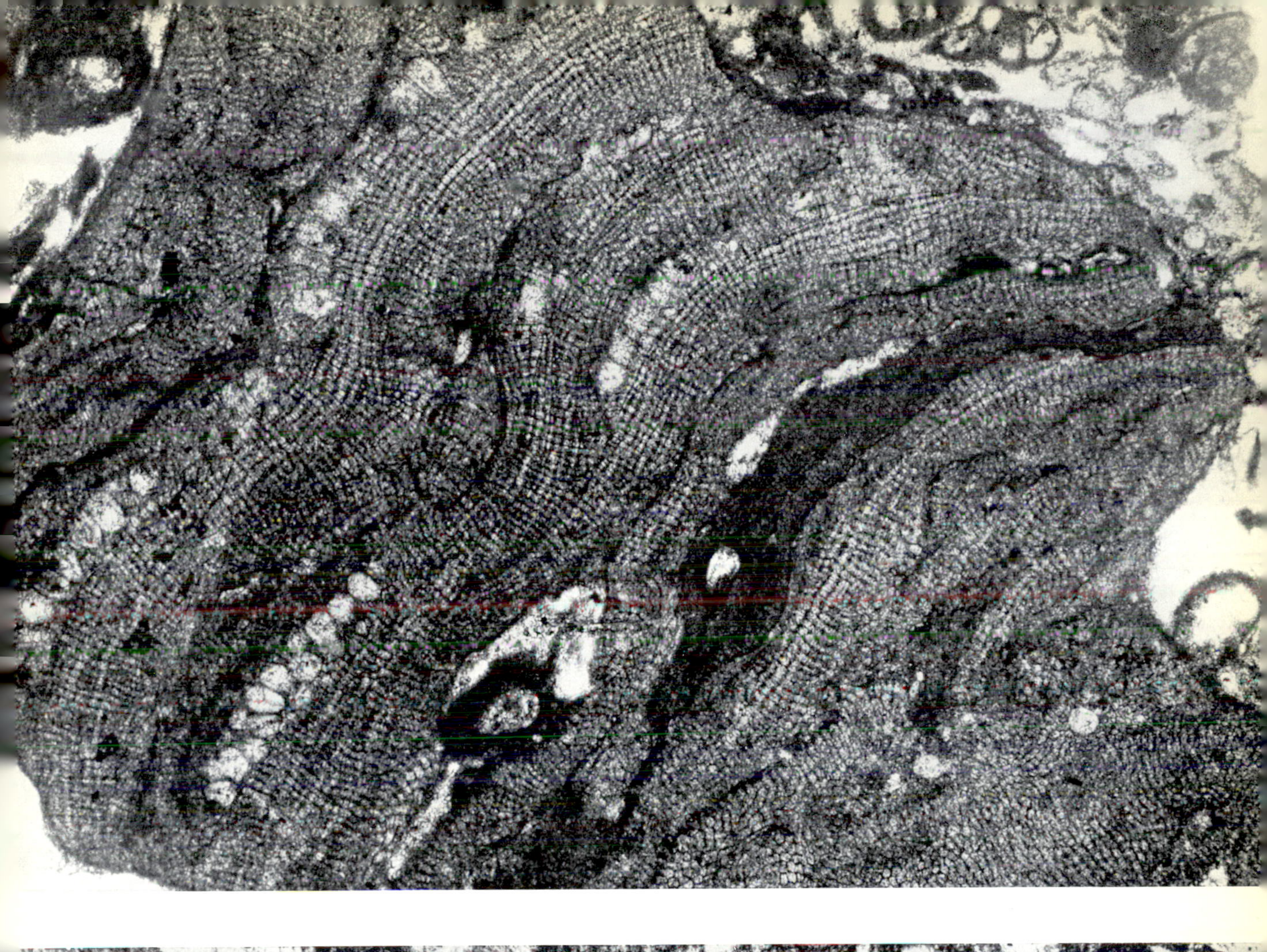

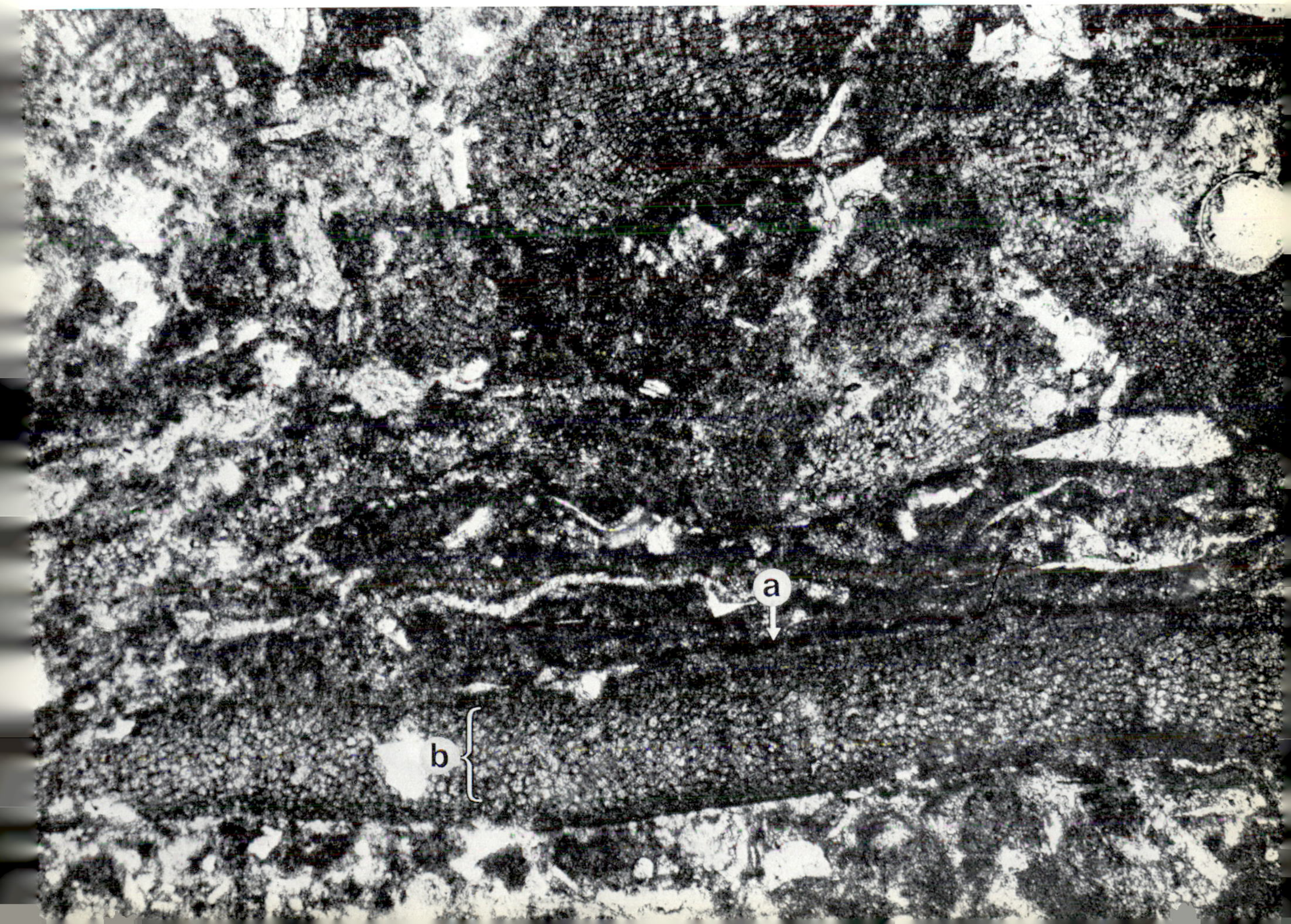

1/10 mm
x 20
x 40
x 60
x 80
x100

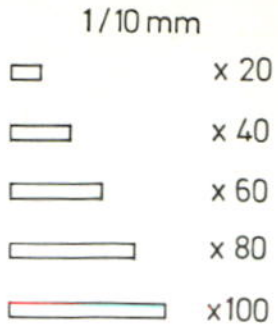

Fig. 1. The codiacean algae *Halimeda* displaying thin outer tubes perpendicular to surface and larger randomly oriented internal tubes. Intertwined internal tubes impart a spaghetti appearance to thin sections of fragments of *Halimeda*. Recent, Chireerete Island, Bikini Atoll, Marshall Islands, Pacific Ocean. IU 8099–476, ×40.

Fig. 2. Cellular calcareous coralline alga at lower center and broken foraminiferal fragment at upper center. Matrix of calcareous mud and some clear quartz grains. Upper Eocene, near highest bed on Lou Cachaou, Biarritz, Basses Pyrénées, France. IU 8099–2B, ×40.

Fig. 3. *Halimeda* and fragment of large foraminifer (a) at right center edge of photomicrograph. As fig. 1, ×20.

Fig. 4. Phylloid (leaflike) algae in cross section. No well-defined internal structures are visible in this photomicrograph, and algal identification is based on structures known from better preserved materials of codiacean or primitive red algae from other localities. Note internal increase in grain size away from walls; this suggests original plant structure left a void subsequently filled by open space sparry calcite cement. Matrix of dark pelletal mud showing small clear fragments, probably from broken algal fronds. Gothic Formation, Middle Pennsylvanian, sec. 21, T. 15 S., R. 84 W., Gunnison County, Colorado, U.S.A. IU 8099–312, ×20.

Fig. 5. Fragment of the codiacean alga *Halimeda*. The intertwined interior tubes become smaller and intersect the surface of the fragments at right angles. Portion of gastropod shell at lower left. As fig. 1, ×20.

Fig. 6. The codiacean alga *Halimeda*. Internal intertwined tubes exhibited, but surface of fragment abraded and does not show tubes intersecting surface at right angles. Very lightly cemented principally along grain contacts. Recent, Rongelap Atoll, Tufa Island, Marshall Islands, Pacific Ocean. IU 8099–473, ×40.

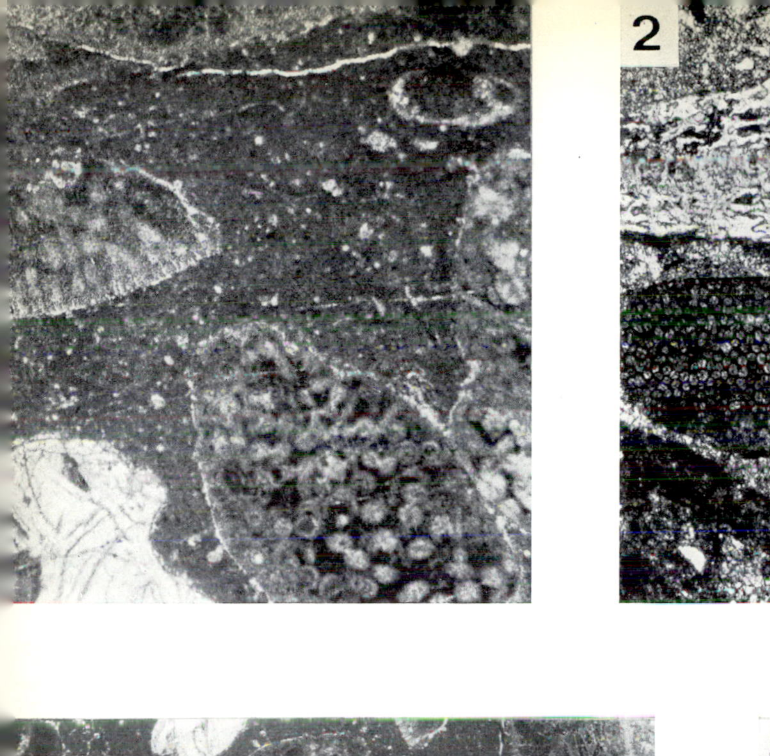

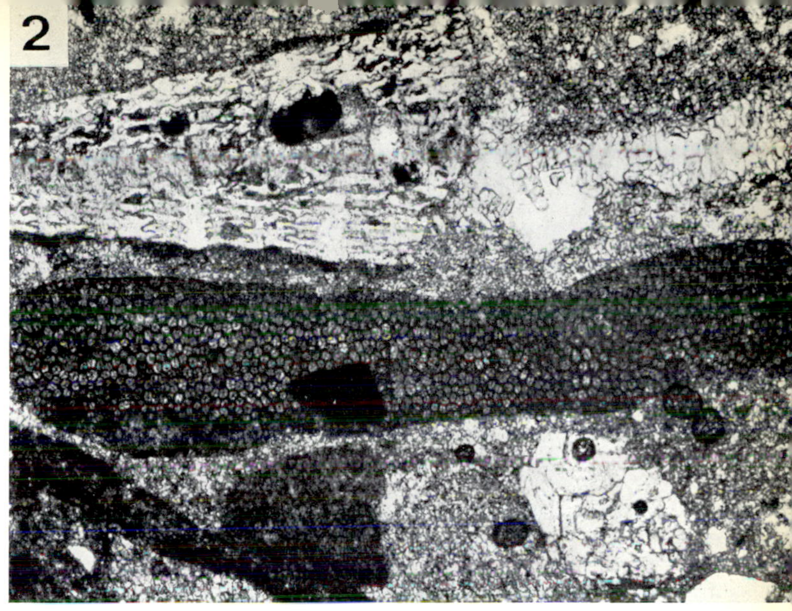
2

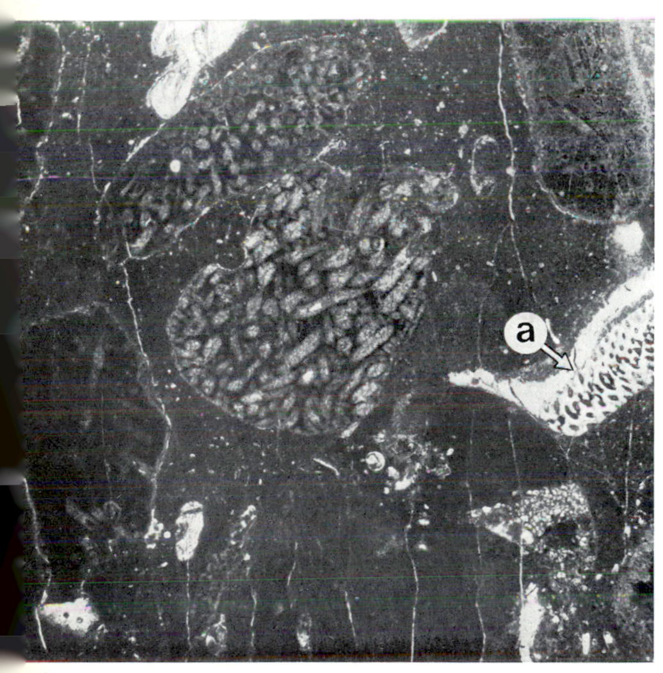
a

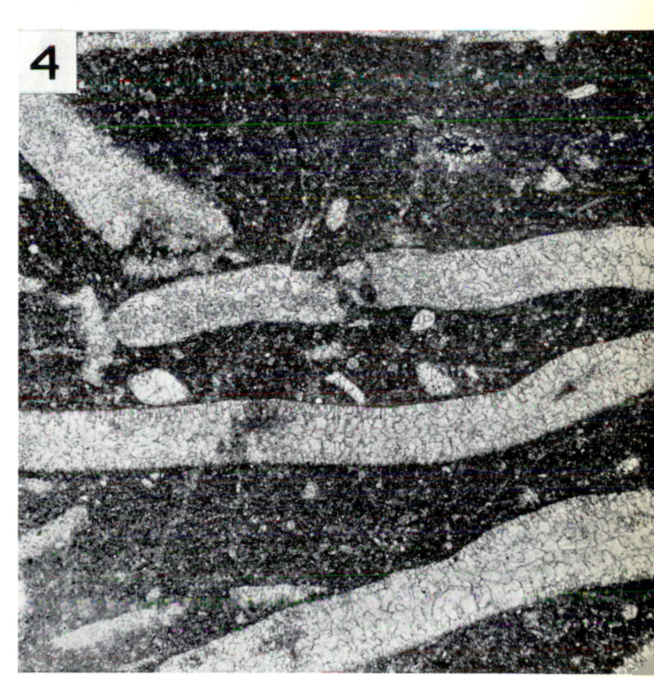
4

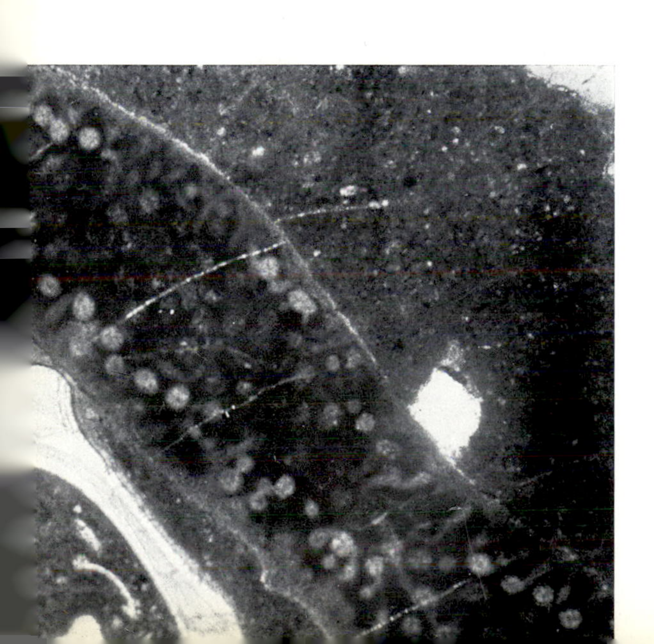

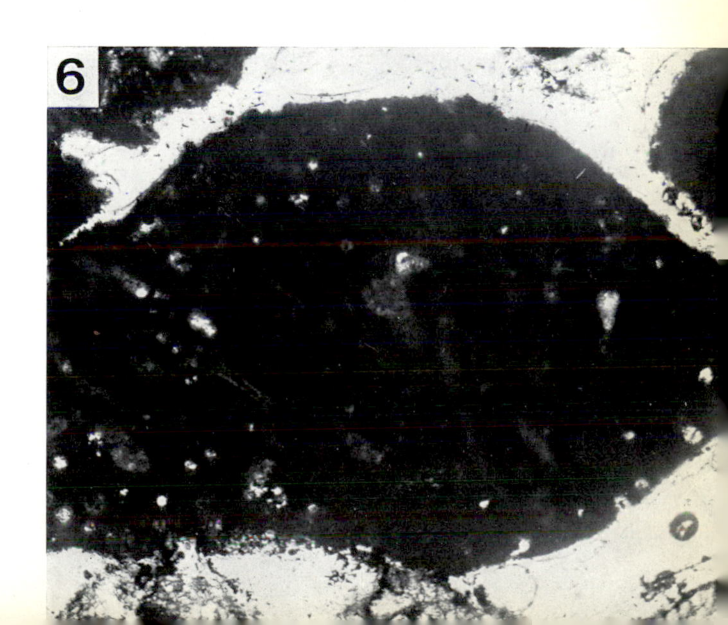
6

1/10 mm
x 20
x 40
x 60
x 80
x 100

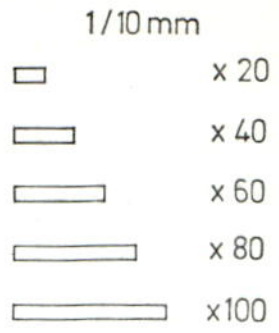

Fig. 1. Longitudinal section of dasycladacean algal fragment and foraminifers, principally fusulinids. Clear calcite cement, probably altered or recrystallized. Nansen Formation, Pennsylvanian, Jugeborg Fjord, northwest Ellesmere Island, District of Franklin, Northwest Territories, Canada. IU 8099–138, ×40.

Fig. 2. Limestone containing *Mizzia*, a dasycladacean alga, which is seen in transverse and longitudinal sections. Note hollow main branch having perforate walls (a) for side branches. Observe rim cement at left center. Tansill Formation, Permian, Guadelupe Mountains, Eddy County, New Mexico. IU 8099–523, ×20.

Fig. 3. Tangential section of dasycladacean alga, possibly *Epimastopora*, exhibiting small circular openings for side branches. Also longitudinal section (a) and small foraminifers. Probably altered cement matrix. As fig. 1.

Fig. 4. *Sphaerocodium* nodule. Note the "spaghetti" aspect of the small intertwined calcareous tubes. See also Pl. 62–1. Middle Silurian, 3 kilometers northeast of Visby, Gotland, Sweden. IU 8099–332B, ×40.

Fig. 5. Charophyte oogonia exhibiting spiral grooving. Note spirals in end section (a) at upper right. Oogonia filled with clear calcite. Matrix of silty argillaceous mud. Morrison Formation, Upper Jurassic, Park Range, near Steamboat Springs, Routt County, Colorado, U.S.A. U.S. Geological Survey loan ×40.

Fig. 6. Single charophyte oogonium associated with codiacean alga, possibly *Cayeuxia* sp. As fig. 5.

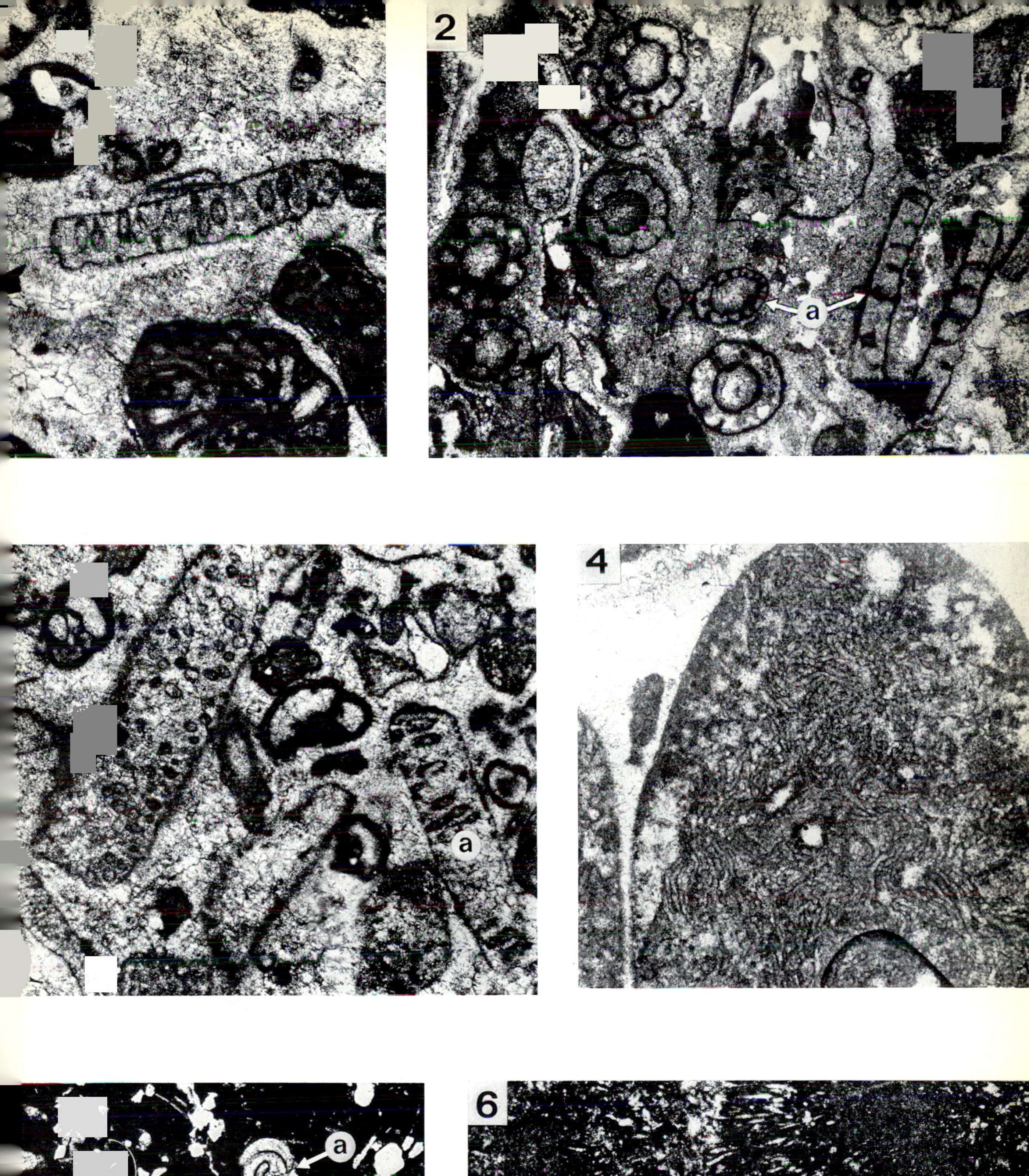
2
a
4
a

a

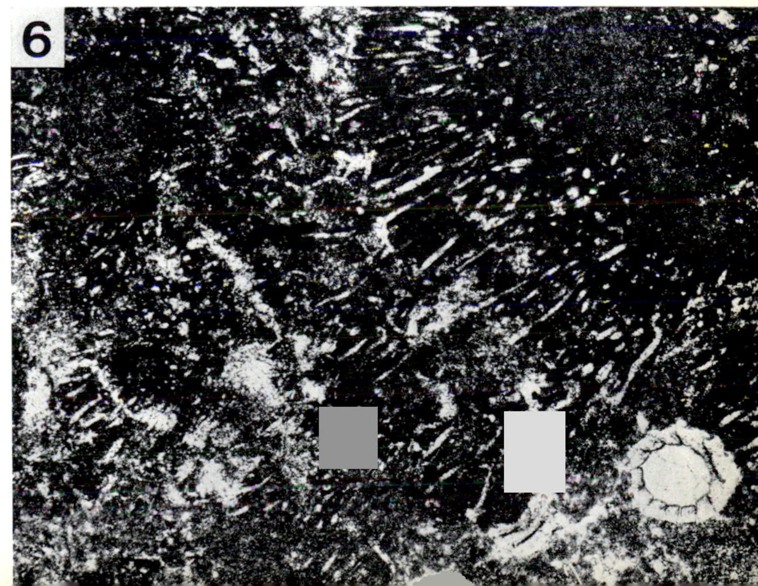
6

1/10 mm
x 20
x 40
x 60
x 80
x 100

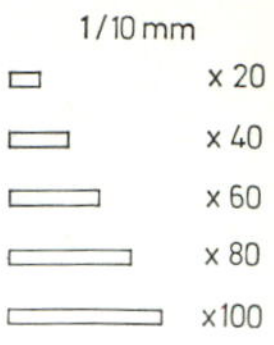

Fig. 1. Dasycladacean fragments in pelletal mud. Porous thinly calcified branches filled with calcareous mud and difficultly distinguishable from matrix. Both tangential and longitudinal views present. Jurassic, Simbrivio Valley, Simbruini Mountains, Lazio, Italy. IU 8099–168, ×80.

Fig. 2. Algal biscuit (oncolite, a type of stromatolite) showing interrupted inner laminae and continuous outer laminae. Flagstaff Limestone, Paleocene, right fork Hobble Creek Canyon, sec. 24, T. 7 S., R. 4 E., Utah County, Utah, U.S.A. IU 8099–571, ×20.

Fig. 3. Stromatolite showing dark lime mud layers and intervening sparry calcite filled irregular layers possibly representing gas bubbles or spar filled areas of former organic material. Tyrone Limestone, Middle Ordovician, Garrard County, Kentucky, U.S.A. IU 8099–910, ×20.

Fig. 4. Large fragment of dasycladacean alga. Clear calcite wall exhibits no distinctive microstructure. Mud-filled perforations represent position of former branches. Dark mud matrix contains difficultly discernible dark-walled quinqueloculine foraminifers and echinoderm plates. Photograph deliberately mounted upside down (geopetal fabric at lower right). Urgonian-Barremian, Lower Cretaceous, Coupe de St. Chamas, Bouches du Rhône, France. IU 8099–777, ×20.

Fig. 5. Cross sections of partially mud-filled broken and/or eroded specimens resembling charophyte oogonia (calcified fruiting structures) but assigned to the foraminifer *Umbellina* on the basis of the aperture in the specimen on the left. Calcispheres lack apertures but sections of *Umbellina* not showing the aperture are difficult to distinguish from some calcispheres. Radiolitid calcispheres, where well-preserved, exhibit radial divisions in the wall that are not present in these specimens. Matrix is pelletal mud. Note properly oriented geopetal fabric. Jeffersonville Limestone, Middle Devonian, active quarry W½ Grant 132, Clark Military Survey, Sellersburg, Clark County, Indiana, U.S.A. IU 8099–986 (SQ–22), ×40.

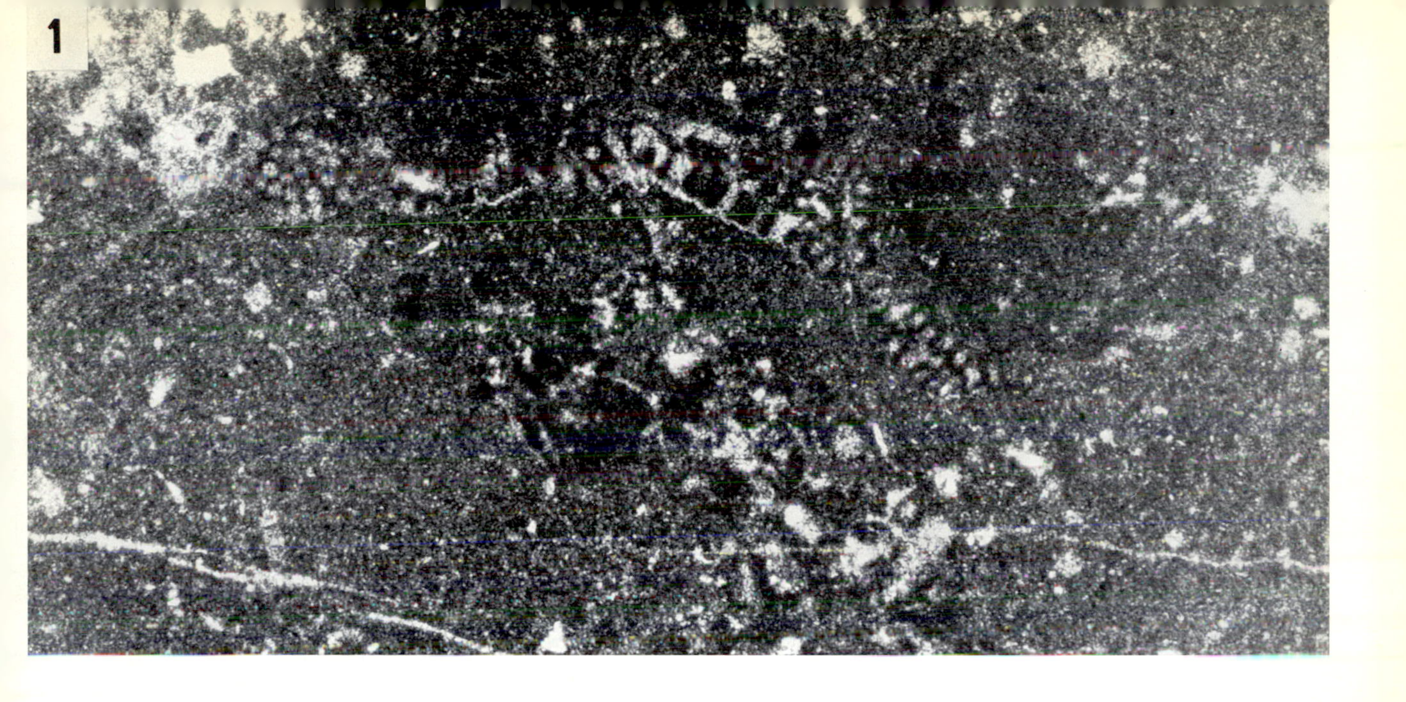
1

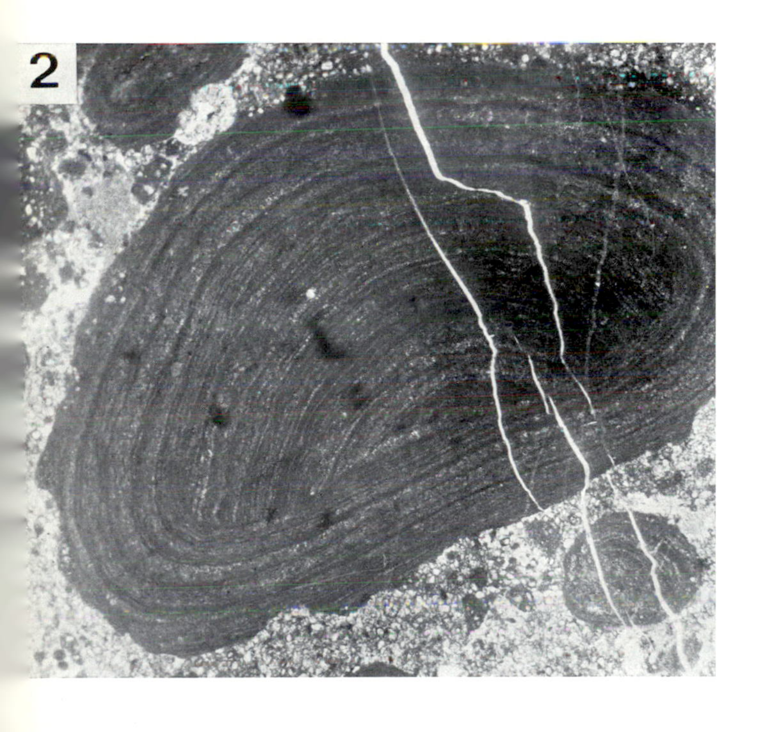
2

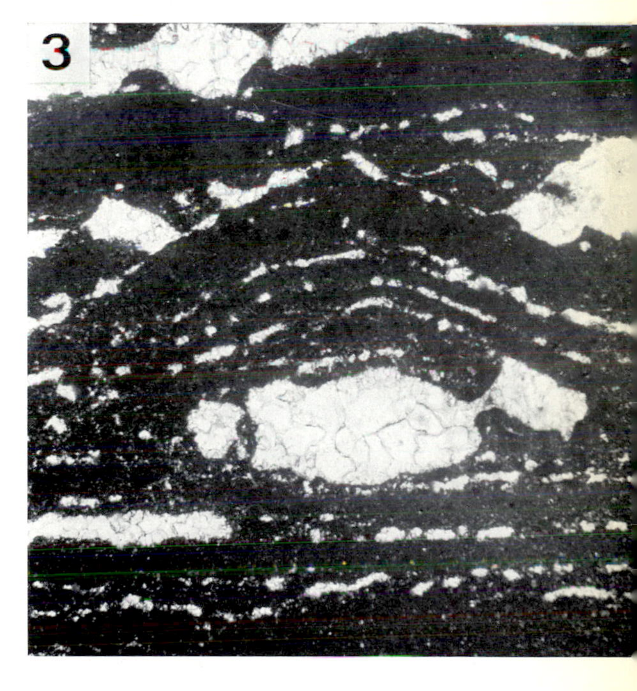
3

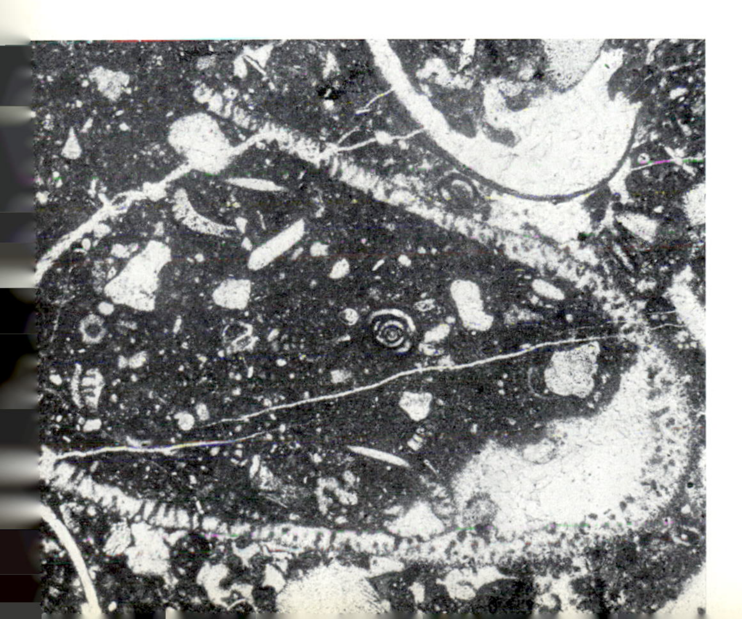

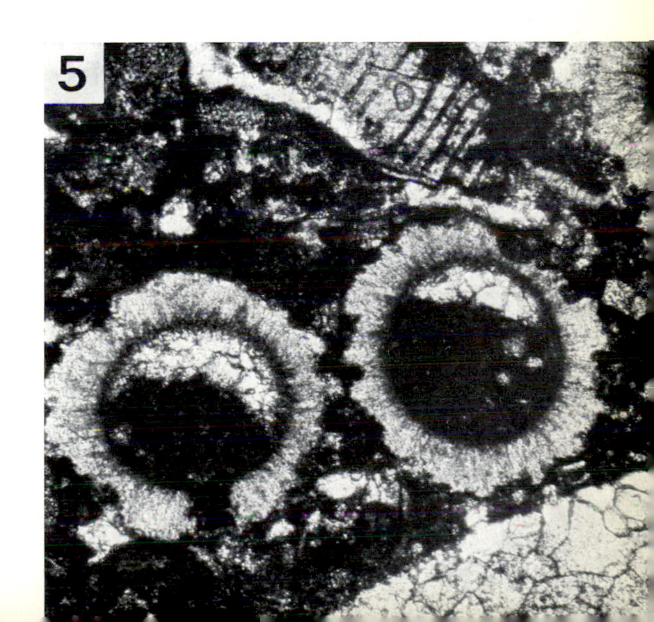
5

1/10 mm
x 20
x 40
x 60
x 80
x100

Fig. 1. Portion of algal ball composed principally of intertwined closely packed tubes of *Sphaerocodium* that contain internal "septa" (equatorial flanges; poorly shown at this magnification) and beaded appearance due to branching of filaments. Large diameter encrusters are foraminifers, *Wetheredella,* which are common associates of *Sphaerocodium* in algal nodules. See also Pl. 60–4. Middle Silurian, 3 kilometers northeast of Visby, Gotland, Sweden. IU 8099-332B, ×80.

Fig. 2. Portion of algal ball composed of *Girvanella* displaying small sinuous tubes. Cambrian, north end of Pensacola Mountains, from moraine close to the base on west side of Spann Mountain, Latitude 81° 45′ South, Longitude 43° West, Antarctica. IU 8099–984, ×100.

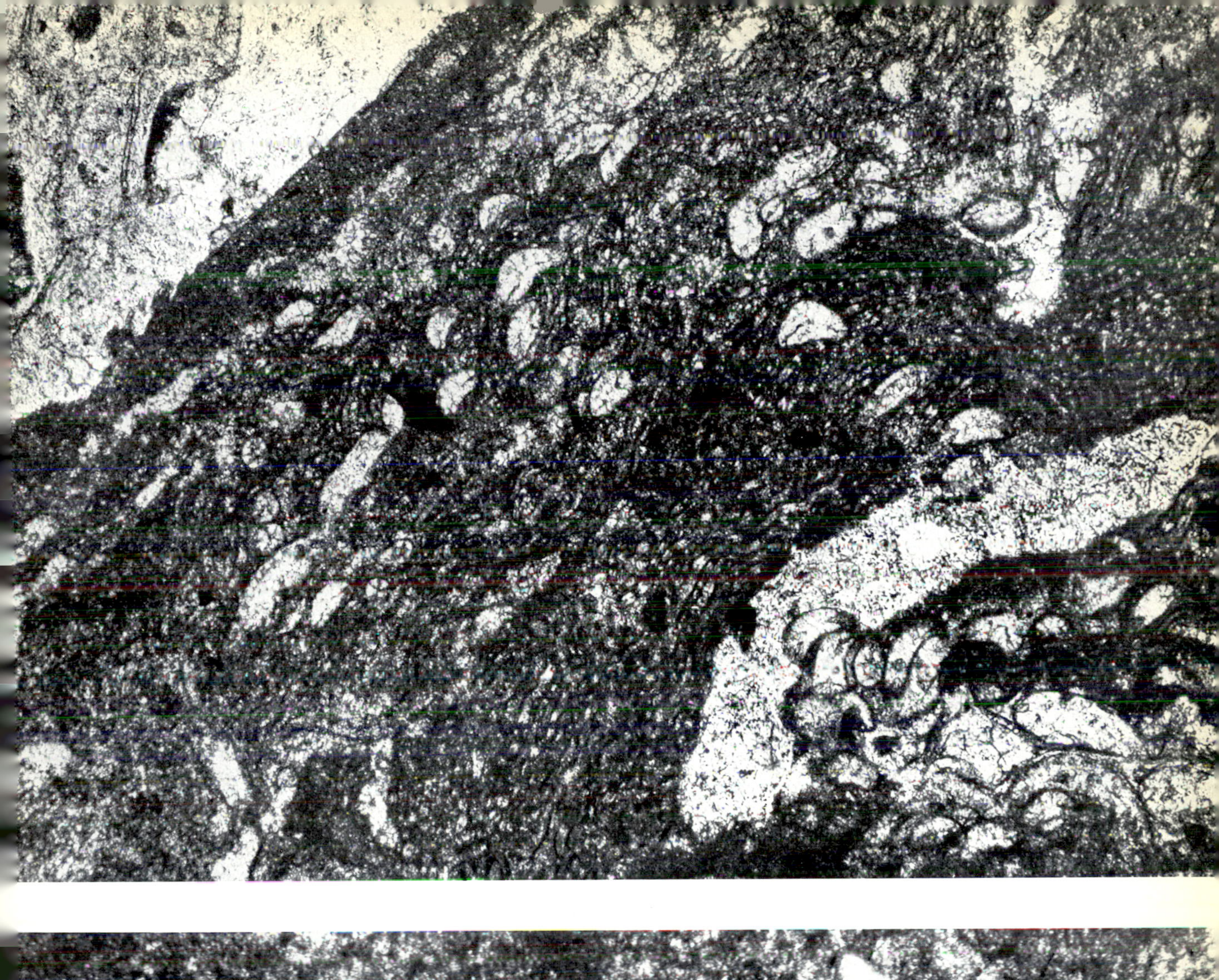

1/10 mm
x 20
x 40
x 60
x 80
x 100

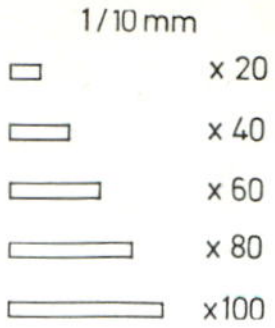

Fig. 1. *Tubiphytes*, a presumed alga showing dark dense fine-grained tissue and broad layering. The calcite filled voids are interpreted by some algologists as the position of the material (subsequently destroyed) to which this encrusting alga attached. These North American forms originally were referred by RIGBY (1958) to the hydrozoans. Wolfcamp Formation, Lower Permian, subsurface well, Ray Smith # 2 Calverly, depth 7881 feet, Glasscock County, Texas, U.S.A. IU 8099–518, ×40.

Fig. 2. *Tubiphytes*, another view. As fig. 1.

Fig. 3. *Tubiphytes* (?), exhibiting dark fine-grained skeleton but pierced by more narrow calcite areas than the Permian forms illustrated in previous figures. *Tubiphytes* has not been reported in post-Paleozoic strata, but forms comparable to those figured here are illustrated by Radiočić (1966, Pls. 11 and 47) who identified them as blue-green algae. See also Pls. 20–3 and 73–2. JOHNSON (1964, Pl. 38, figs. 3, 4; Pl. 39, figs. 1–3) illustrated similar forms that are referred to the encrusting codiacean algal genus *Lithocodium*. This is the way, dear reader, that one attempts to identify unknowns by means of the matching method. Jurassic, Marsica Mountains, L'Aquila, Abruzzo, Italy. IU 8099–165. ×40.

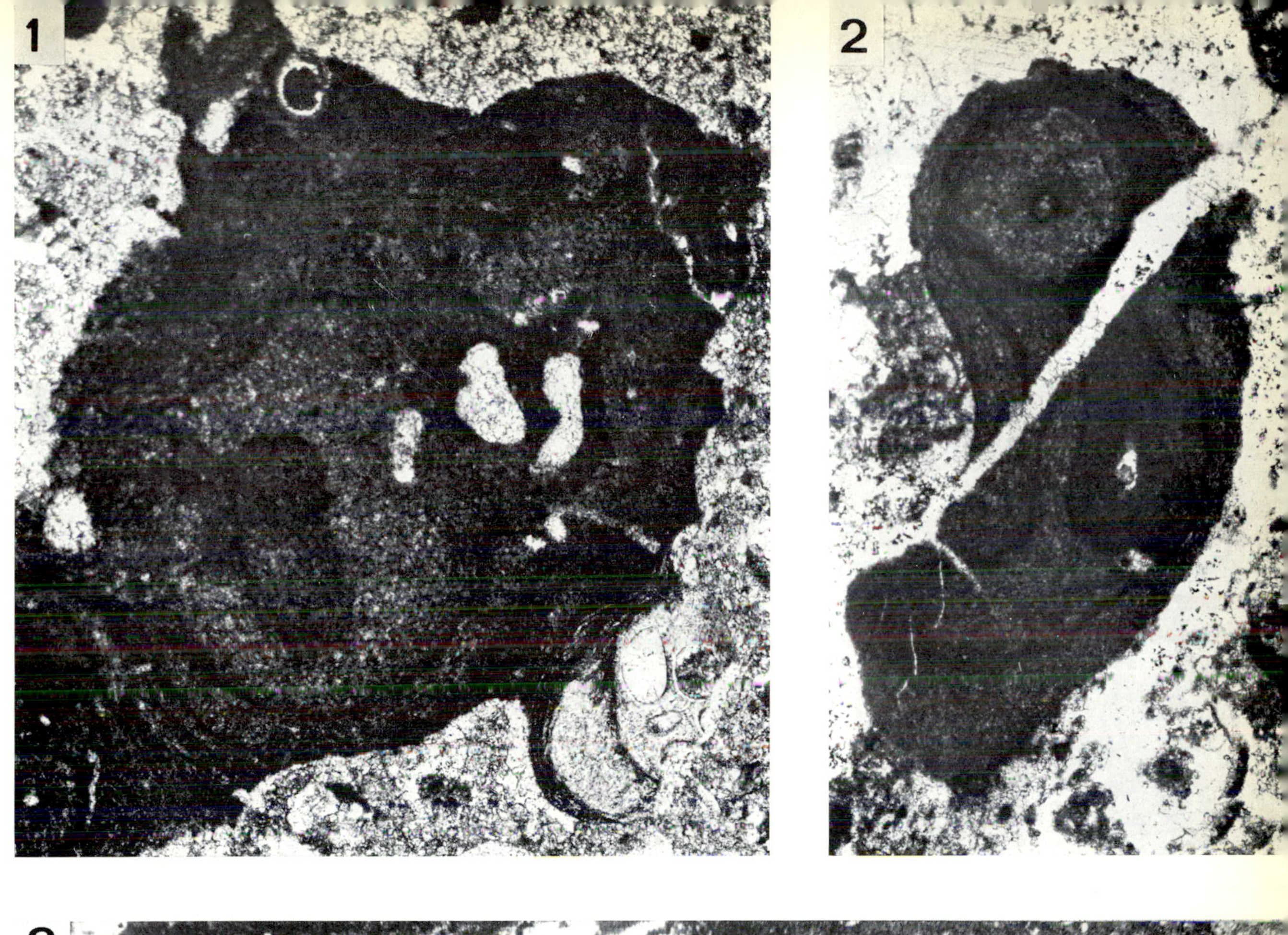
1
2

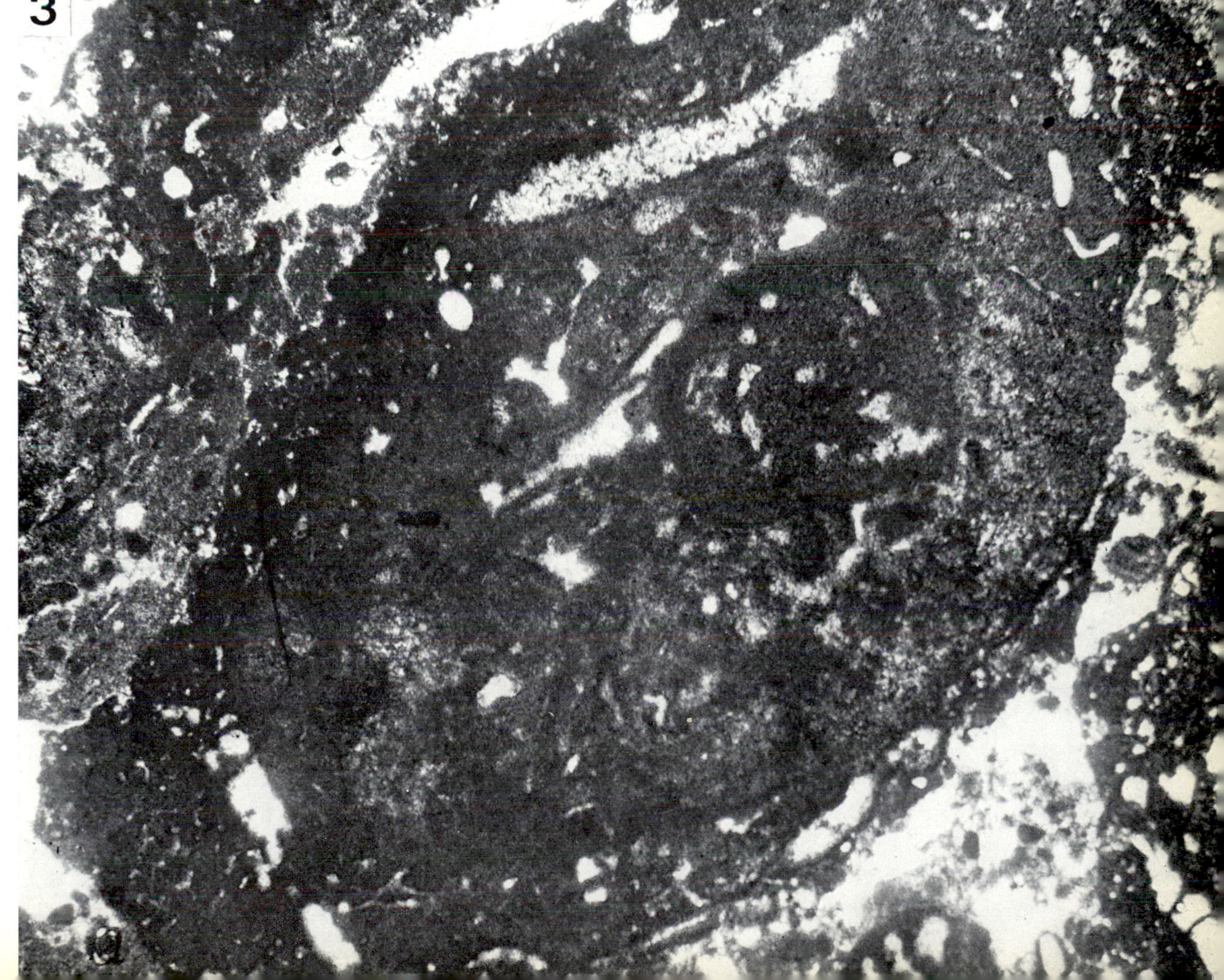
3

PLATE 64. CENOZOIC

1/10 mm

x 20

x 40

x 60

x 80

x100

Very porous calcarenite showing thin rims of initial cement. Cement rims have a drusy ("marching men", "picket fence") surface. Cement completely infills only the smallest voids between grains. Solution packing is absent. Compare this open packing with the tight packing of Pl. 68. Framework includes grains of molluscan debris (note well-developed crossed-lamellar structure of large grain in center), miliolid foraminifers (a) cut in various directions, algae (b), *Halimeda* (c), and several altered grains of either rock fragments or unidentifiable fossil debris. Belmont Limestone, Pleistocene, Devonshire Bay, Bermuda, Atlantic Ocean. IU 8099–S142, ×50.

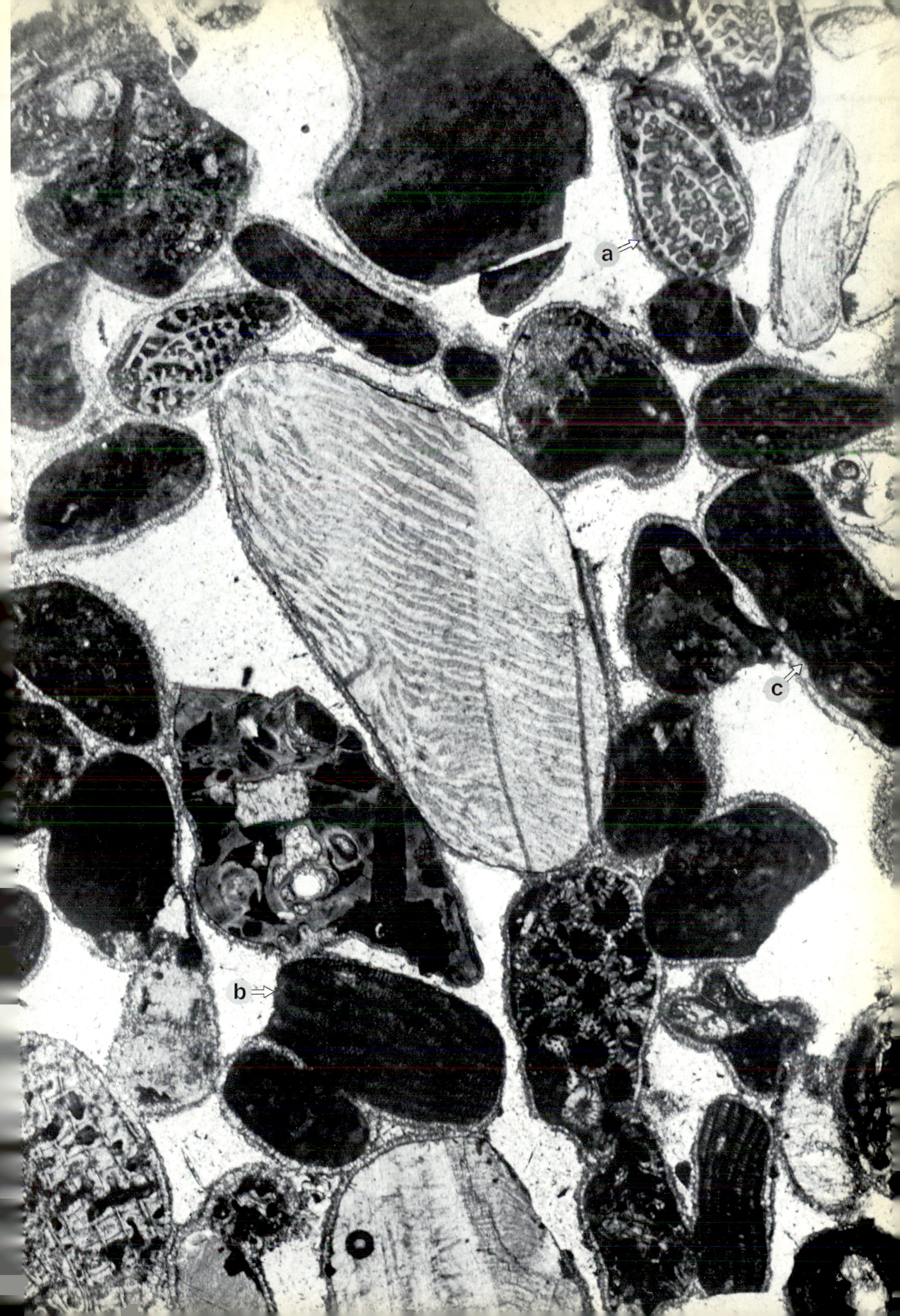
a
b
c

1/10 mm
x 20
x 40
x 60
x 80
x100

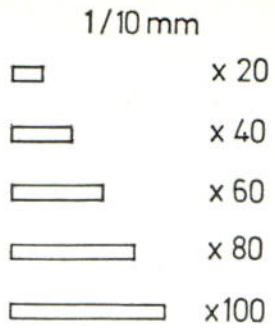

Grainstone, well-sorted, well-washed (poorly cemented, impregnated with plastic). Prominent molluscan fragments (a) showing crossed-lamellar structure, algae, echinodermal plate, detrital carbonate rock fragments of various kinds, reworked bryozoan, and superficial ooliths. Lack of matrix and lack of solution packing make this an easy rock to point count because constituents are clearly defined. Compare with Pl. 68. Typical of late Mesozoic or Tertiary limestones. Miocene, near Latitude 30° 25′ North, Longitude 19° 35′ East, Libya. IU 8099–826, ×50.

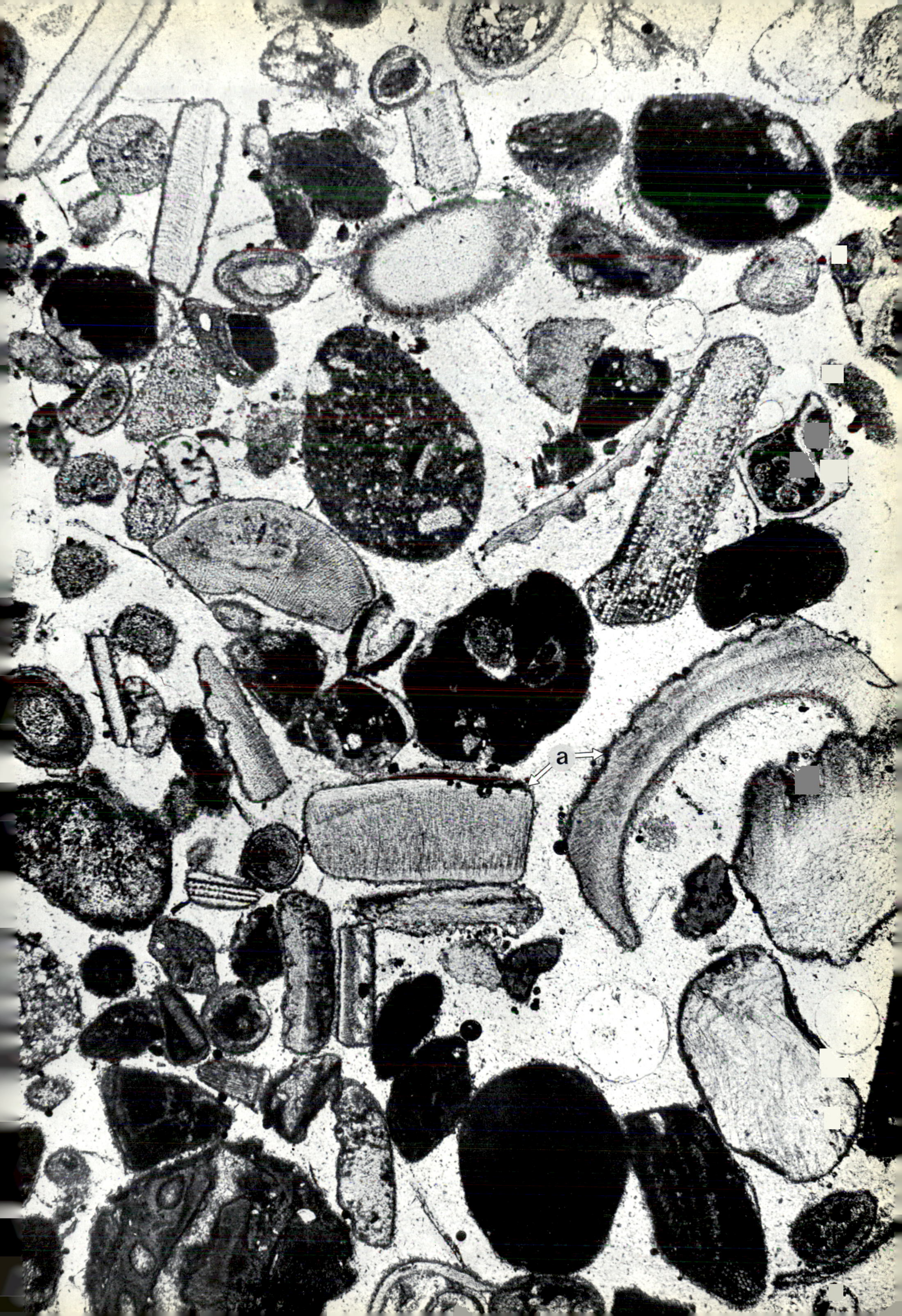
a

1/10 mm
x 20
x 40
x 60
x 80
x100

Fig. 1. Foraminiferal-algal limestone with sparry calcite cement. Some pelletal mud (a), encrusting algal fragment, *Lithoporella* sp. (b), has rather large cell size; spar-filled voids between cell layers represent position of sporangia (fruiting areas). Mariana Formation, Pleistocene, karenfeld, 1500 feet southwest of Pati Point, Guam, Mariana Islands, Pacific Ocean. IU 8099–475, ×80.

Fig. 2. Foraminiferal-algal limestone in micrite matrix. Echinoderm plate (a), rotalid foraminifers (b), sparry calcite (c), articulating coralline algae (d), some quartz silt. Upper Eocene, near highest bed on Lou Cachaou, Biarritz, Basses Pyrénées, France. IU 8099–2, ×40.

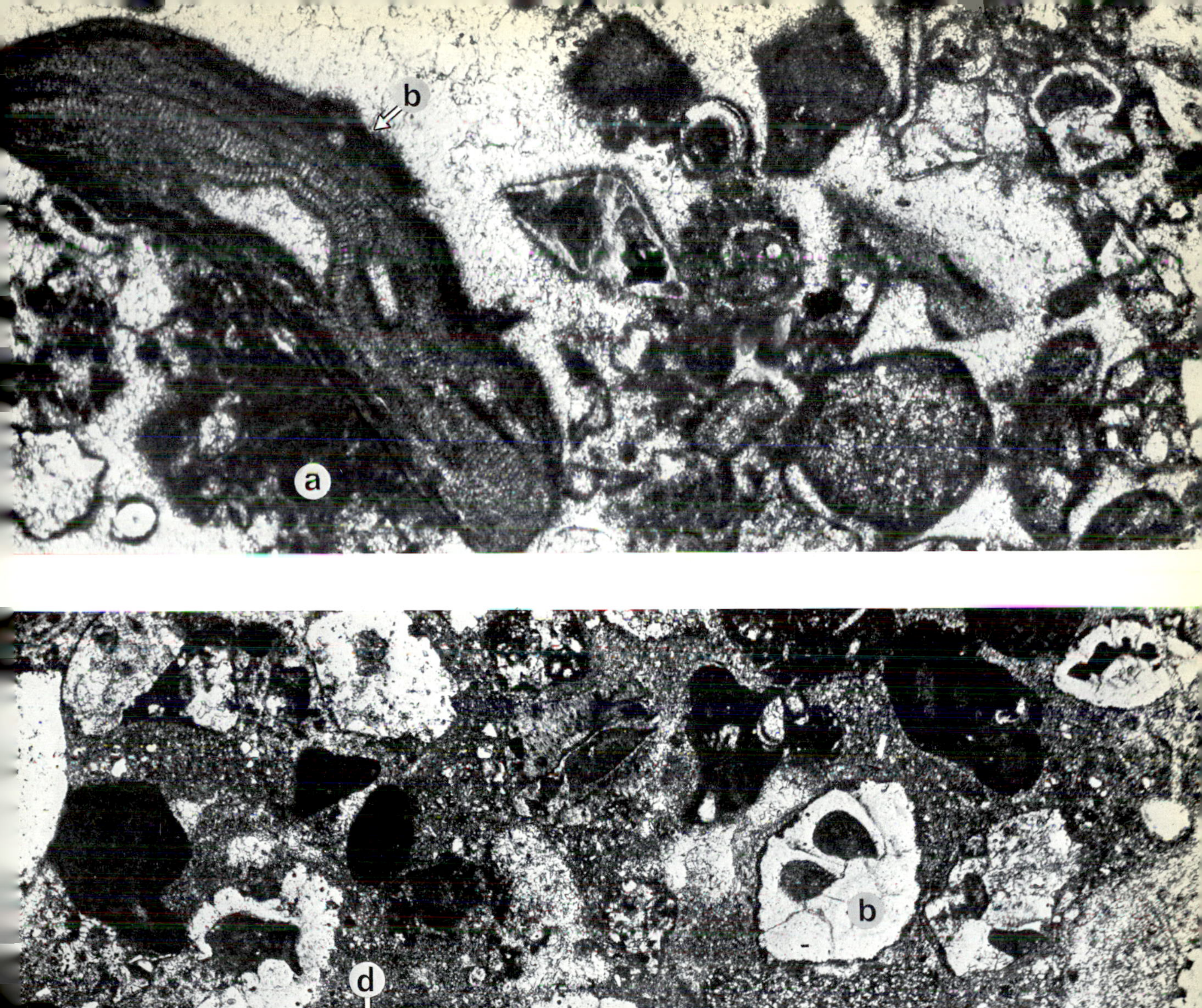
b
a

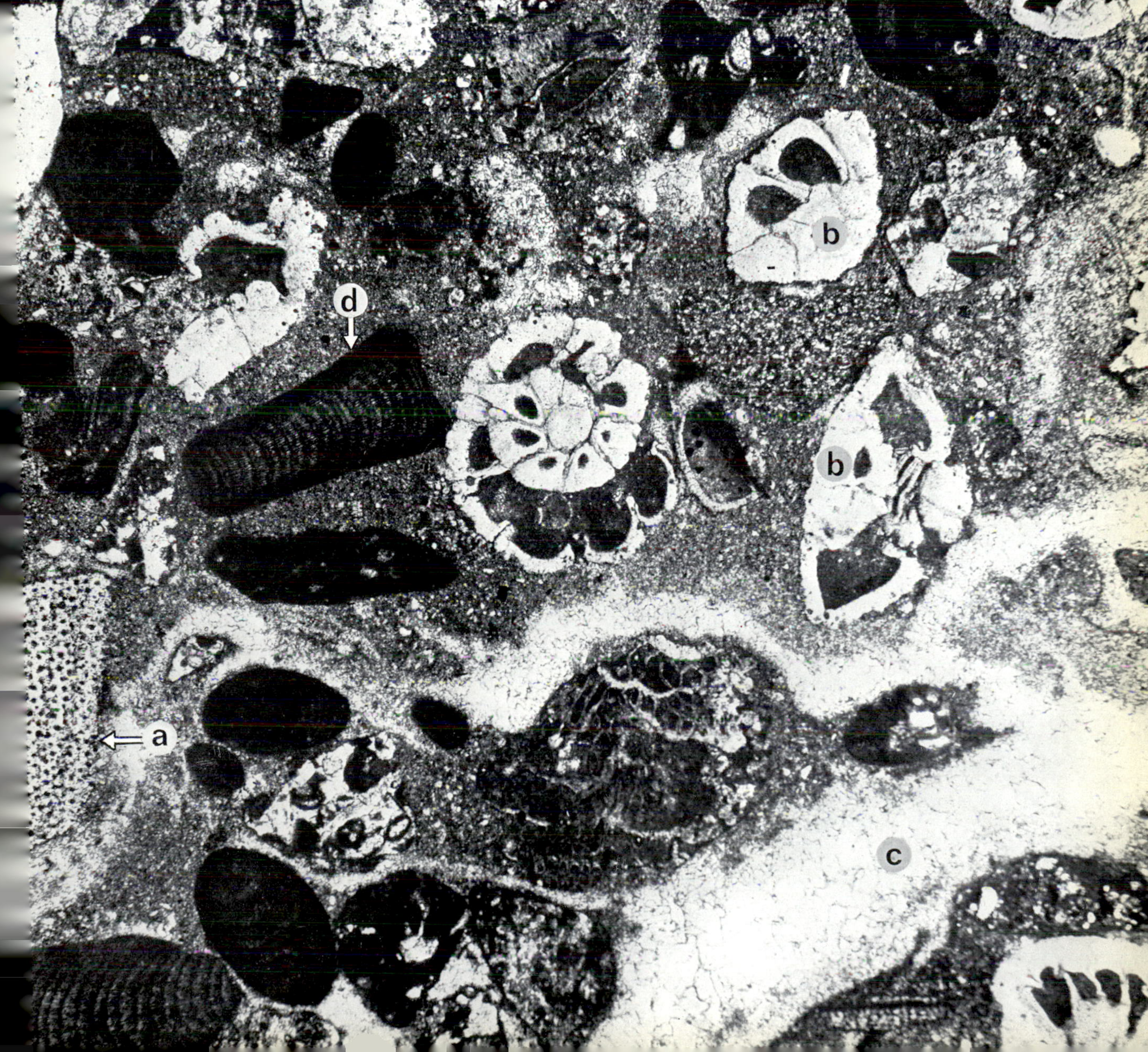
b
d
b
a
c

1/10 mm

▭	x 20
▭	x 40
▭	x 60
▭	x 80
▭	x100

Globigerinid limestone having dark mud matrix. Coiled shells, cut in numerous different planes, show one to five or more globular chambers. The perforate and finely spinose (hispid) shells are filled internally by both mud and calcite cement. Note that many small shell fragments are embedded in the matrix. Oligocene, Pietraroia, Benevento, Italy. IU 8099–367, ×100.

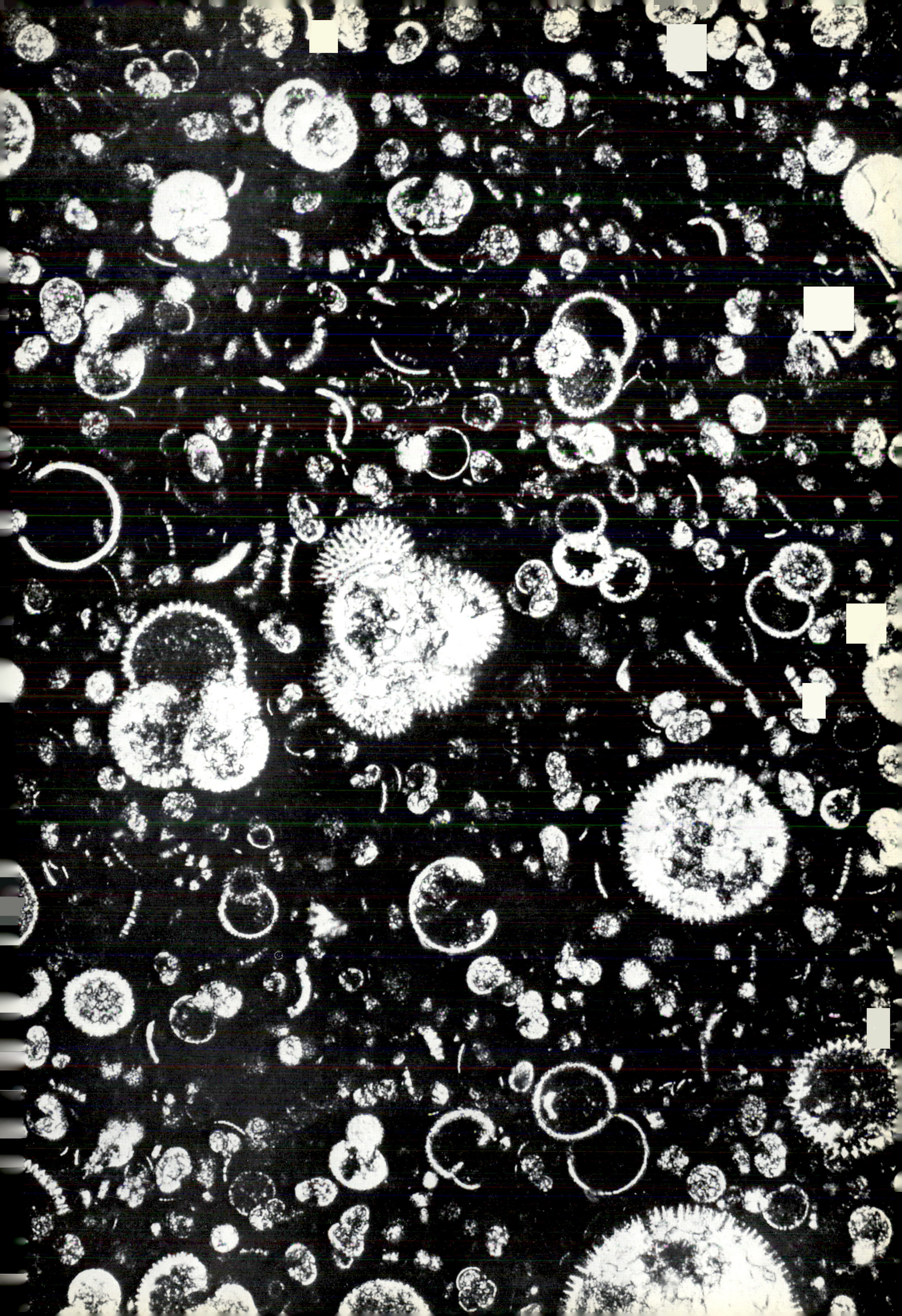

1/10 mm

x 20

x 40

x 60

x 80

x100

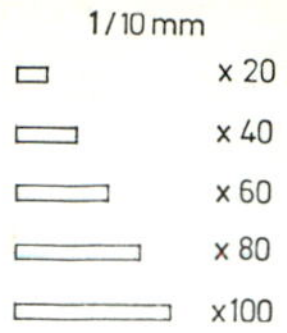

Closely packed, poorly sorted, quartz-bearing skeletal limestone. Bryozoan (a), foraminiferal (b), algal (c) and echinodermal (d) debris is present. Compare solution packing with Pls. 64 and 65. Ebishima Formation, Pliocene, Nagasaki Prefecture, Japan. IU 8099–967, ×50.

b
c
a
d

1/10 mm

x 20

x 40

x 60

x 80

x100

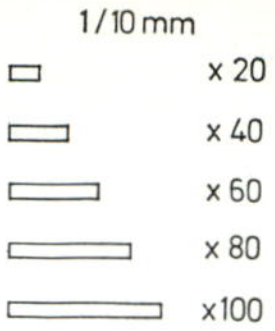

Poorly sorted, algal-echinodermal-bryozoan-foraminiferal-molluscan limestone. Large dasycladacean alga in center is cut obliquely and shows longitudinal section at right and tangential section at left. Perforations are mud filled. Probable echinoid spine in right center. Prominent mud-filled bryozoan at lower left. Coiled foraminifers (a) have dark, fine-grained walls and are difficult to distinguish from matrix. Mollusk fragment at upper left. Dark, dense pelleted matrix is difficult to study. Lower Cretaceous, Coupe de Chamas, Bouches du Rhône, France. IU 8099-777, ×50.

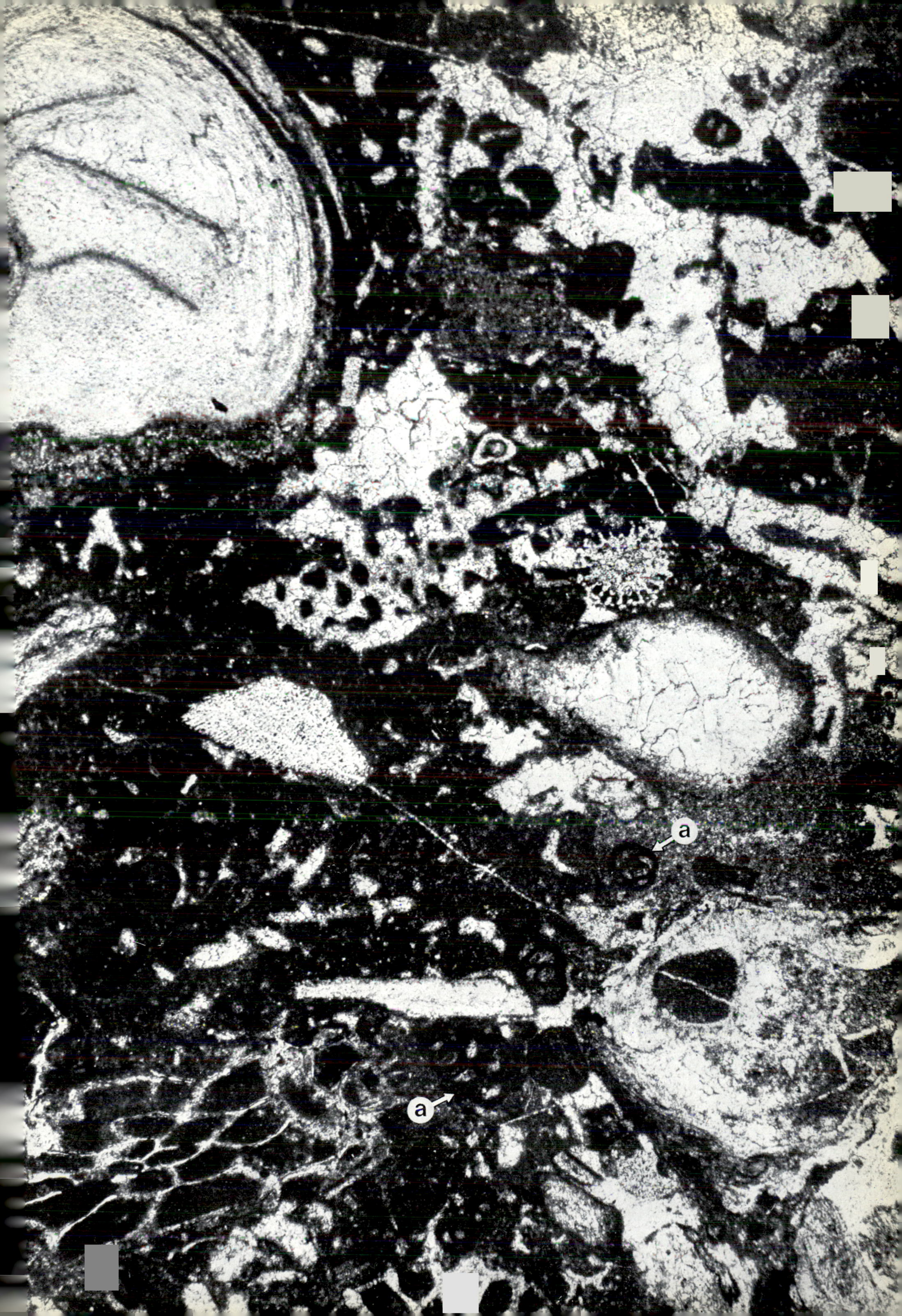
a
a

1/10 mm

x 20

x 40

x 60

x 80

x 100

Bryozoan limestone. Spar and mud matrix. Bryozoans exhibit transverse, longitudinal, and tangential sections. Zoaria (colonies) are bifoliate (a) and ramose (b); many zoaria appear broken and fragmented. Upper Cretaceous, Faxe Limestone Quarry, Sjaelland, Denmark. IU 8099–446, ×50.

a
b

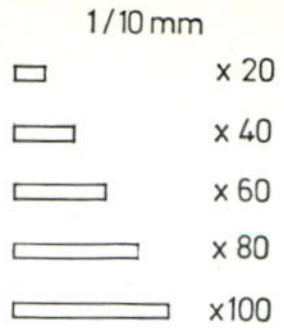

1/10 mm

x 20

x 40

x 60

x 80

x100

Poorly sorted, silty, micritic foraminiferal limestone. Biserial foraminifers (a) are cut in different planes. Note angular quartz grains. Foraminiferal section in lower-right corner appears uniserial (b). Note segmented debris (c), possibly foraminifers (*Frondicularia* sp.). Barra do Dande Limestone, Cretaceous, 50 kilometers northwest of Luanda, Luanda, Portuguese Angola. IU 8099-749, ×50.

a
c
a
b

1/10 mm
x 20
x 40
x 60
x 80
x100

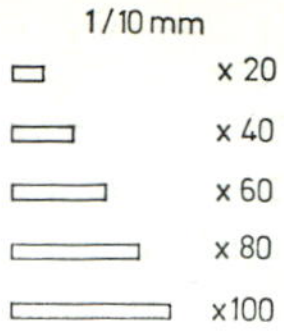

Foraminiferal-molluscan limestone exhibiting a large vertical fracture filling at right. Foraminifers are quinqueloculines shown in transverse and longitudinal sections. Recrystallization obscures some details, but limestone probably had an original well-sorted framework with some lime mud. Some foraminifers were abraded and filled with mud prior to lithification. Much solution packing. Upper Cretaceous, hillslopes 0.95 miles northeast of Lamuda, Barrio Tortugua, San Juan, Puerto Rico. IU 8099–758, ×50.

1/10 mm
x 20
x 40
x 60
x 80
x100

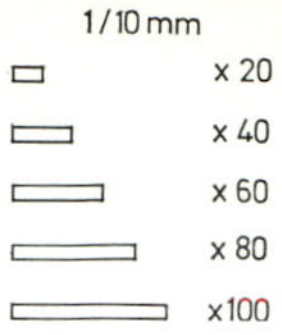

Fig. 1. Well-sorted foraminiferal limestone cemented by sparry calcite: foraminifers, *Kurnubia* sp. (a), porous echinodermal plates (b), quartz grains (c). Beersheva Formation, Oxfordian, Upper Jurassic, Makhtesh Ramon, Israel. IU 8099–635, ×40.

Fig. 2. Micritic, arenaceous algal-molluscan limestone. Note the calcite-filled voids in the large dark fragments of encrusting algae (a), ?*Lithocodium* sp. See Pls. 20–3 and 63–3. Additional biotic debris includes: molluscan fragments (b), echinoderm plate (c), and an obscure foraminifer (d). Quartz (e). Pliensbachian, Lower Jurassic, Coupe de la Ste. Baume, Bouches du Rhône, France. IU 8099–798, ×40.

a
b
c

e
c
d
a
b
a
b
e

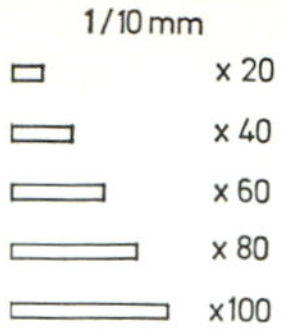

Fig. 1. Pellet-foraminiferal limestone. Coiled and uniserial foraminifers display diverse cross sections. Middle Jurassic, Montagna, Spaccata Dam, south of Alfedena, L'Aquila, Italy. IU 8099–373, ×40.

Fig. 2. Framework of loosely packed echinodermal debris and other skeletal debris in black calcareous mud. One foraminifer (a) is conspicuously ornamented. Some echinodermal debris has unusually coarse pores (b). Echinodermal arm ossicle (c). Rotwandel Limestone, Toarcian, Lower Jurassic, Steinernes Meer Mountains, Salzburg Province, Austria. IU 8099–554, ×80.

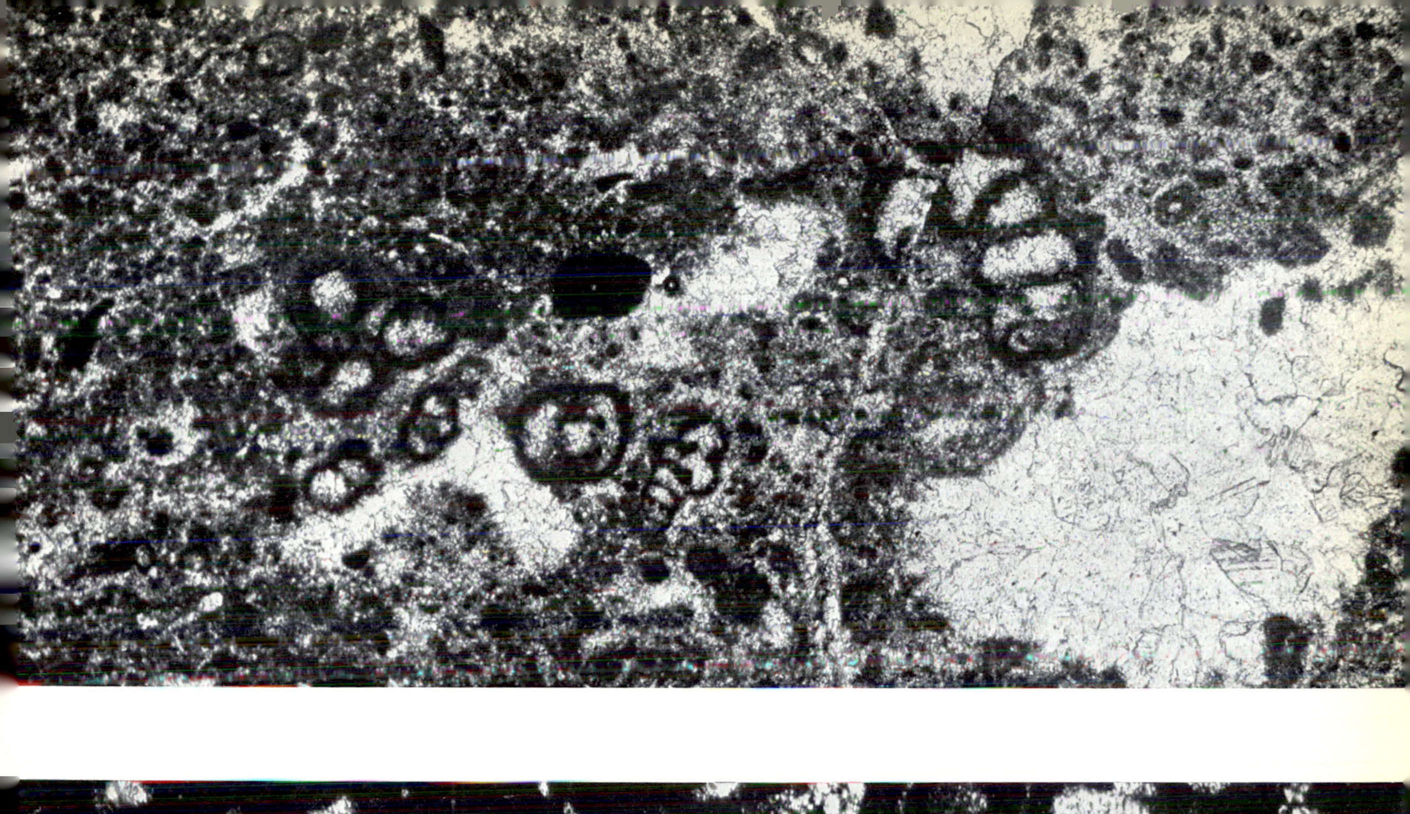

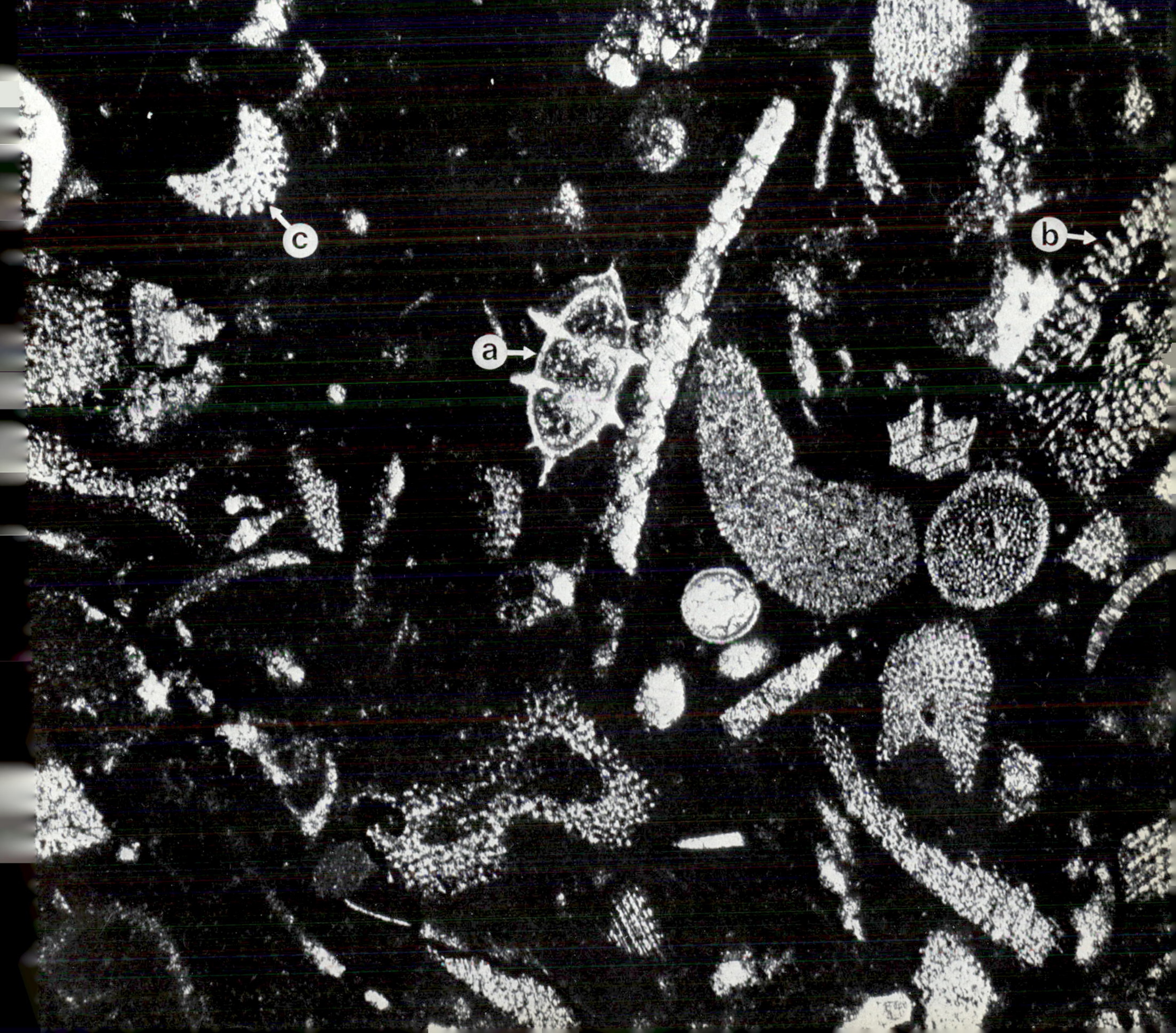
c
a
b

1/10 mm
x 20
x 40
x 60
x 80
x100

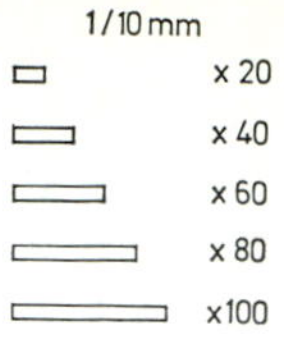

Well-sorted gastropod limestone. Matrix is calcite spar and micrite. Longitudinal and transverse sections of gastropods many of which are filled with dark ferruginous mud and small debris. Gastropod shells are recrystallized. Base of Triassic Limestone Group, Emarat, Heraz Valley, northeast of Tehran, Iran. IU 8099–307, ×50.

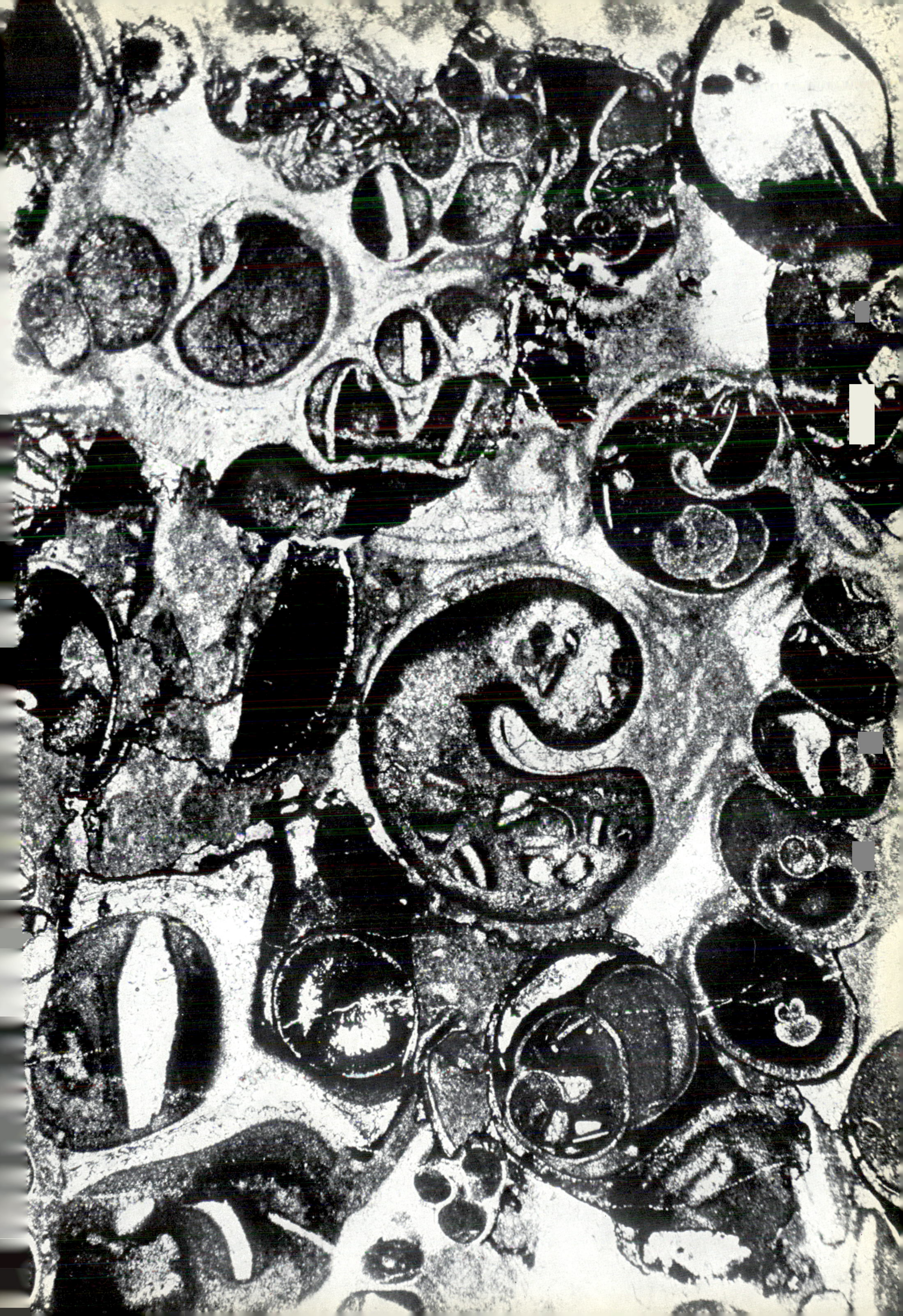

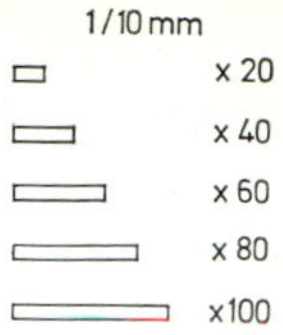

1/10 mm
x 20
x 40
x 60
x 80
x100

Poorly sorted molluscan-echinodermal limestone. Note variability in porosity of large echinodermal plate. Molluscan debris is either recrystallized or the leached shells were filled by calcite, silica, and pyrite. Pellets are a prominent part of the dark matrix. Whitehorse Formation, Upper Triassic, Llama Mountain, west-central Alberta, Canada. IU 8099–204, ×50.

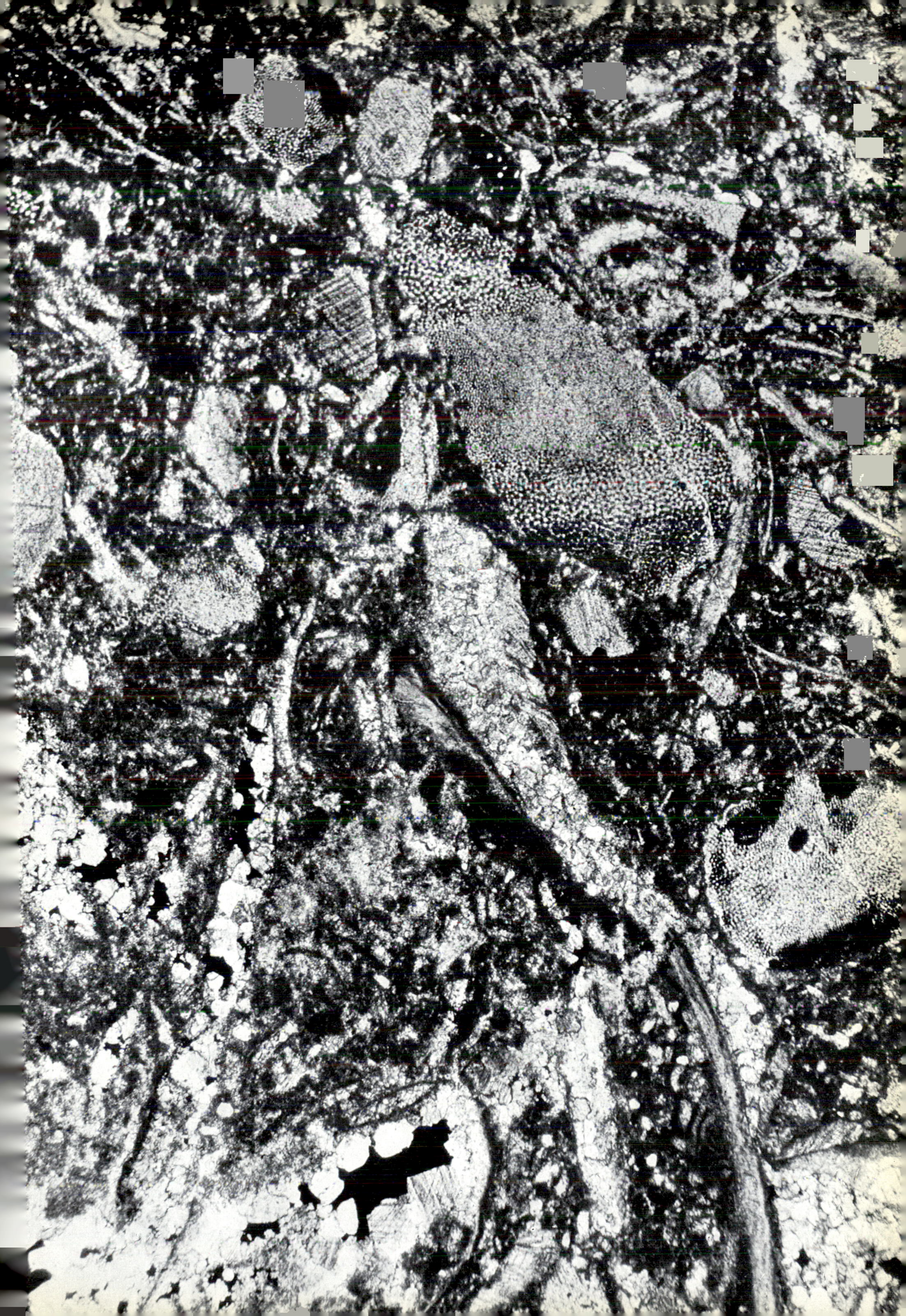

1/10 mm
x 20
x 40
x 60
x 80
x100

Brachiopod-ostracode-echinoderm limestone: large punctate brachiopod (a), ostracodes (b) frequently with both valves together, porous echinodermal plates (c), ? nubecularid (encrusting) foraminifers (d). Note dense skeletal framework in partially recrystallized lime mud. Some ostracodes exhibit dark borders of unknown origin. Wolfcamp Formation, Permian, Texaco, Inc. 2 Curry B, depth 7861 feet, Glasscock County, Texas, U.S.A. IU 8099–516, ×50.

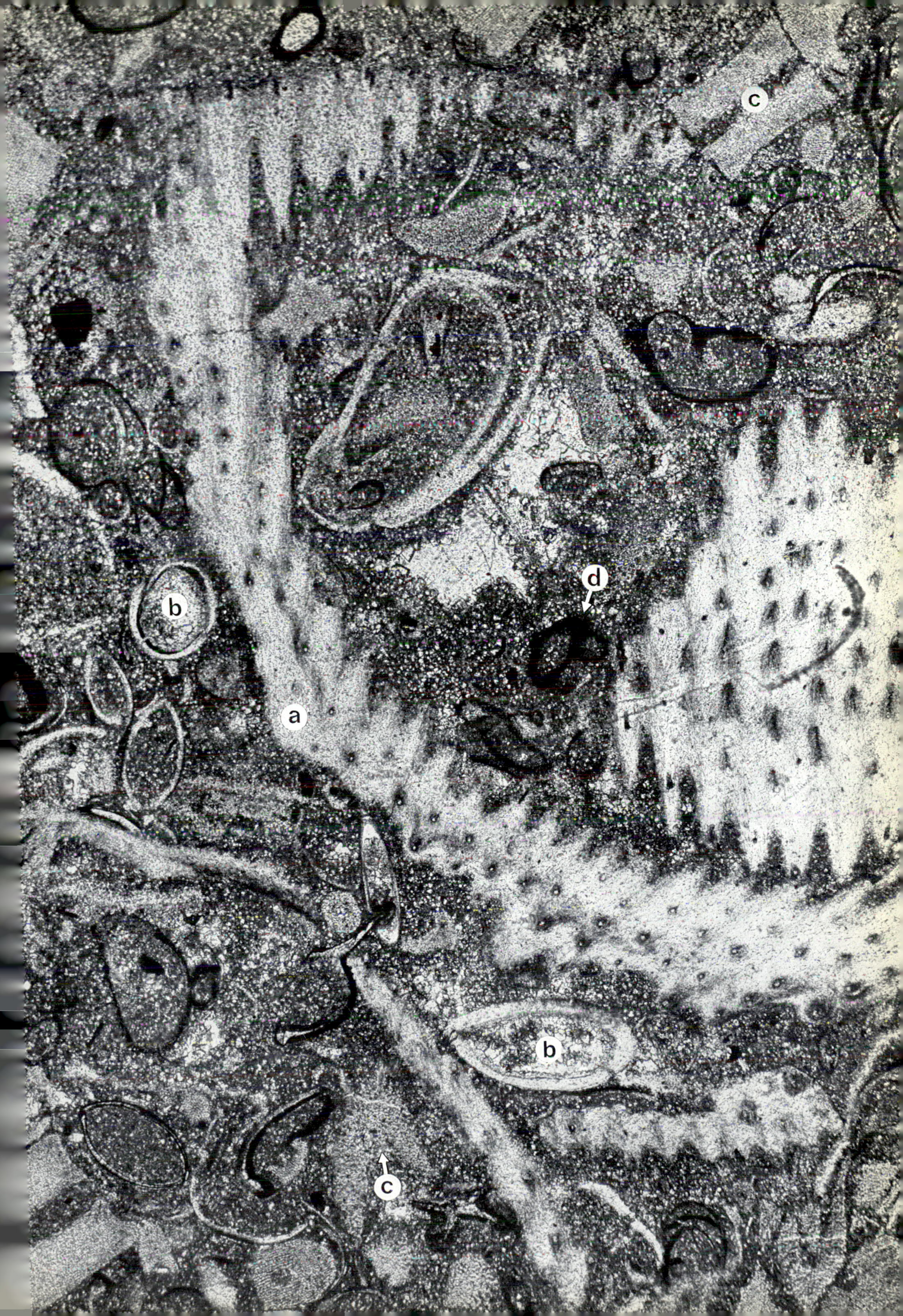

c
b
d
a
b
c

1/10 mm
x 20
x 40
x 60
x 80
x 100

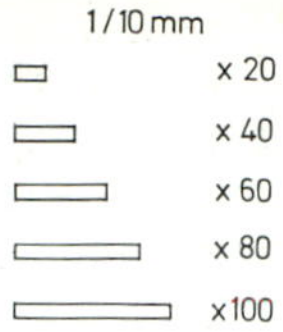

Bryozoan-echinodermal limestone. Quartz silt and fine pelletal matrix. Echinodermal plates exhibit typical gray color due to very finely porous texture. Bryozoans are bifoliate and polyporid (fenestrate) types. The polyporid (a) clearly shows the central clear granular wall layer surrounded by a darker laminated outer wall. The bifoliate fronds exhibit a dark fine-grained inner wall and a lighter outer wall whose laminated structure is poorly shown partially due to silicification. Smaller fragments are probably brachiopods and ostracodes. Riepe Spring Formation, Early Permian, "The Loop" in Carlin Canyon, sec. 22, T. 33 N., R. 53 E., Elko County, Nevada, U.S.A. IU 8099–563, ×50.

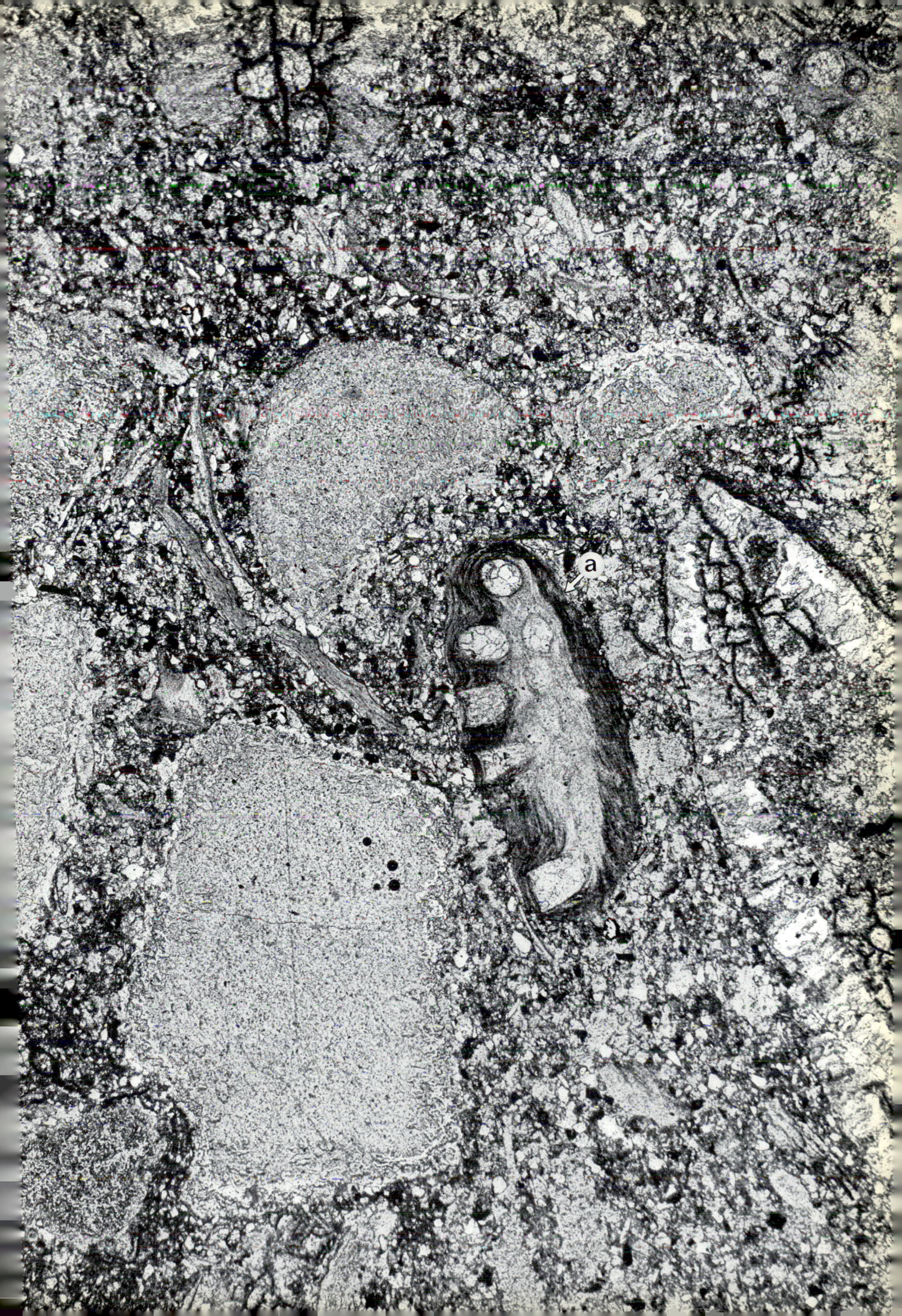
a

Well-sorted foraminiferal limestone having sparry calcite cement. Coiled foraminifers show diverse cross sections. Few small echinoderm grains and molluscan fragments, usually having micritic coats. Foraminiferal walls appear altered. Interiors of foraminifers exhibit both clear calcite cement and cloudy interiors. The foraminiferal shapes in this illustration are comparable to those in many Cretaceous and Tertiary limestones (see Pl. 11–2). Fusulinid limestone in Kenosha Shale Member, Tecumseh Formation, Shawnee Group, Virgilian, Upper Pennsylvanian, one half mile west of Weeping Water, center south line, sec. 25, T. 11 N., R. 11 E., Cass County, Nebraska, U.S.A. IU 8099–854, ×50.

1/10 mm
x 20
x 40
x 60
x 80
x100

Typical well-washed Paleozoic bryozoan-brachiopod limestone: punctate (a) and impunctate (b) brachiopod shells, fenestrate bryozoans (c), echinoderm plate (d), molluscan fragment (e), broken brachiopod spine (f), quartz grains (g), and sparry calcite cement (h). Note that in punctate brachiopod fragment (a) holes (punctae) are cut in longitudinal (left) and transverse (right) directions. Negli Creek Limestone, Chester Series, Mississippian, northwest side of Peach Hill, 3 miles northeast of Tobinsport, Perry County, Indiana, U.S.A. IU 8099–280, ×50.

e
a
c
f
h
d
g
c
b

PLATE 81. MISSISSIPPIAN

1/10 mm

x 20

x 40

x 60

x 80

x100

Very well-washed Paleozoic echinodermal-bryozoan limestone: ramose bryozoan (a), echinodermal plates (b), echinoid spines (c), brachiopods (d), trilobite fragment (e), grapestone (f), ooliths (g), quartz-bearing mud filling gastropod (h), and quartz grains (i). Note pelletal mud (j) defining geopetal fabric. Mud filling of some fragments indicates that they were reworked prior to final deposition and lithification. Open packing of framework is filled by sparry calcite. Note micritic rims. Bethel Formation, Chester Series, Mississippian, Hardin County, Kentucky, U.S.A. IU 8099–P211, ×50.

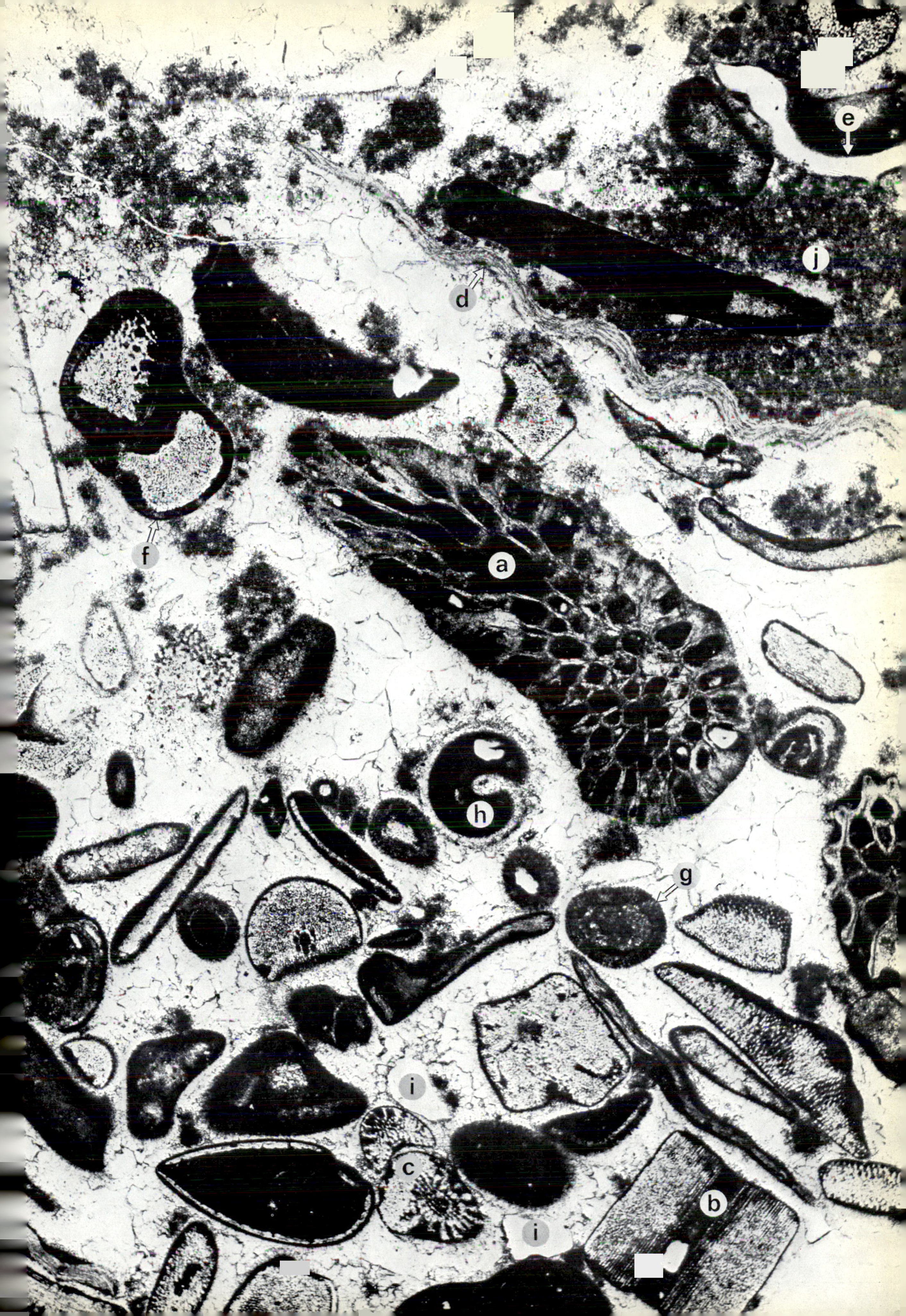
e
j
d
f
a
h
g
i
c
i
b

PLATE 82. MISSISSIPPIAN

1/10 mm
x 20
x 40
x 60
x 80
x100

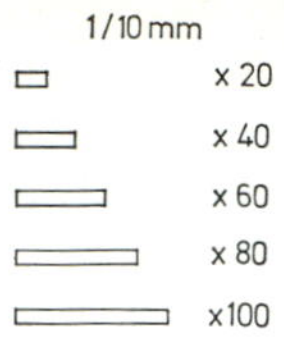

Bryozoan and echinodermal debris washed from a shale has been mounted in plastic to form an artificial skeletal limestone. Note that many grains do not exhibit contacts with adjacent grains and "float" in the plastic cement. Borden Group, Mississippian, Sugar Creek, $W\frac{1}{2}$ $SW\frac{1}{4}$ sec. 29, T. 19 N., R. 4 W., Montgomery County, Indiana, U.S.A. IU 8099–1070, $\times 50$.

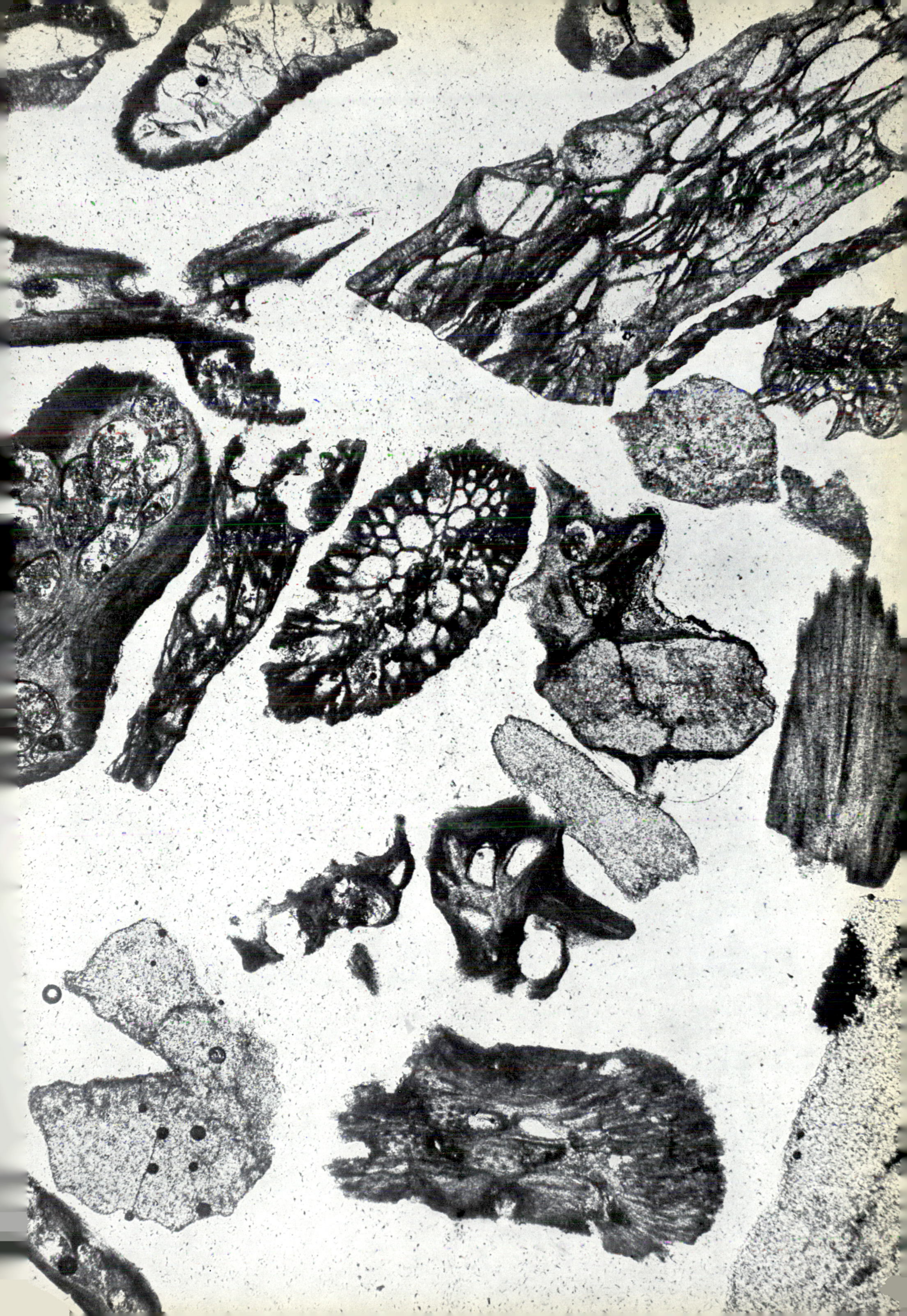

PLATE 83. MISSISSIPPIAN

1/10 mm
x 20
x 40
x 60
x 80
x100

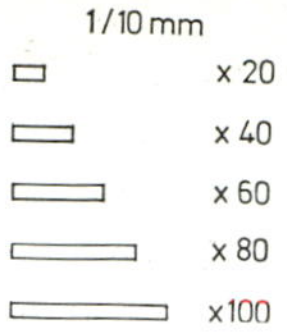

Well-sorted Paleozoic fenestrate bryozoan-echinodermal limestone. Fenestrate bryozoans shown in transverse and longitudinal section. Note thin inner lighter walls and laminated darker thick outer walls. Zooecia commonly mud-filled and suggest redeposition because matrix contains little lime mud. Echinodermal plates show prominent syntaxial rim cement and twinning. A few brachiopod fragments exhibit fibrous internal microstructure. Solution packing (arrow) shown by stylolitic boundary between grains. Akiyoshi Limestone, Lower Carboniferous, Yamaguchi Prefecture, Japan. IU 8099–999, $\times 50$.

Silty, poorly sorted bryozoan-ostracodal limestone. Ostracodes present as articulated and disarticulated valves. Some articulated valves filled with clear calcite cement. Thin dark coatings in valves probably finely disseminated pyrite. Note ostracodes are ornamented (irregular valve outlines). Ostracodal debris provides much of matrix. Solution packing and some recrystallization in matrix. Oblique section of trepostomatous bryozoan exhibits thin-walled interior region and barely perceptible laminated exterior walls. Outer zooecia filled with matrix to produce an obscure boundary with matrix. Helderberg Limestone, Lower Devonian, active quarry one half mile south of Andreas, north side of Blue Mountain, West Penn Township, Schuylkill County, Pennsylvania, U.S.A. IU 8099–671, ×50.

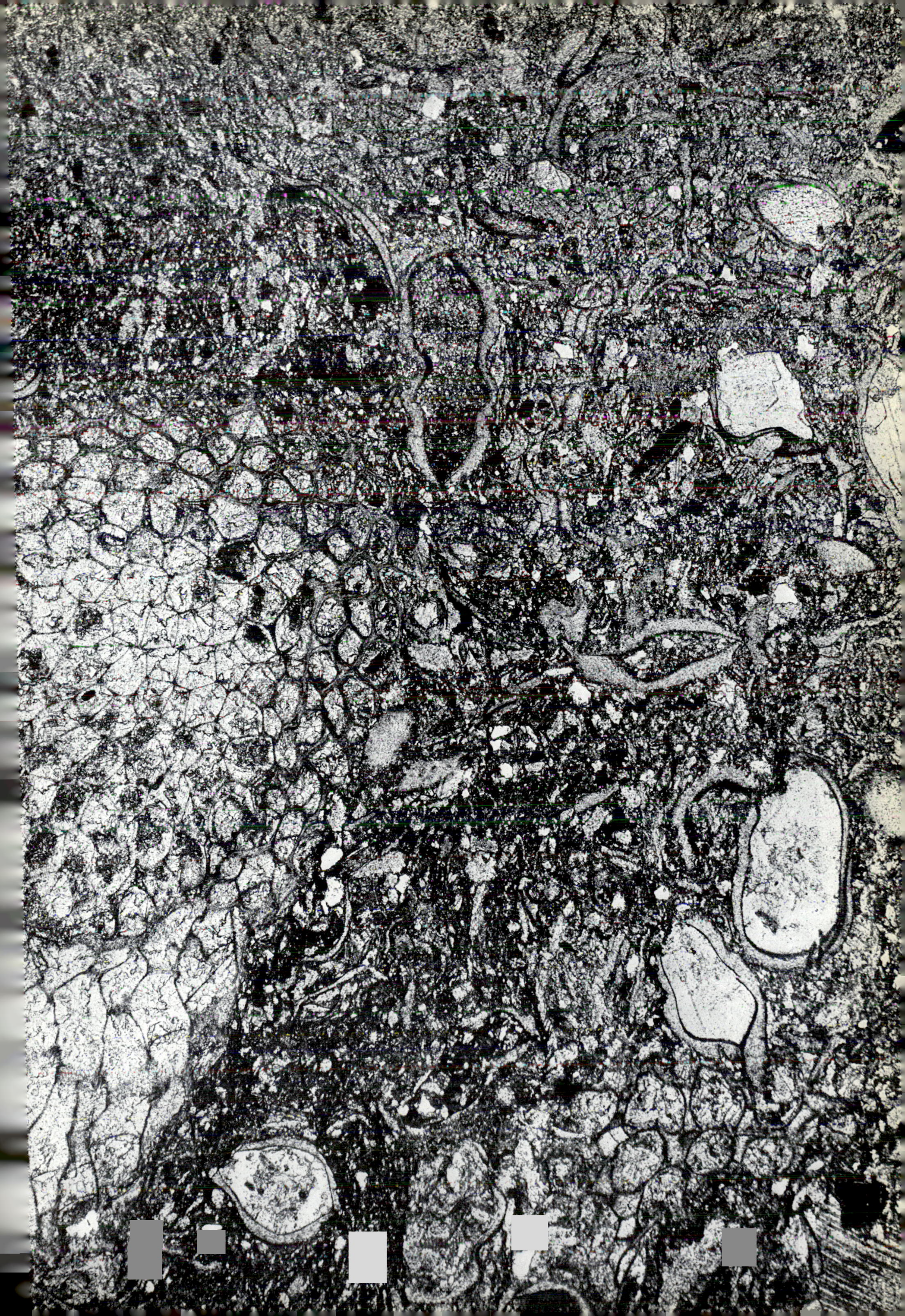

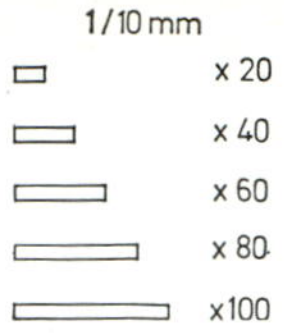

1/10 mm

x 20

x 40

x 60

x 80

x100

Echinodermal limestone containing a large fistuliporoid bryozoan at bottom and some fragmentary punctate brachiopod debris. Fistuliporoid bryozoans are characterized by vesiculose (cystose) skeletal plates or walls between the zooecial (living chambers) tubes. Upper portion of zooecia filled with mud and fine debris; lower portions filled with clear calcite cement. Diaphragms uncommon in zooecial tubes. Calcareous sand covers *in situ* bryozoan. Dark phosphatic grain (a), either conodont or vertebrate fragment. North Vernon Limestone, Middle Devonian, Jennings County, Indiana, U.S.A. IU 8099–525B, $\times 50$.

a

1/10 mm
x 20
x 40
x 60
x 80
x100

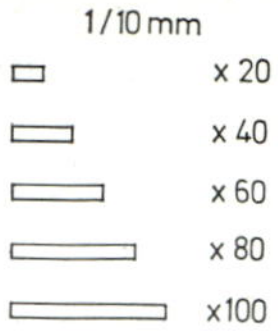

Poorly sorted skeletal limestone containing pelletal (?) dark muddy matrix. Some sparry calcite areas in protected spaces beneath (a) or within (b) skeletal debris. Echinodermal plates exhibit meshwork filled with dark mud. Other constituents include ostracodes, trilobite fragments (c), and a large bryozoan colony at right center. Mud contains finely comminuted fossil debris. Murrindal Limestone, Middle Devonian, Rocky Camp, four miles north of Buchan, Victoria, Australia. IU 8099–459, ×50.

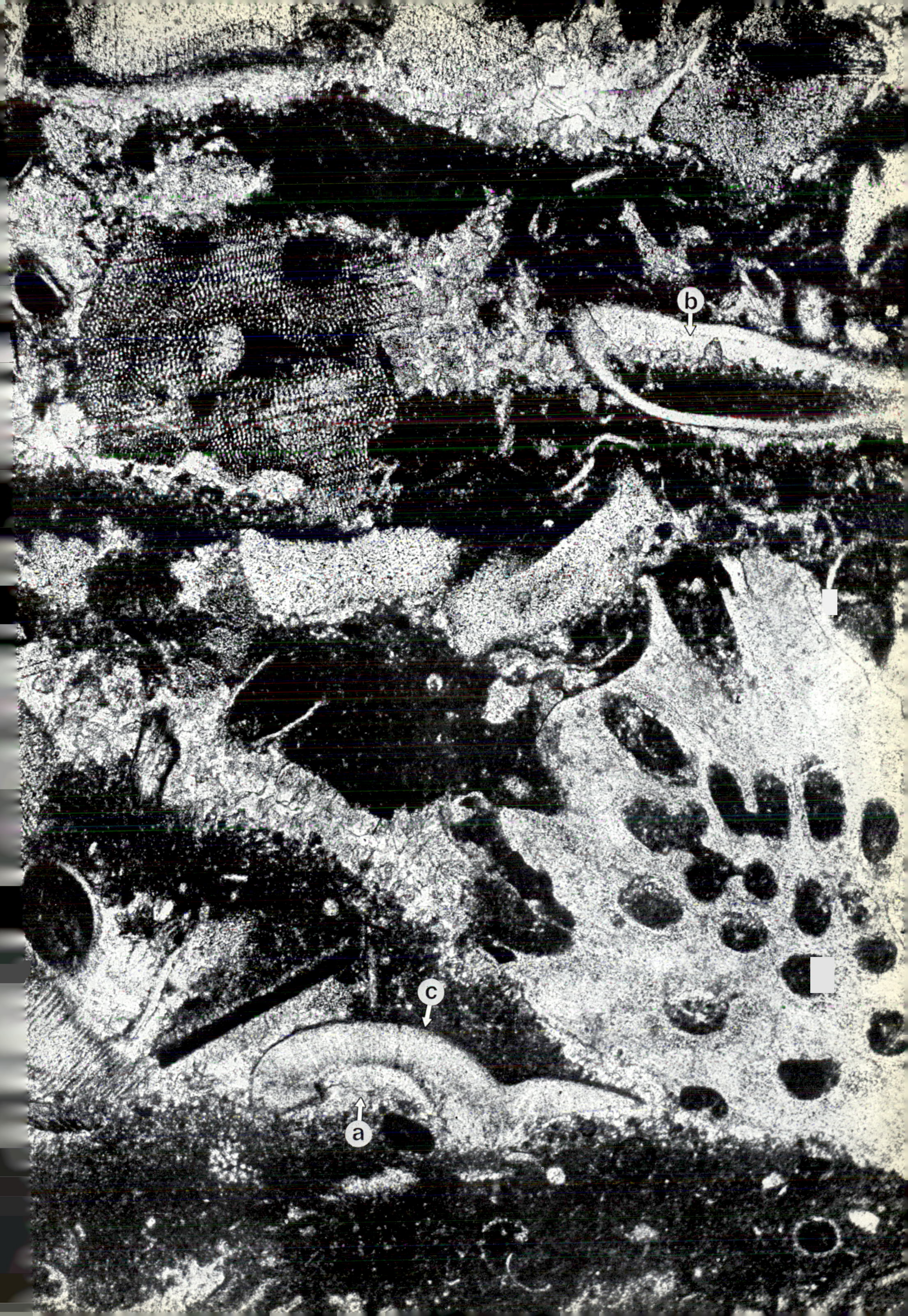
b
c
a

1/10 mm

x 20

x 40

x 60

x 80

x100

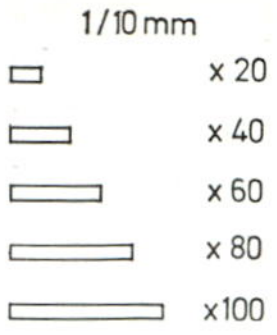

Brachiopodal-echinodermal-coral limestone containing dark mud matrix and finely comminuted fossil debris. Brachiopod (a) has fibrous and somewhat altered prismatic shell microstructure. Coral wall recrystallized and identified on basis of cross sectional shape. Note accumulation of insolubles along stylolites (b) in mud matrix. Loyola Limestone, Lower Devonian, quarry at side of Howes Creek Road, Loyola, Victoria, Australia. IU 8099–464, ×50.

b
a

PLATE 88. SILURIAN

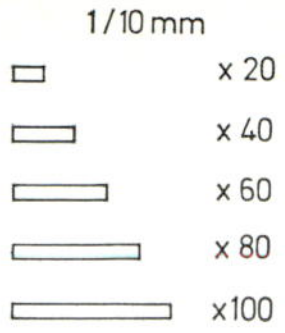

1/10 mm

x 20

x 40

x 60

x 80

x100

Two views of same thin section, a well-washed grainstone. Much of debris may be fragments of pentamerid brachiopods (coarsely prismatic shell microstructure as viewed under crossed nicols; compare with Pl. 29–3). Because thin sections do not show any complete valves, our identification is not conclusive. Fragment at left center of upper photomicrograph may be a coral. The sample from which thin section was prepared reveals large brachiopod shells. Conclusion: sometimes a single thin section is not sufficient for satisfactory identifications. Fossil debris partially pyritized. Silurian, Morcha Well 3, 703 meters depth, Latitude 26° 11′ North, Longitude 38° 30′ West, Spanish Sahara. IU 8099–424, ×30.

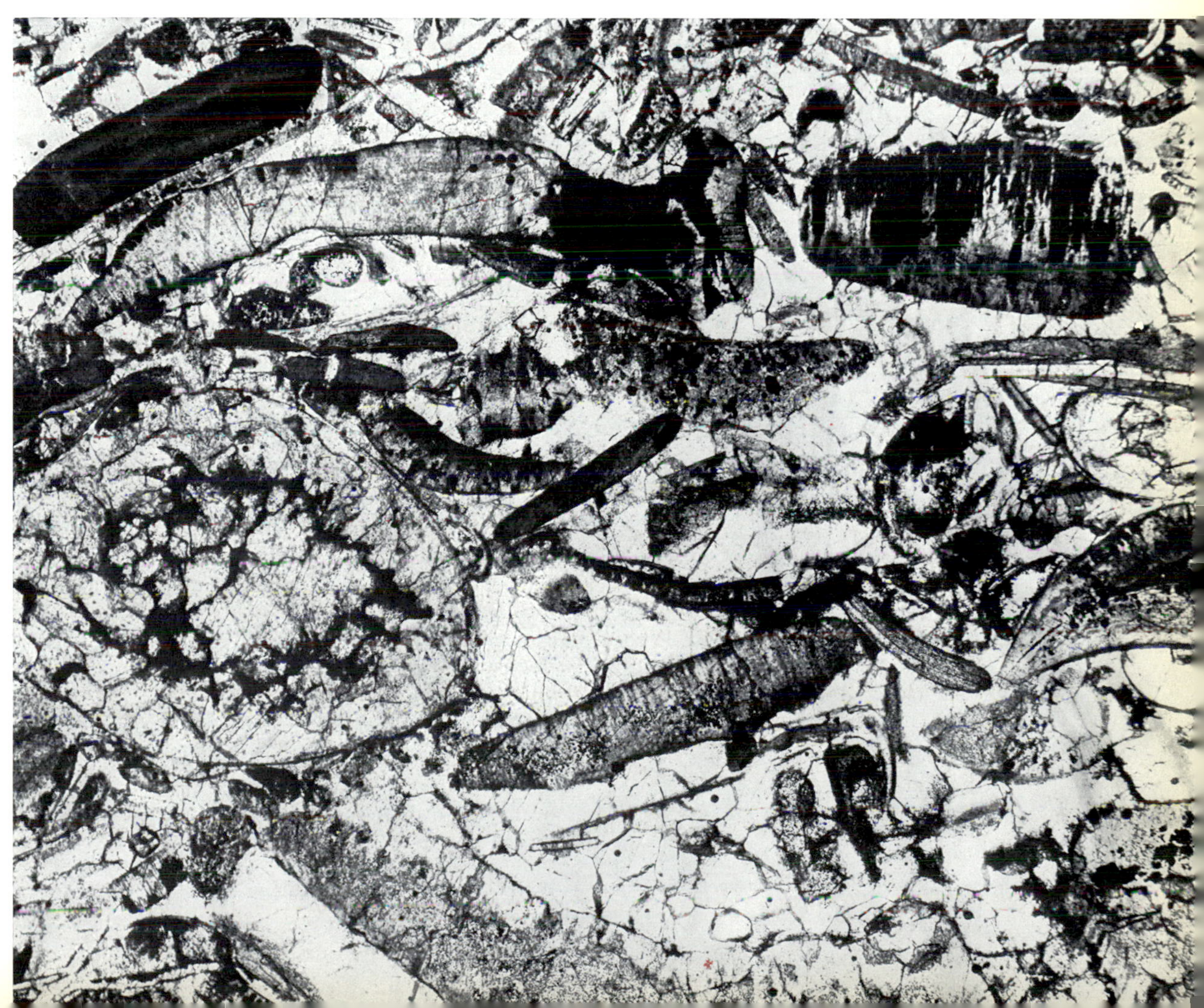

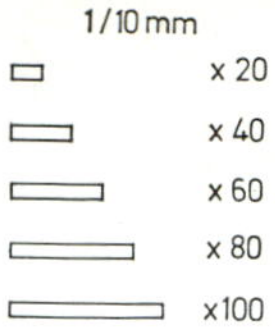

1/10 mm
x 20
x 40
x 60
x 80
x100

Echinodermal plates, ramose (dark transverse section at center of plate), encrusting (lower center), and fenestrate bryozoan debris in clear sparry calcite cement. Other constituents include coral (a), encrusted by bryozoan, possible recrystallized pelecypod (b) fragment at lower left, and small ostracode valve (c). Well-washed and well-sorted skeletal framework. Note difference in diameter between corallites and zooecia. Some solution packing between larger skeletal grains. Middle Silurian, 7 kilometers north of Visby, Gotland, Sweden. IU 8099–329, ×50.

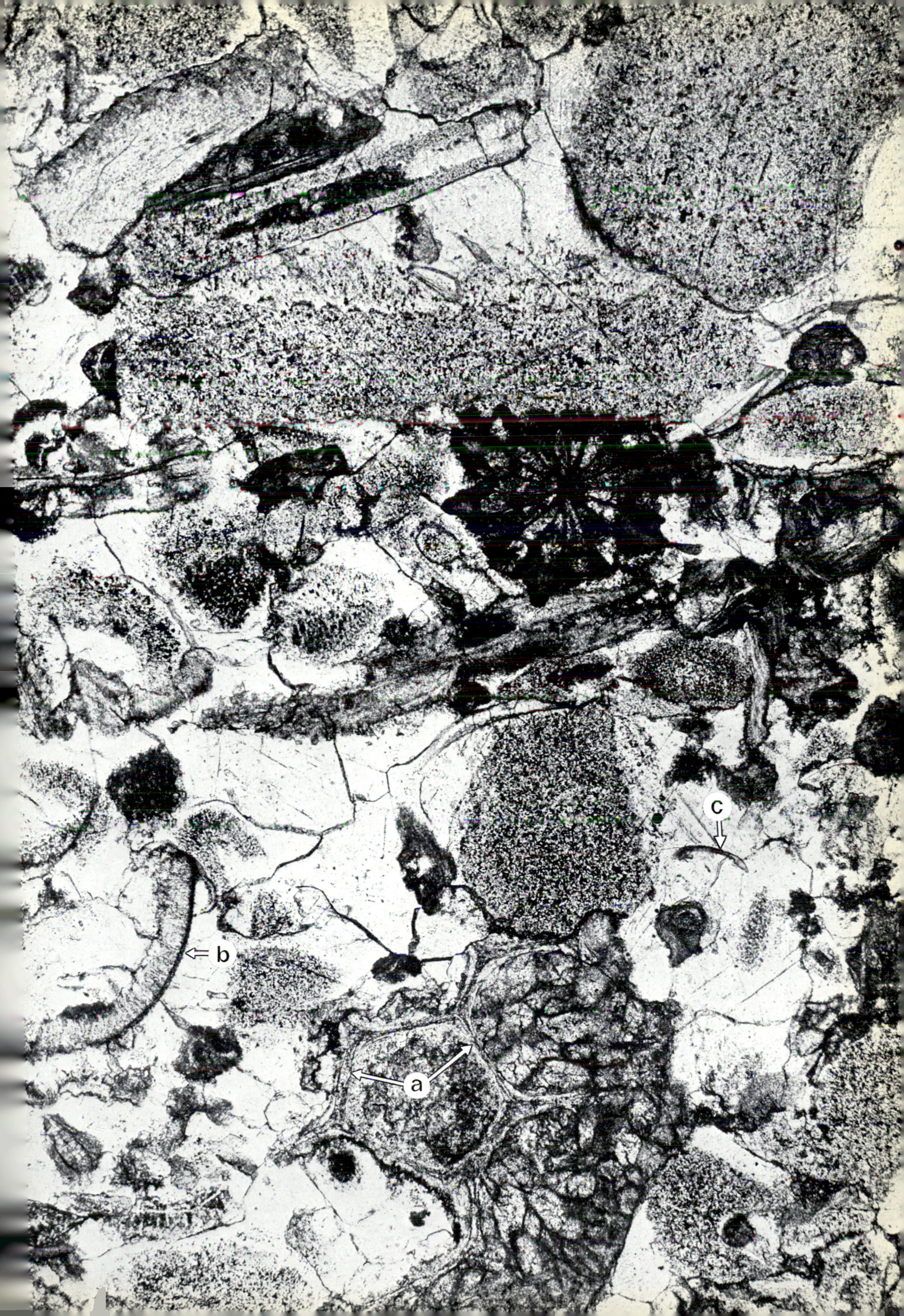

c
b
a

1/10 mm

x 20

x 40

x 60

x 80

x100

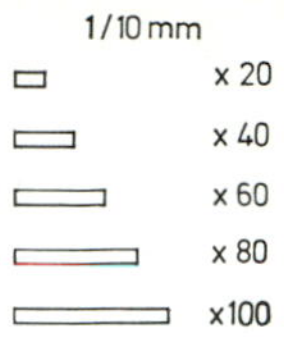

Fig. 1. Peel of pentamerid brachiopod limestone. These brachiopods have thick prismatic shell layers. Two brachiopods show prominent internal septum and are filled by echinodermal debris. Note chain coral (a). Compare this peel with thin section on Pl. 29–3. Zone 9a, Upper Wenlockian, Middle Silurian, road section between Vik and Sundvollen, Ringerike region, northwest of Oslo, Norway. IU 8099–810AP, ×20.

Fig. 2. Peel of chain coral (halysitid) cut oblique to corallites and showing internal tabulae. Corallites filled with sparry calcite which contrasts with dark mud matrix. As fig. 1.

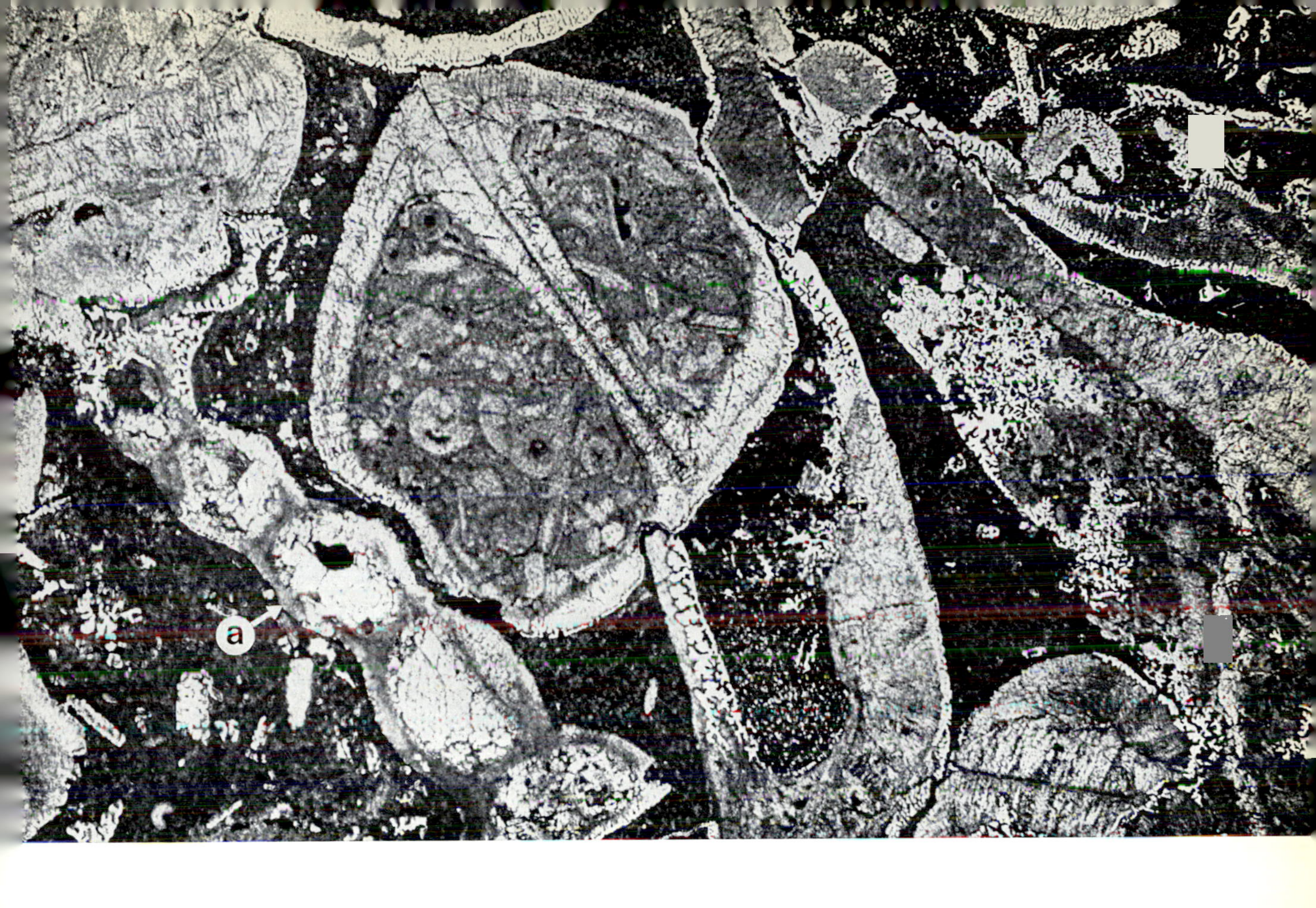
a

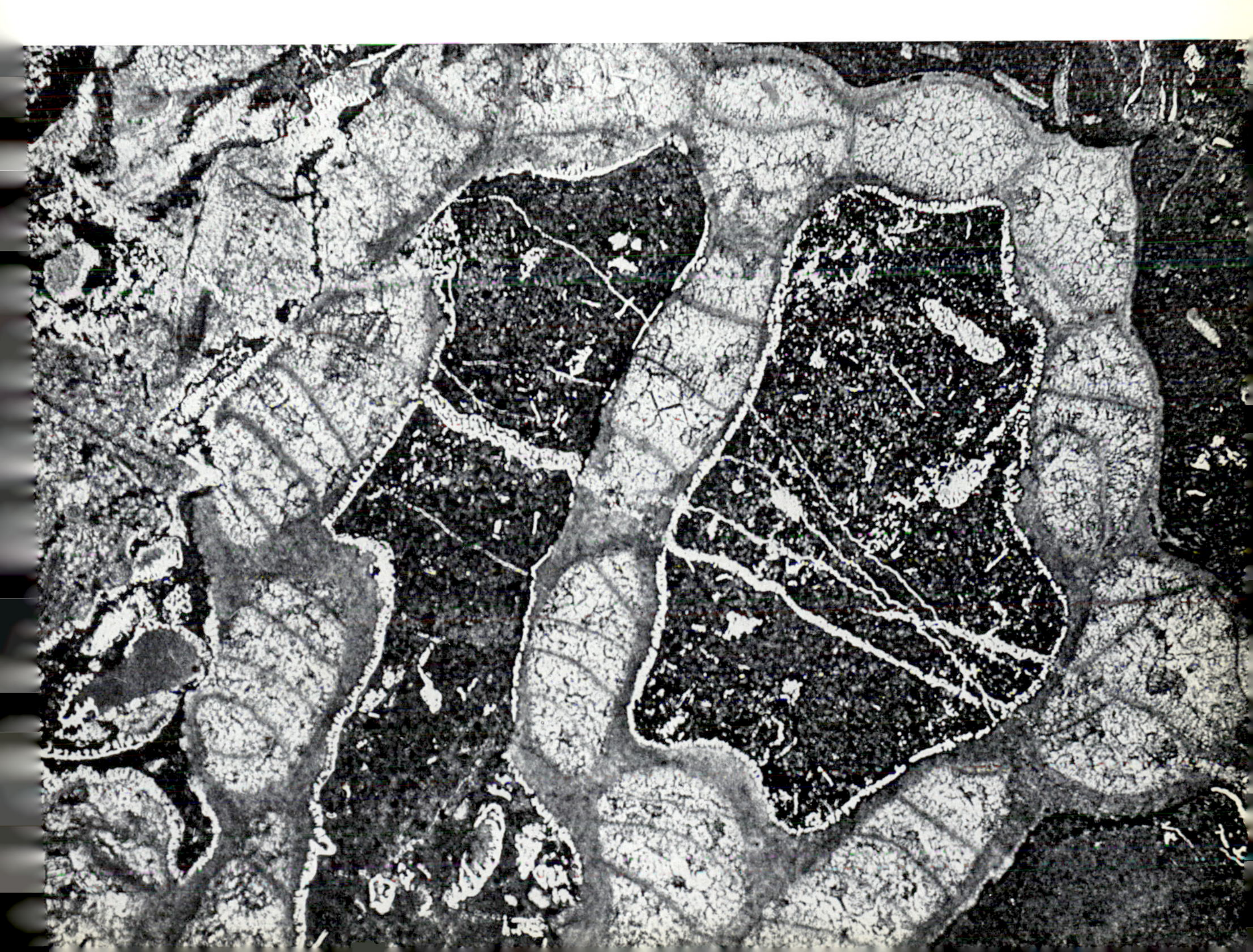

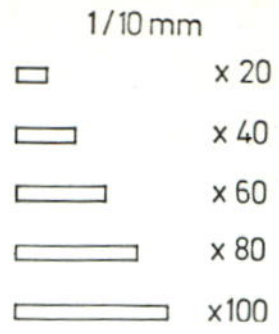

1/10 mm

x 20

x 40

x 60

x 80

x100

Echinodermal plates and fenestrate bryozoan framework in a sparry calcite cement. Fenestrate bryozoans are cut in both longitudinal (a) and transverse (b) sections. Brachiopod fragment at right center. Solution packing between many skeletal grains produces stylolitic boundaries. Middle Silurian, 7 kilometers north of Visby, Gotland, Sweden. IU 8099-329, ×50.

b
a

1/10 mm

x 20

x 40

x 60

x 80

x100

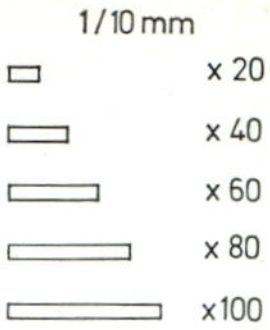

Bryozoan-brachiopod-echinodermal limestone. Prominent punctate shell in center of figure is a trilobite fragment. Ribbed (corrugated in cross section) brachiopod shells exhibit fibrous structure. Finely porous echinodermal plates (a) are difficult to distinguish from muddy or finely pelleted matrix. Bryozoan wall structure (b) is shown. Some shells, especially brachiopods, form a type of geopetal texture by trapping fine matrix above the shell and exhibit clear calcite cement below the shell (c) where matrix was not deposited. Upper Ordovician, near Monroe, Butler County, Ohio, U.S.A. IU 8099–915, ×50.

c
a
b
a

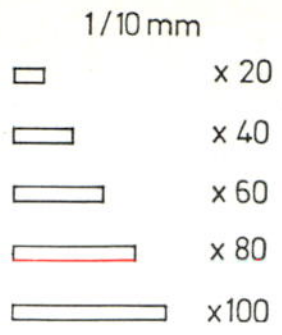

1/10 mm
x 20
x 40
x 60
x 80
x100

Fig. 1. Thin section of disarticulated brachiopod shells in partially dolomitized (a) lime sand. Brachiopod at upper right is pseudopunctate. Note thin ramose bryozoan at upper left. Unknown encrusting organism on brachiopod in center. Compare quality of thin section with peel below. Peel wins. Tanglewood Member, Lexington Limestone, Middle Ordovician, railroad cut $1^1/_2$ miles south of Winchester, Clark County, Kentucky, U.S.A. IU 8099–1044, ×20.

Fig. 2. Peel of large brachiopods and bryozoans in partially dolomitized (a) lime sand. Brachiopods show clear transverse pillars of pseudopunctae. Fossil debris in lime sand is difficult to identify in peel at this magnification. Detail in this peel compares favorably with thin section above. As fig. 1, IU 8099–1044AP, ×25.

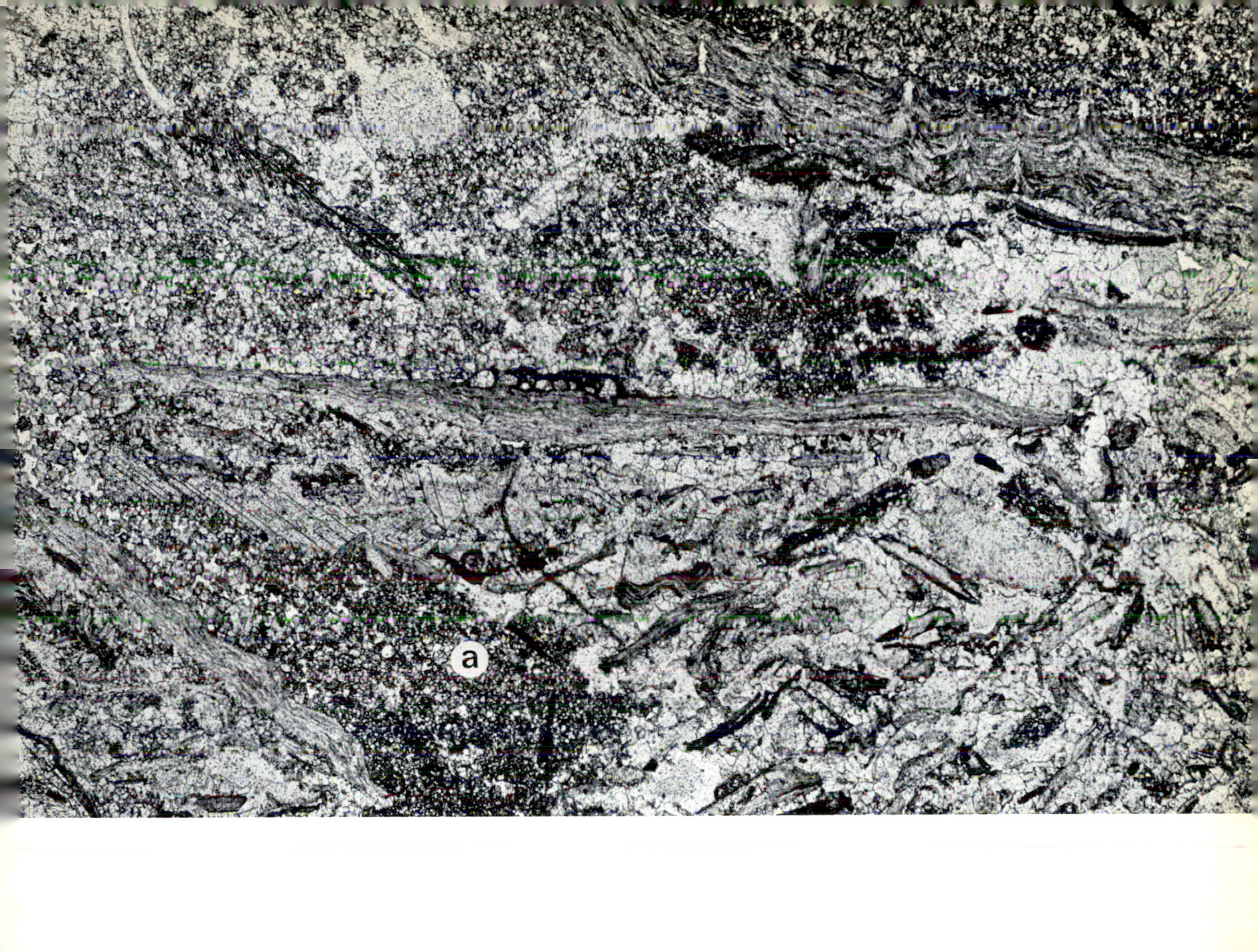
a

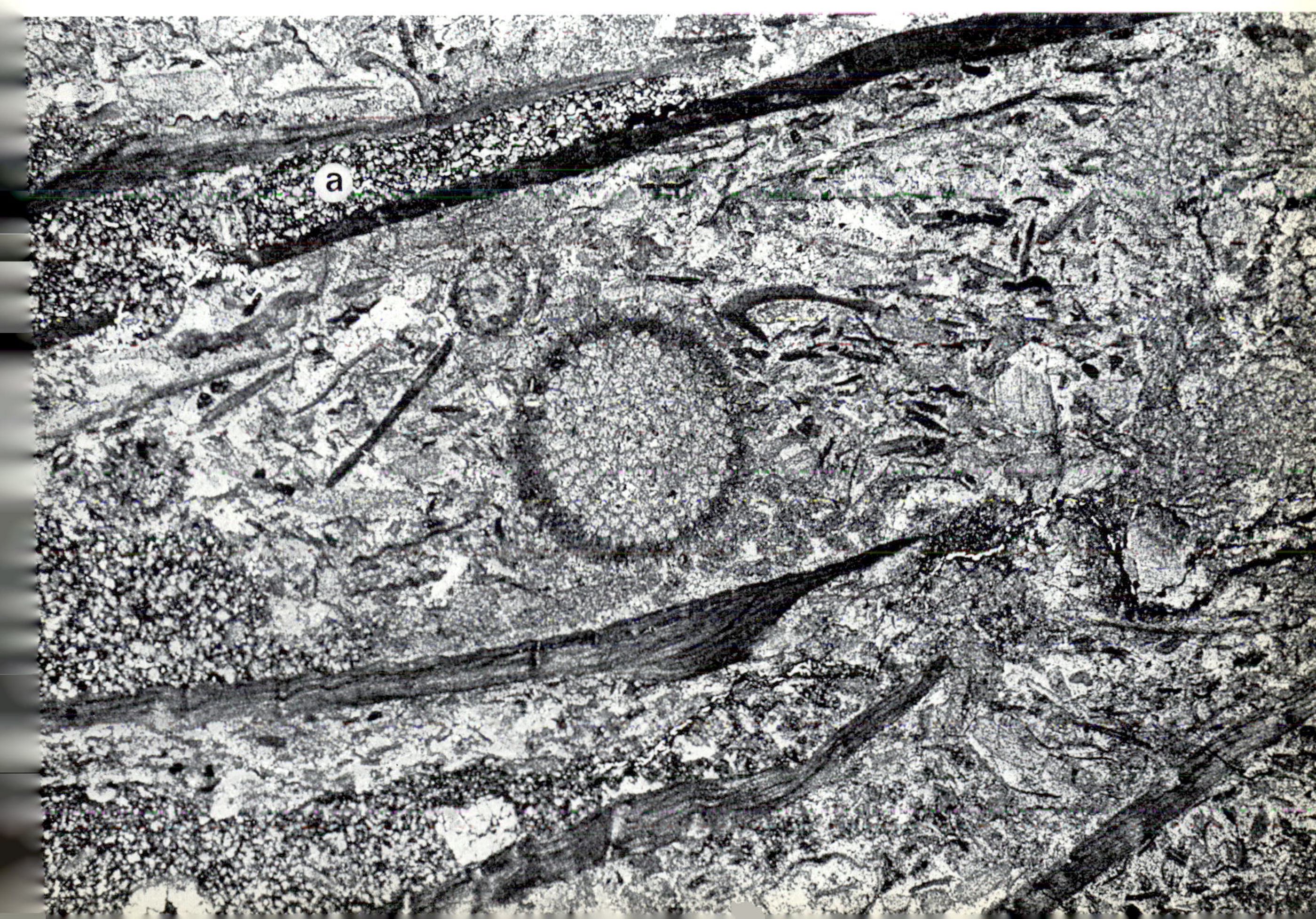
a

PLATE 94. CAMBRIAN

1/10 mm

x 20

x 40

x 60

x 80

x100

Recrystallized limestone with archaeocyathids and trilobites. Transverse section of archaeocyathid shows septa, but walls are recrystallized. Trilobite debris is recrystallized and poorly preserved (a). Some geopetal fabric due to mud caught on shells. Note twinned sparry calcite cement. Recrystallization and calcite twinning combine to make recognition of biotic debris difficult. Mural Formation, Lower Cambrian, Latitude 53° 32′ North, Longitude 121° 13′ West, Cariboo Mountains, British Columbia, Canada. IU 8099–218, ×50.

a
a

1/10 mm
x 20
x 40
x 60
x 80
x100

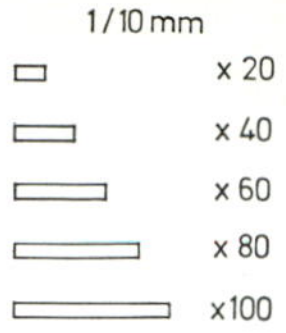

Archaeocyathid limestone. Note porous outer wall (a) and septa (b) in longitudinal section at center. Drusy cement lines septa and additional large crystals fill center of specimen. Large crystals are twinned as befits calcite from a tectonic area. Other archaeocyathid at lower left. Ghosts of ooliths (c). Angular quartz silt and dark fine-grained intraclasts in coarsely altered sparry calcite matrix. Mural Formation, Lower Cambrian, Latitude 53° 54′ North, Longitude 120° 46′ West, western ranges of Rocky Mountains, British Columbia, Canada. IU 8099–219, ×50.

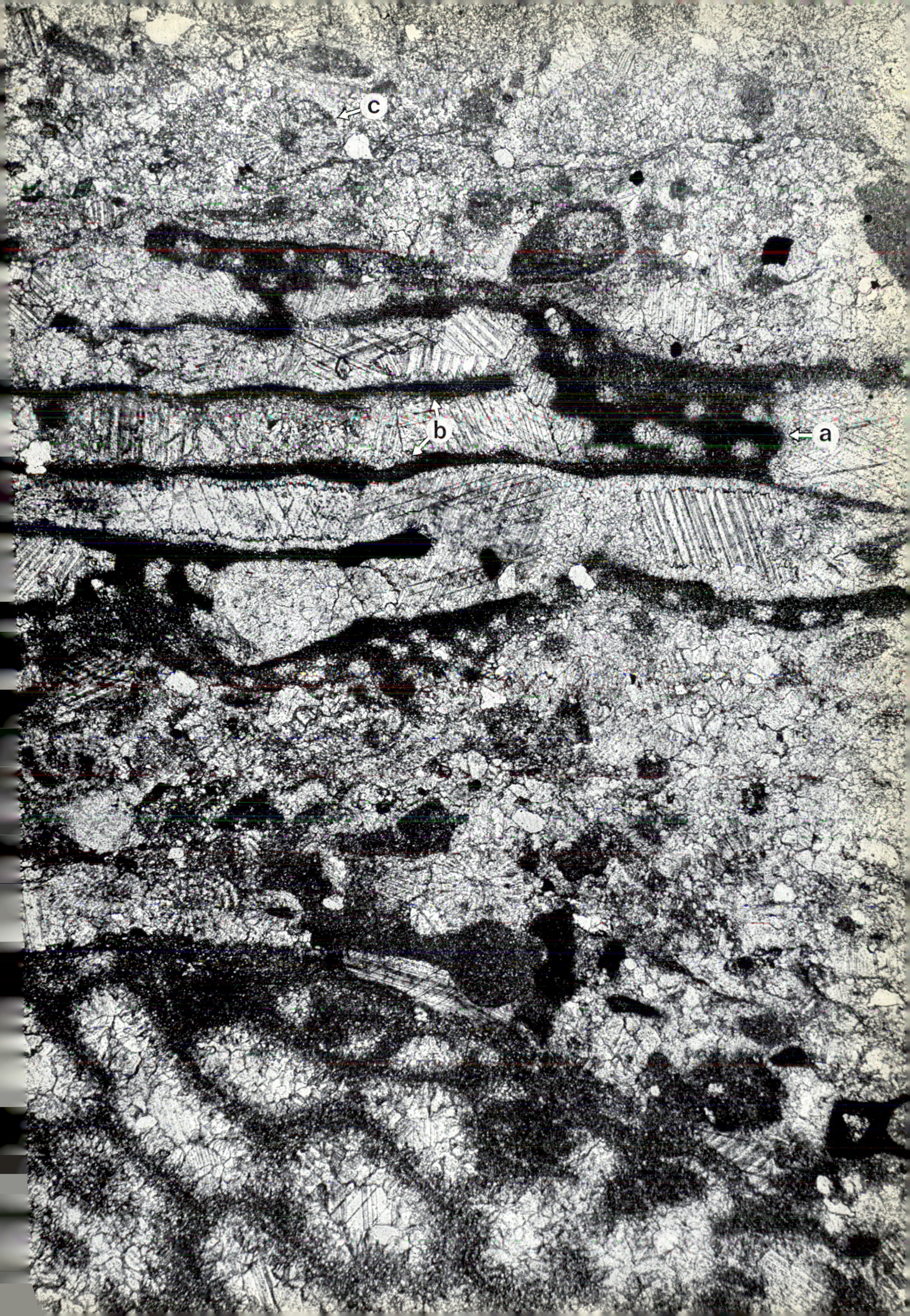
c
b
a

1/10 mm
x 20
x 40
x 60
x 80
x100

Trilobite-echinodermal limestone. Trilobite microstructure altered, but some shapes are characteristic as in the recurved fragment ("shepherd's crook") in the center of the photomicrograph. Solution packing is conspicuous (compare with Pls. 83 and 89), and the matrix consists of fine quartz silt and dark pelletoid grains. Eighteen meters above base of Wolsey Formation, Middle Cambrian, NE$\frac{1}{4}$ NW$\frac{1}{4}$ sec. 9, T. 1 S., R. 2 W., Madison County, Montana, U.S.A. IU 8099-1072, ×50.

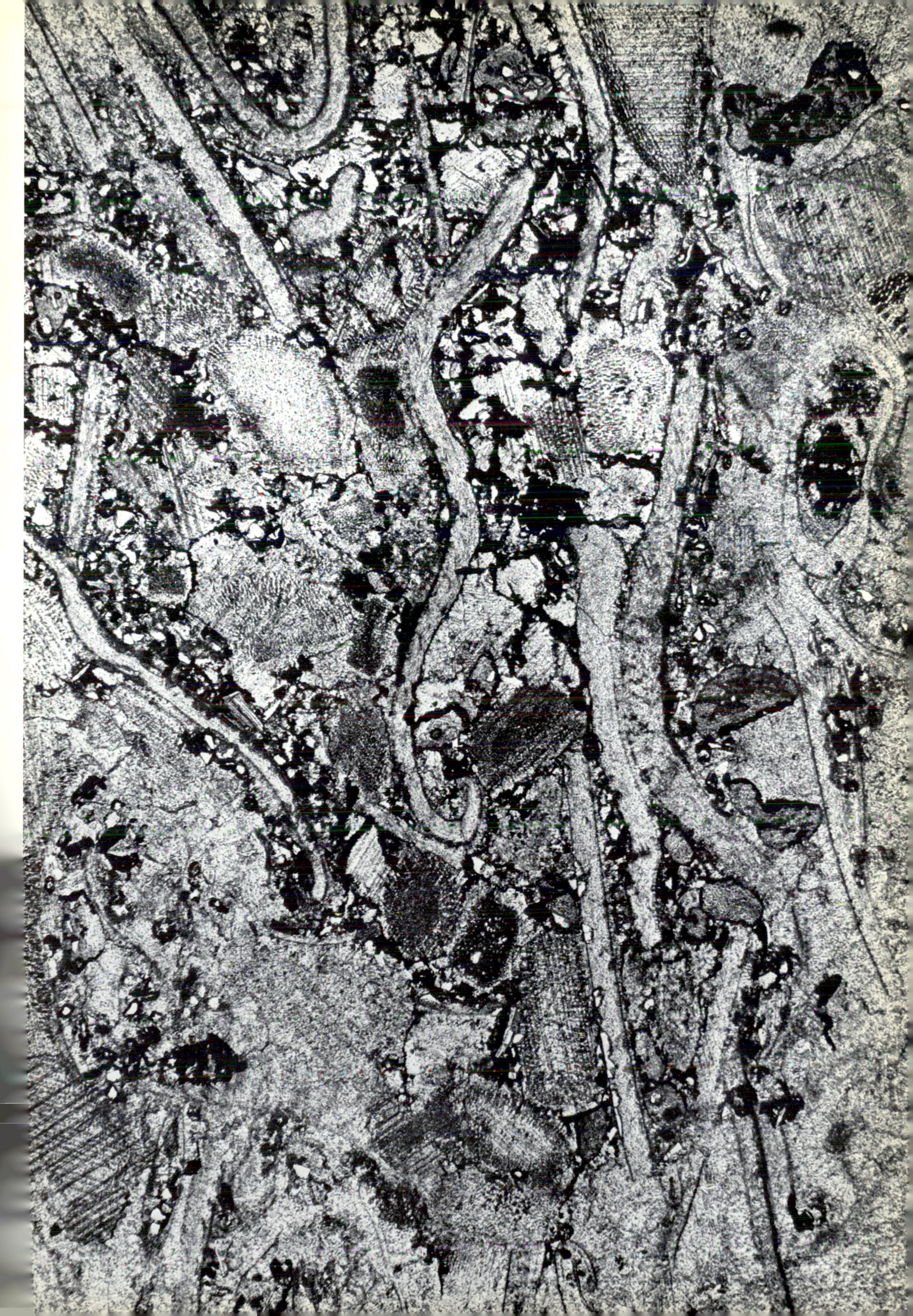

PLATE 97. PLEASANT POTPOURRI

1/10 mm
x 20
x 40
x 60
x 80
x100

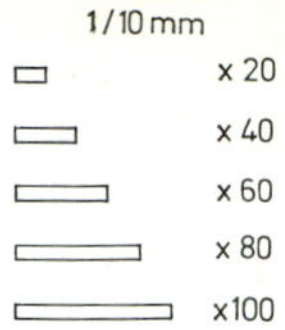

Fig. 1. Echinodermal-brachiopod-bryozoan limestone. Well-sorted, sparry calcite cement. Tanglewood Member, Lexington Limestone, Middle Ordovician, Kentucky, U.S.A. U.S. Geological Survey loan, W–11, ×40.

Fig. 2. Laminated shell microstructure (a) of pelecypods comparable to fibrous microstructure of brachiopods (fig. 1 above) as viewed in thin section. Identifications based on age of associated fossils rather than on shell microstructure. Dark grains are phosphate probably from vertebrate teeth and bones. Matrix of inoceramid prisms, which are cut in transverse and longitudinal directions, and sparry calcite. Graneros Shale, Upper Cretaceous, cut bank on a small tributary of Wolf Creek, approximately $7^1/_2$ miles north-northwest of Holyrood, NE$\frac{1}{4}$ sec. 5, T. 16 S., R. 10 W., Ellsworth County, Kansas, U.S.A. Hattin Collection, KG–AD=18E, ×20.

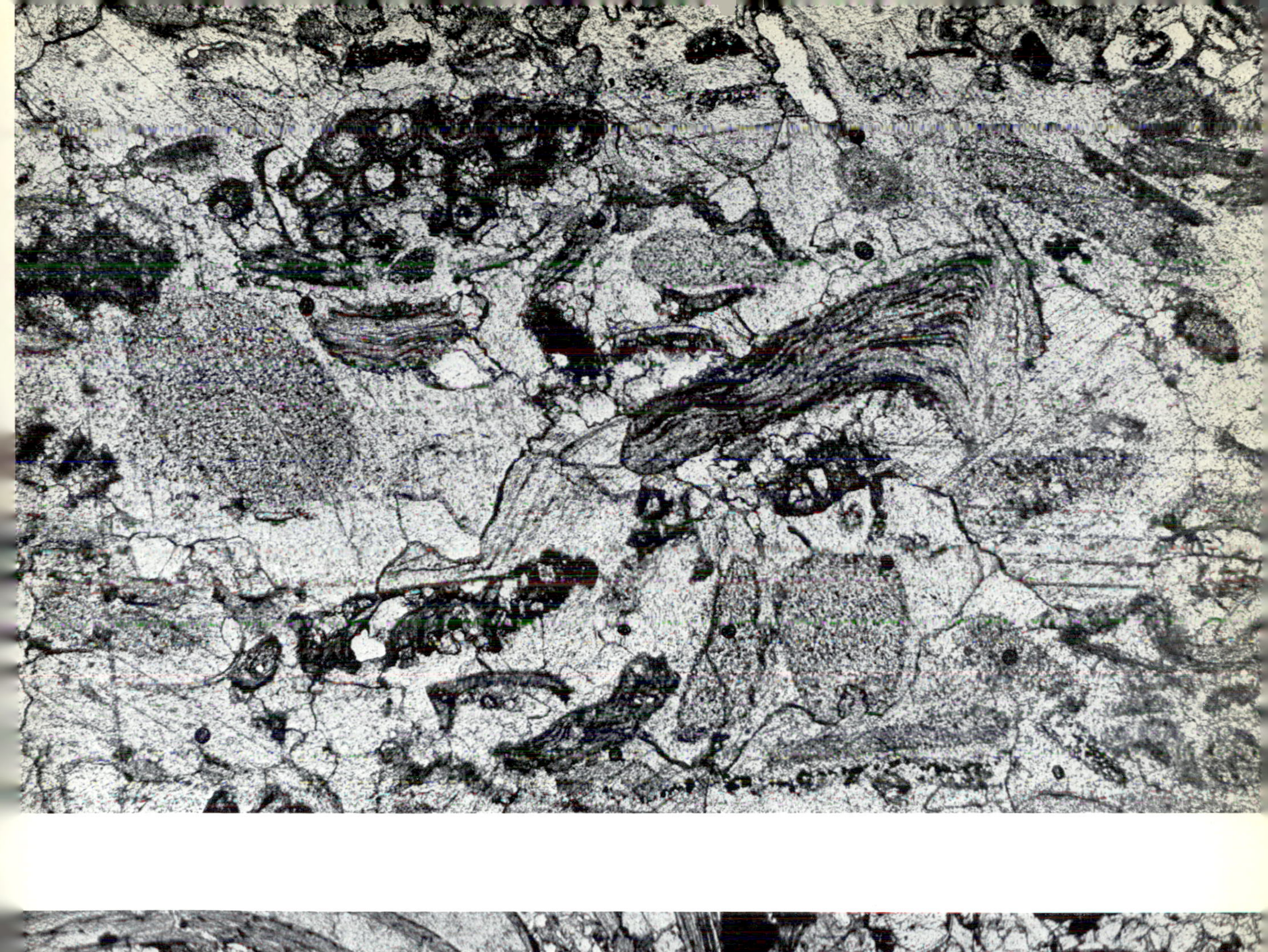

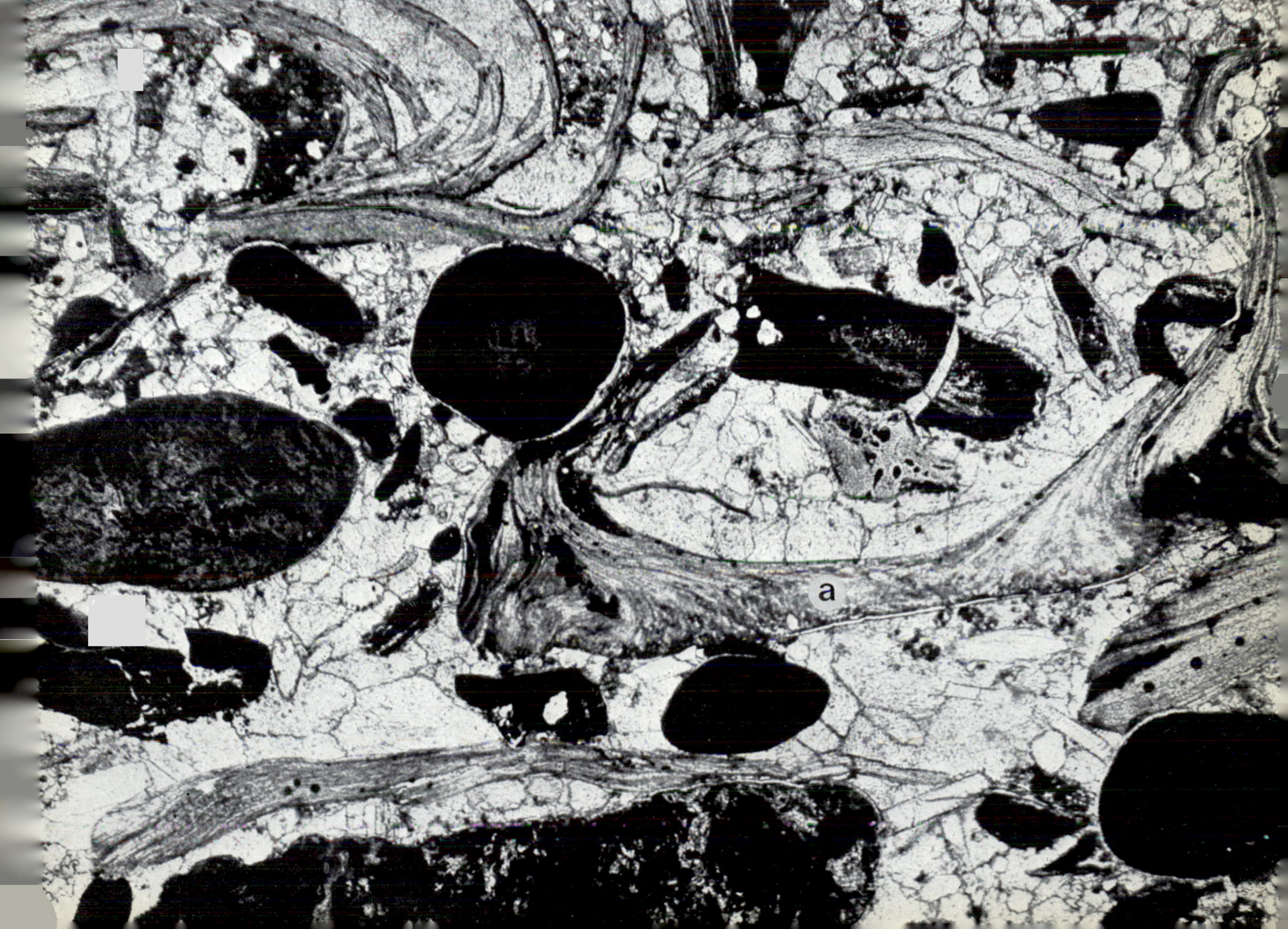
a

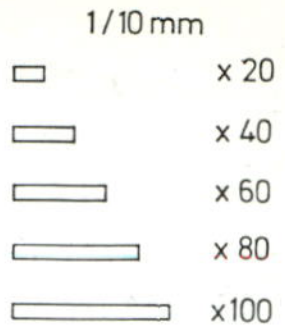

1/10 mm

x 20

x 40

x 60

x 80

x100

Fig. 1. Peel of large nummulitic foraminifers in a matrix of well-sorted silt-sized biotic debris, most of which is unidentifiable but which may be of nummulitic origin. Compare with Pl. 5–2. Wadi Tamet Formation, Middle Eocene, subsurface eastern Sirte Basin, Cyrenaica, Libya. IU 8099–698AP, ×10.

Fig. 2. Peel of altered molluscan debris in oolitic matrix. Some quartz grains are centers of ooliths. Corallian, Oxfordian, Upper Jurassic, Osmington Mills, Dorset, England. IU 8099–57AP, ×15.

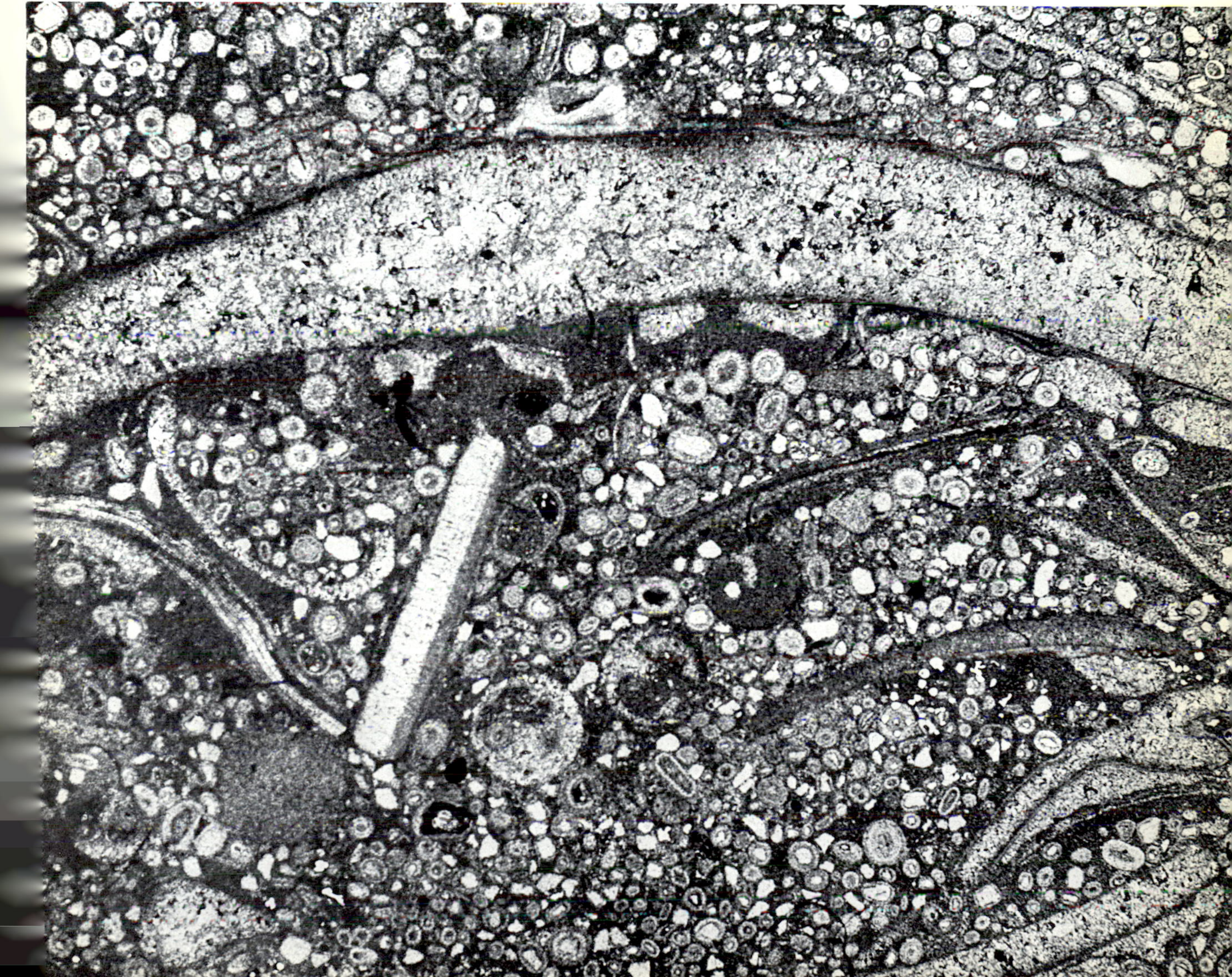

PLATE 99. SATISFACTORY SUCCOTASH

1/10 mm
x 20
x 40
x 60
x 80
x100

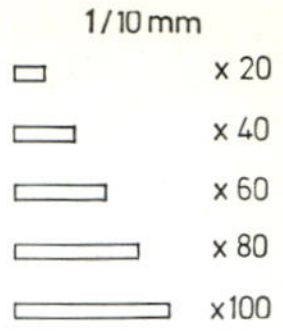

Fig. 1. Finely ribbed (corrugated) brachiopod containing internal spiralia (elongate "grains" having micritic coats) and well-rounded dark grains some of which originated as skeletal debris. Gastropod (left) and bryozoan (right) both in contact with brachiopod shell. Debris inside and outside of shell displays micritic coats. Tighter packing outside of shell contrasts with open packing within shell where grains were protected from solution packing. Sparry calcite cement. Salem Limestone, Middle Mississippian, active quarry, NW$\frac{1}{4}$ SW$\frac{1}{4}$ sec. 34, T. 12 N., R. 2 W., Morgan County, Indiana, U.S.A. IU 8099–81, $\times$20.

Fig. 2. Pellet limestone having sparry calcite cement. Finer texture in upper right represents intraclast. Juarez Limestone, Jurassic, Plomosas, Chihuahua, México. Beales Collection, EP8(18), $\times$40.

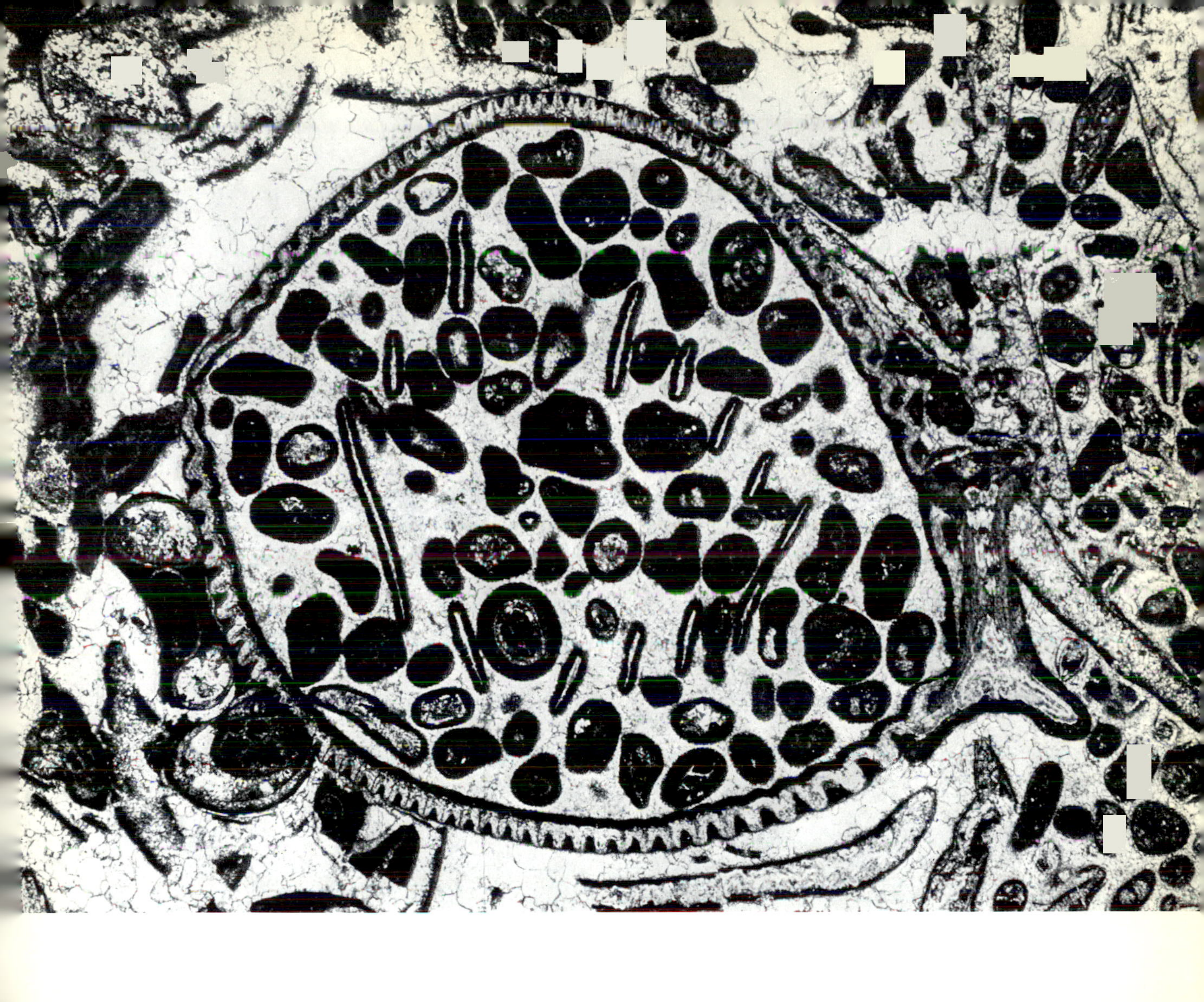
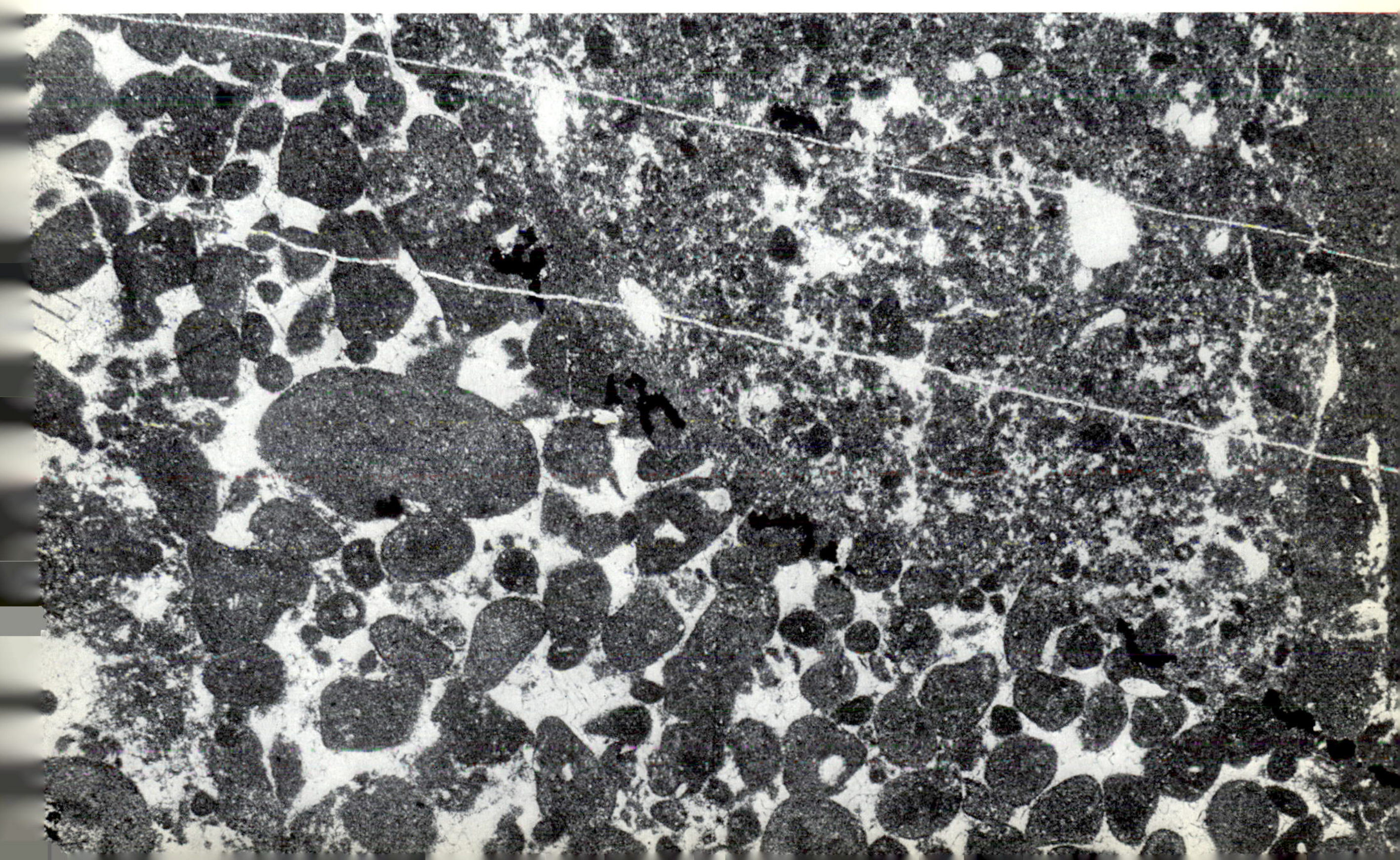

1/10 mm
x 20
x 40
x 60
x 80
x100

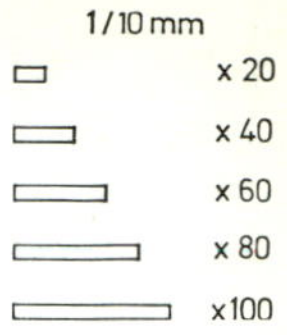

Two views of pellet limestone cemented by sparry calcite. Both figures show intraclasts of pellet limestones defined by tighter packing densities. Palliser Limestone, Devonian, Whiternan's Pass, Alberta, Canada. Beales Collection, WPT3(11), ×40.

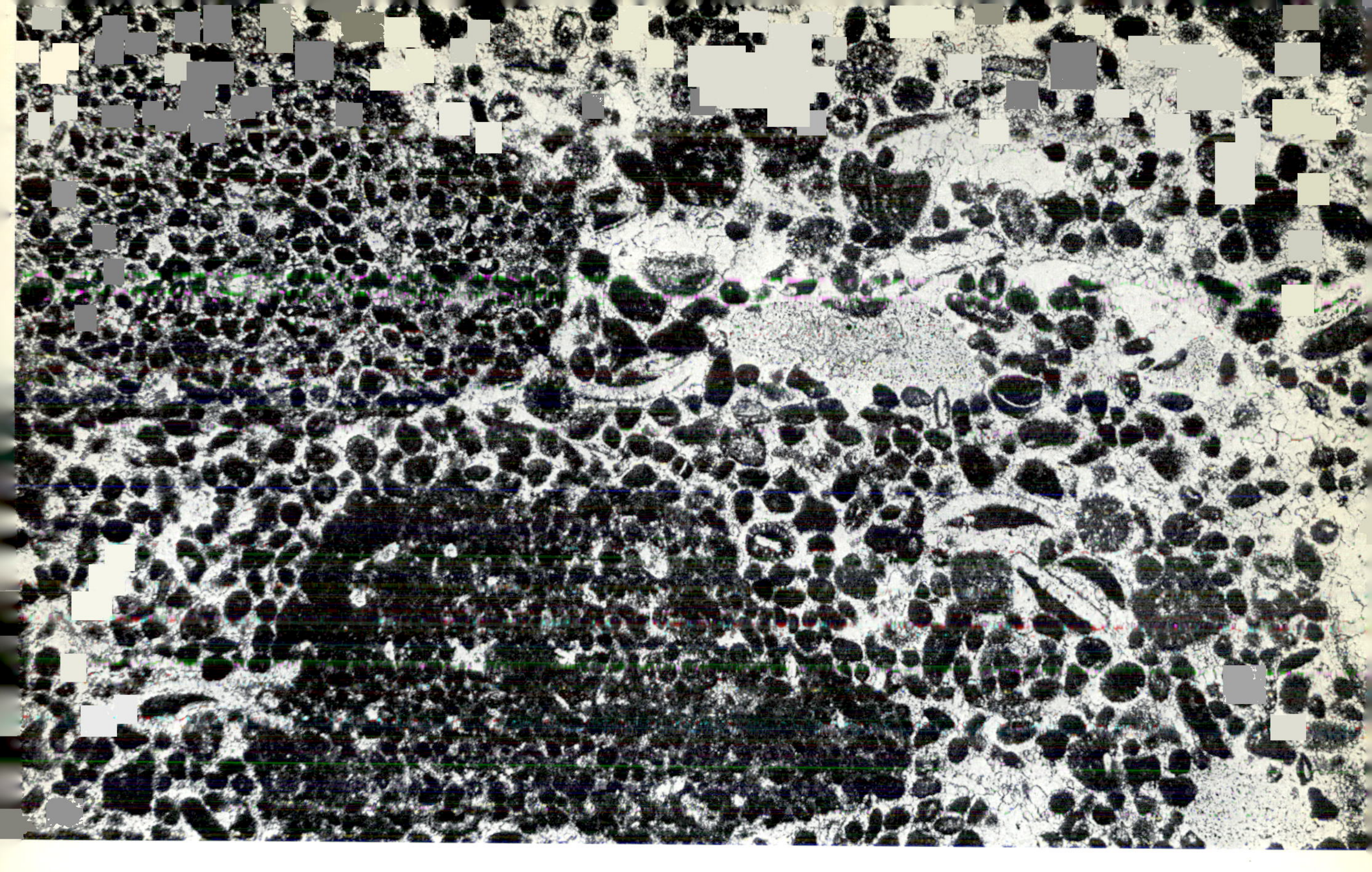

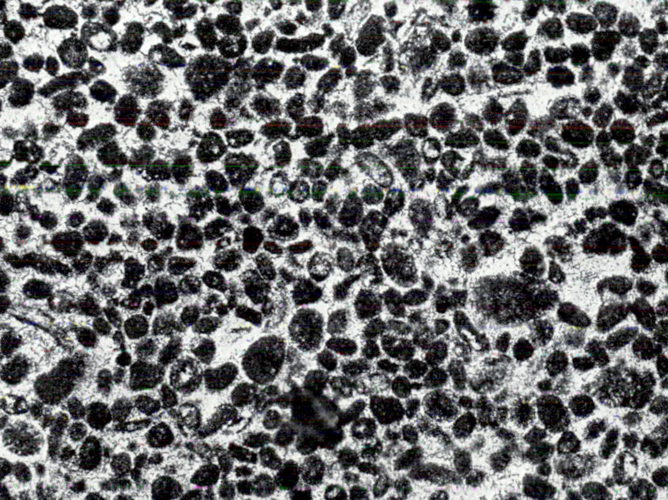

Index to Plates

Bold numbers indicate plates devoted entirely to individual fossil groups

LANTIC OCEAN · *TECUMSEH FORMATION*, PENNSYLVANIAN, WEEPING WATER, CASS CO., NEBRASKA, U.S.A. · *NEGLI CREEK LIMESTONE*, MISSISSIPPIAN, PEACH HILL, PERRY CO., INDIANA, U.S.A. · *WHITEHORSE FORMATION*, TRIASSIC, LLAMA MOUNTAIN, ALBERTA, CANADA · *SILURIAN*, VISBY, GOTLAND, SWEDEN · *PLIENSBACHIAN*, JURASSIC, STE. BAUME, BOUCHES DU RHÔNE, FRANCE · *MANLIUS FORMATION*, DEVONIAN, HELDERBERG MOUNTAINS, RENSSELAER CO., NEW YORK, U.S.A. · *CRETACEOUS*, TIRUCHCHIRAPPALLI, TIRUCHCHIRAPPALLI DISTRICT, MADRAS, INDIA · *HUNGRY HOLLOW FORMATION*, DEVONIAN, WEST WILLIAMS TOWNSHIP, MIDDLESEX CO., ONTARIO, CANADA · *DENTON CLAY MEMBER*, CRETACEOUS, FORT WORTH, TARRANT CO., TEXAS, U.S.A. · *BEERSHEVA FORMATION*, JURASSIC, MAKHTESH RAMON, ISRAEL · *BRASSFIELD LIMESTONE*, SILURIAN, JEFFERSON CO., INDIANA, U.S.A. · *GERSTER FORMATION*, PERMIAN, LEACH MOUNTAINS, ELKO CO., NEVADA, U.S.A. LOWER ORDOVICIAN, BORNHOLM, DENMARK · *BETHEL FORMATION*, MISSISSIPPIAN, HARDIN CO., KENTUCKY, U.S.A. · *LOWER JURASSIC*, COMANA VALLEY, PERSANI MOUNTAINS, ROMANIA · *JURASSIC*, POLIWNA-ON-THE-